Introduction to Pharmaceutical Biotechnology, Volume 2

Enzymes, proteins and bioinformatics

Introduction to Pharmaceutical Biotechnology, Volume 2

Enzymes, proteins and bioinformatics

Saurabh Bhatia

Amity Institute of Pharmacy, Amity University, Gurgaon, Haryana, India

IOP Publishing, Bristol, UK

ISBN 978-0-7503-1302-5 (ebook)
ISBN 978-0-7503-1303-2 (print)
ISBN 978-0-7503-1304-9 (mobi)

DOI 10.1088/978-0-7503-1302-5

Version: 20180901

IOP Expanding Physics
ISSN 2053-2563 (online)
ISSN 2054-7315 (print)

British Library Cataloguing-in-Publication Data: A catalogue record for this book is available from the British Library.

Published by IOP Publishing, wholly owned by The Institute of Physics, London

IOP Publishing, Temple Circus, Temple Way, Bristol, BS1 6HG, UK

US Office: IOP Publishing, Inc., 190 North Independence Mall West, Suite 601, Philadelphia, PA 19106, USA

Dedicated to my Mother (Mrs Vinod Bala Bhatia), my Father (Mr Ram Lal Bhatia) and my Brother (Mr Sanjay Bhatia).

Contents

Preface

There has been a long-felt requirement for a textbook representing current and recent developments in the field of pharmaceutical biotechnology. The few available books have become out-of-date, in particular due to the development of new biotechnology from advances in concepts and techniques. This volume has been written with a view to provide background knowledge on the state-of-the-art of the subject and provide a practical view of the developments to date in genes, enzymes and proteins. The book covers several different facets of evolutional progress and achievements in each line of inquiry. Attention is focused on newer perspectives on the roles of genes, enzymes and proteins in the field of pharmaceutical biotechnology. The current volume is organized into three important areas—enzymes, genomics and proteomics—and is comprised of ten chapters. The book begins with the general properties, mechanisms, applications, production, immobilization and purification of enzymes, with some applications in the form of biosensors and biotransformation reactions. This information is followed by characterizations in genomics, genomes sequencing, comparative genomics and genomic evolution. Other highlights pertain to protein engineering. This book will be of interest to biochemists, biologists, microbiologists, biotechnologists, food technologists and others involved in research on the biotechnological applications of proteins and enzymes.

Acknowledgements

It is a moment of gratification and pleasure to look back with a sense of contentment at the long-traveled path, to be able to recapture some of the fine moments, to be able to thank the infinite number of people, some who were with me from the beginning and some who joined me later, whose love and blessing has made this day possible for me.

Here, I pay tribute to my beloved family, especially Shri. Ram Lal Bhatia, Smt. Vinod Bhatia and my dear brother Sanjay Bhatia. I am grateful to them for their love, trust, patience and support, and acceptance of lost evenings and weekends; and of course for bearing all the kinds of stresses they could, to make me what I am. I will be forever indebted to them; I owe everything to them. Words cannot express my deep sense of gratitude to my guide, mentors and supervisor for their time-to-time encouragement and their excellent guidance, which always kept me on the right track. I am grateful to Amity institute of Pharmacy, Gurgaon, for providing the excellent facilities required for completion of my book project.

Words defeat me to express my heartfelt thanks to all my colleagues for their constant encouragement and support throughout the tenure of this work. Their constant assistance, affection, fidelity, vibrant encouragement and conviction in me played a key role in accomplishing this project. I thank the Almighty for showering infinite bounties and grace upon me and for being my constant companion.

Dr Saurabh Bhatia
Bengaluru

Author biography

Saurabh Bhatia

Dr Saurabh Bhatia graduated from Kurushetra University, India, in 2007, followed by a post-graduate degree in 2009 from Bharati Vidyapeeth University, India. He successfully completed a PhD program in pharmaceutical technology at Jadavpur University, Kolkata, India (2015). Currently, he is working as an Associate Professor at Amity institute of Pharmacy, Amity university. Gurgaon, Haryana in India. He has nine years of academic experience and his areas of interest include nanotechnology (drug delivery), biomaterials, natural products science, biotechnology (plant and animal), microbiology, modified drug delivery systems, analytical chemistry, parasitology (leishmaniasis) and marine science. He has promoted several marine algae and their derived polymers throughout India and has published more than five books and thirty articles in many areas of pharmaceutical science. He has participated in more than thirty national and international conferences.

Chapter 1

Introduction to enzymes and their applications

1.1 Introduction

The cell is the structural and functional unit of life—the basic building block of living systems. Cells have the capability to effectively utilize biocatalysts, known as enzymes, which have outstanding catalytic efficiency and both substrate and reaction specificity. Enzymes have amazing catalytic power and their high level of specificity for their substrate makes them suitable for biological reactions. They are crucial for cellular metabolism. Each and every chemical reaction that takes place in plants, micro-organisms and animals proceeds at a quantifiable rate as a direct result of enzymatic catalysis. Most of the history of biochemistry is directly or indirectly related to the history of enzyme research. Catalysis in biological systems was initially reported in the early 1800s based on research into the digestion of meat. In this report the catalytic activity of secretions from the stomach, the conversion of starch into sugar by saliva, and various plant extracts were reported.

In 1837, Berzelius documented the catalytic nature of fermentation. In the 1850s Louis Pasteur reported that fermentation was a process initiated by living organisms. During this study it was reported that the fermentation of sugar into alcohol by yeast was catalyzed by ferments. He also hypothesized that these ferments are close to the structure of yeast. These ferments were later called enzymes (in yeast). The key breakthrough in the history of enzymes came in 1897 when Edward Buchner isolated, from yeast cells, the soluble active form of the set of enzymes that catalyzes the fermentation of sugar to alcohol. Emul Fischer reported the first systematic studies on enzyme specificity in the early twentieth century [1]. Later, in 1926, James Sumner extracted *urease* in pure crystalline form from jack beans [2]. He also recognized the protein nature of *urease*. In 1930, John Northrop and his co-workers crystallized pepsin and trypsin and established them as proteins [3]. In subsequent years enzymology developed rapidly (table 1.1). The important developments during

doi:10.1088/978-0-7503-1302-5ch1

Table 1.1. Chronology of enzyme studies.

Name	Year	Work
Anfinsen	1956–8	The sequence of an amino acid regulates the folding pattern and activity of a ribonuclease.
Beatle and Tatum	1940	'One gene one enzyme' hypothesis.
Bertrand	1896–7	Co-enzyme or co-ferment (currently known as co-factors).
Berzelius	1835	Concept of catalysis.
Berzelius	1837	Exploration of biological catalysis.
Briggs and Haldane	1925	Derivation of enzyme rate equations using the steady-state approximation.
Buchner	1897	Isolation of the soluble active form of enzymes from yeast cells.
Chances	1943	Application of spectroscopic techniques for studying enzymes.
Cori and Cori	1937–9	Muscle phosphorylase.
Duclaux Henri	1898	Nomenclature: substrate plus suffix 'ase'.
Fischer	1894–5	'Lock and key' hypothesis of enzyme specificity.
Harden and Young	1901–3	Methods for the derivation of kinetic rate laws; principle of enzyme–substrate complex. 1906 Co-ezymase (NAD).
Jacob, Monod and Changeux	1961	Allosterism.
Koshland	1953	'Induced fit' hypothesis.
Kuhne	1878	Explored trypsin catalyzed reactions; introduction of word 'enzyme'.
Michaelis and Menten	1913	Extension of the kinetic theory of enzyme catalysis.
Northrop and Kunitz	1930–3	Crystallization of proteolytic enzymes.
Pasteur	1850	Fermentation of sugar into alcohol by yeast.
Payen and Persoz	1833	Alcohol precipitation of thermolabile 'diastase' from malt.
Phillips, Johnson and North	1965	Three-dimensional structure of lysozyme obtained at 1.5 A resolution.
Sumner	1926	Crystallization of urease.
Sutherland	1956	Cyclic AMP adenyl cyclase.
Umbarger, Yates and Pardee	1956	Regulation of enzyme activity via feedback inhibition.
Wilhelmy	1850	Quantitative evaluation of the rates of sucrose inversion.

this period are: the elucidation of major metabolic pathways, such as the glycolysis and tricarboxylic acid cycle; the detection of numerous biochemical events of digestion, coagulation, muscular contraction and endocrine function, and their roles in the maintenance, control and integration of complex metabolic processes; the kinetic backgrounds to explain the observations of enzyme action and inhibition; and the development of protocols for examining the structures of functionally sensitive proteins. There has been exhaustive research on enzyme-catalyzed reactions and enzymes involved in cell metabolism. At present, 2000 different enzymes have been recognized, each of which catalyzes a different chemical reaction. Currently, more focus is being directed towards the application of enzymes. The high efficiency of enzymes makes them commercially valuable and their specificity of action is offering diverse advantages in clinical medicine.

1.2 Properties of enzymes

Enzymes are the complex protein molecules, often called biocatalysts, which are produced by living cells. They are highly specific both in the reactions that they catalyze and in their choice of reactants, which are known as substrates. An enzyme typically catalyzes a single chemical reaction or a set of closely related reactions [4]. Side reactions resulting in the wasteful formation of by-products are rare in enzyme-catalyzed reactions, in comparison to uncatalyzed ones. Enzymes can also be defined as soluble, colloidal and organic catalysts that are produced by living cells, but are capable of acting independently of the cells [4]. Enzymes are currently being used in diverse areas in the food, feed, paper, leather, agriculture and textiles industries, resulting in major cost reductions. Simultaneously, rapid scientific progress is now encouraging the chemistry and pharmacological industries to embrace enzyme technology, a trend supported by concerns regarding energy, raw materials, health and the environment. One of the most common advantages of enzymes is their ability to function continuously even after their removal or separation from the cells. This means that even after the separation of cells from *in vivo* environments, they continue to work efficiently under *in vitro* conditions; we can conclude that these biocatalysts remain in an active state even after their isolation. Principally, enzymes are non-toxic, biodegradable and can be produced in ample amounts by micro-organisms for industrial applications. In this chapter, the isolation, production, purification, utilization and application of enzymes (in soluble and immobilized or insoluble form) are discussed in detail. Procedures such as recombinant DNA technology and protein engineering are frequently used to produce more efficient and beneficial enzymes. The industrial production and utilization of enzymes is an important part of industry. Interdisciplinary collaboration between areas such as chemistry, process engineering, microbiology and biochemistry is required to develop the best possible enzyme technology, and eventually to achieve increased production and maintain the enzyme's physico-chemical properties under *in vitro* environments.

For catalytic action, small quantities of an enzyme are sufficient, where this quantity of enzyme is much smaller in comparison to its substrates. The overall concentration of substrate transformed per mass of enzyme is often very large. Without exception, all enzymes are proteinaceous and exhibit all the properties of a protein. The treatment of enzymes by extreme temperature or extreme pH, or by treatment with other denaturing agents, results in the complete loss of catalytic activity. Structural configurations such as the primary, secondary, tertiary and quaternary structures of enzyme proteins are essential for their catalytic activity. The degree of catalytic activity chiefly depends on the integrity of the enzyme's structure as a protein. As per reports, enzymes have molecular weights ranging from about

12 000 to over 1 million Da. A number of enzymes consist only of polypeptides and contain no chemical groups other than amino acid residues, e.g. pancreatic ribonuclease. Numerous enzymes require a specific, heat stable, low molecular weight organic molecule, known as a co-enzyme. Moreover, a number of enzymes require both a co-enzyme and one or more metal ions for activity. A complete biochemically active compound is formed by the combination of a catalytically active enzyme (also called the protein part) with a co-enzyme or a metal ion—this is called a holoenzyme. The protein part of a holoenzyme is called an apoenzyme. In this arrangement a co-enzyme may bind covalently or noncovalently to the apoenzyme. In certain enzymes the co-enzyme or metal ion is only loosely and transiently bound to the protein. However, in others it is tightly and permanently bound, in which case it is known as a prosthetic group. A prosthetic group signifies a covalently bound co-enzyme. According to reports, co-enzymes and metal ions are stable under heating, while the protein part of an enzyme (the apoenzyme), is denatured by heat.

$$\underset{\text{(Total enzyme)}}{\text{Holoenzyme}} = \underset{\text{(Protein)}}{\text{Apoenzyme}} + \underset{\text{(Non-protein)}}{\text{Prosthetic group}}$$

Prosthetic groups may be classified functionally into two major classes: co-enzymes and co-factors. Co-enzymes may be considered to be biosynthetically related to the vitamins, such as the co-enzyme nicotinamide adenine dinucleotide (NAD) which is vital for cellular energy metabolism and integrates the vitamin niacin into its chemical makeup. Moreover, a co-enzyme may be considered as a co-substrate, experiencing a chemical transformation throughout the enzyme reaction (NAD is reduced to NADH), the reversal of which requires a separate enzyme, perhaps from a different cellular site. Co-enzymes might thus travel intra-cellularly between apo-enzymes and, by transferring chemical groupings, integrate several metabolic processes. Table 1.2 shows a list of the more common co-enzymes and their functions. In contrast to co-enzymes, co-factors, such as pyridoxal phosphate or hem groups, remain with one enzyme molecule and in conjunction complete a cycle of a chemical change brought about by one enzyme turnover [5]. Other enzymes, such as carboxypeptidase, require metal ions as co-factors, the divalent cations Mg^{2+}, Zn^{2+} and Mn^{2+} being the most common; these are often called enzyme activators [6]. Table 1.3 lists several enzymes and their respective co-factors.

Table 1.2. Several co-enzymes employed in the transfer of specific atoms or functional groups.

Co-enzyme	Entity transferred
Thyamin pyrophosphate	Aldehydes
Tetrahydrofolate	Other one carbon groups
Pyridoxal phosphate	Amino groups
Nicotinamide adenine dinucleotide	Hydrogen atoms (electrons)
Flavin adenine dinucleotide	Hydrogen atoms (electrons)
Co-enzyme A	Acyl groups
Biocytin	CO_2
3′-deoxyadenorylcohalamine (co-enzyme B12)	H atoms and alkyl groups

Table 1.3. Several enzymes and their co-factors.

Enzyme	Co-factor(s)	Enzyme	Co-factor(s)
Pyruvate kinase	K^+ and Mg^+	Urease	Ni^{++}
Nitrate reductase	Mo	Peroxidase	Fe^{++} or Fe^{+++}
Glucose 6-phosphatase		Hexokinase	Mg^+
DNA polymerase	Zn	Glutathione peroxidase	Se
Cytochrome oxidase	Cu**	Cytochrome oxidase	Fe^{++} or Fe^{+++}
Carbonic anhydrase	Zn^{++}	Catalase	Fe^{++} or Fe^{+++}
Arginase	Mn	Alcohol dehydrogenase	Zn^{++}

1.3 Catalysis

The role of a catalyst is to increase the speed of a chemical reaction. When the rate of a chemical reaction is governed by a soluble catalyst, which may result in a further increase in the rate of chemical reaction, it is called homogeneous catalysis. In this case catalysis occurs in a solution. When the catalyst is in a separate phase from the reactants, or when catalysis occurs on a insoluble surface or an immobilized matrix, it is known as heterogeneous catalysis. Enzymes are also called biological catalysts. These biological catalysts generally have the properties of homogeneous catalysts, however, a number of enzymes present in membranes are insoluble, and thus are called heterogeneous catalysts. Enzyme specificity is the absolute specificity of protein catalysts to identify and bind to only one or a few molecules. In this process the enzyme carries a defined arrangement of atoms in their active site to bind with the substrate. This active site on the enzyme should have a shape that accurately matches the substrates. Thus specificity is achieved when an enzyme with an active site binds with the chemical reactants (the substrates) at their active sites via weak bond interactions. To undergo a chemical reaction, this active site carries certain residues that form a temporary bond with the chemical reactants, termed the binding site, whereas the catalytic site carries the residues that are responsible for catalysis. Specificity is achieved when a substrate binds to an enzyme that has a defined arrangement of atoms in the active site. An enzyme always catalyzes a single type of chemical reaction, which involves the formation and breakdown of covalent bonds. Since they are specific to one particular reaction, this feature of enzymes is called reaction specificity, also known as absolute reaction specificity, i.e. no by-products are formed.

1.4 The structure of enzymes

Enzymes always act as catalysts and small quantities compared to their substrate are required to considerably increase the rate of chemical reactions, wherein the enzymes themselves experience no overall change [7, 8]. In contrast to all true catalysts, an enzyme does not alter the ultimate equilibrium position of a reaction, which is thermodynamically determined, thus merely the rate of completion of equilibrium of a feasible reaction is augmented. In addition to catalytic properties, enzymes exhibit the physico-chemical behavior of proteins: their solubility, electrophoretic properties, electrolytic behaviors and chemical reactivity [7, 8]. The primary structural configuration and catalytic action of enzymes is determined by the linear chain of amino acid residues linked via peptide bonds, which constitute a protein molecule. Localized folding of the primary structure is called a secondary structure, whereas the complete folding of the molecule is known as a tertiary structure. In contrast to these structural configurations, a quaternary structure is the agglomeration of several folded chains. The structural features of enzymes are shown in figures 1.1 and 1.2. In contrast to traditional chemical catalysts, e.g. hydrogen ions, heavy metals or metal oxides, which are most effective in organic solvents, at very high temperatures or at extreme pH values, enzymes operate most efficiently under very mild conditions. When using enzymes, there are certain issues that require attention, such as deviation from

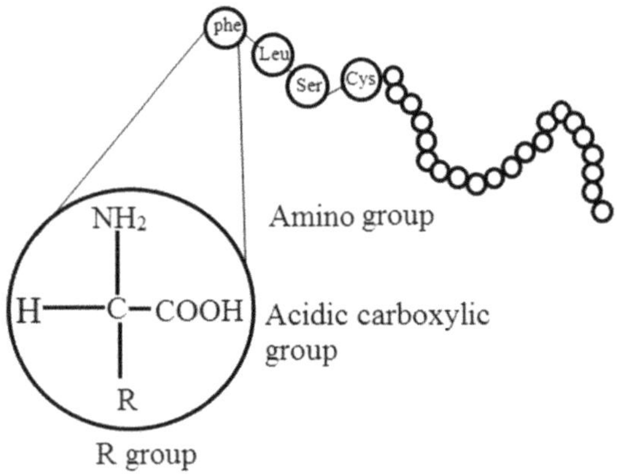

Figure 1.1. Structural features of enzyme.

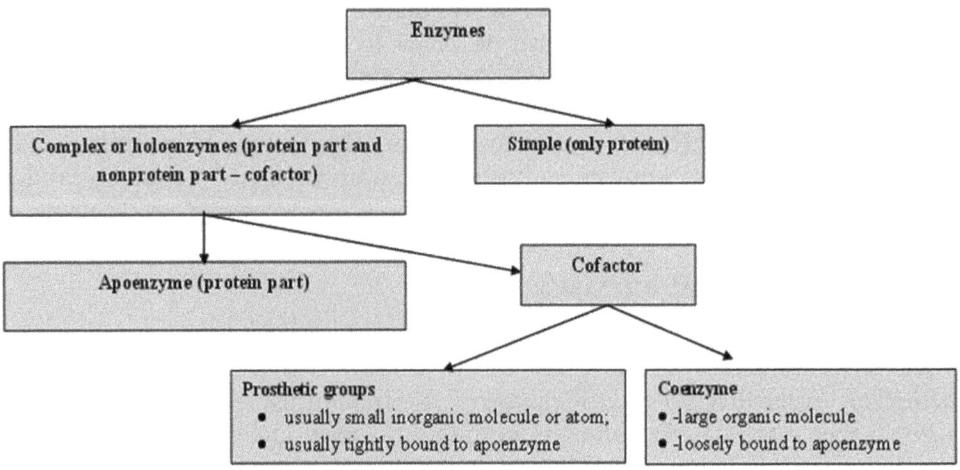

Figure 1.2. Principle components of an enzyme.

homogeneous aqueous solutions, physiological pH and temperature, which can rapidly destroy enzyme activity. However, under normal conditions the increase in reaction rate is rarely matched by their non-protein counterparts.

1.5 Structural features: primary and secondary structures

Three-dimensional analysis of the amino acid sequence of lysozyme of hen's egg white has demonstrated some features essential for primary structure [9, 10]. These are:

- Molecules derived from a similar source have a similar order of amino acid residues and appear to be random with no obvious predictability.
- Even though numerous enzymes are intramolecularly crosslinked via disulfide bridges of cysteine, no branching occurs.

Current databases suggest that a small number of amino acids are extra and most are 'functional', i.e. the majority of them co-operatively control the higher orders of structural organization and therefore the catalytic activity. When comparing the primary structures of enzymes performing similar functions, wide structural homologies are detected in their sequence, mainly in the patterns of their nonpolar residues. For example, pancreatic juice contains five inactive precursors (zymogens), namely chymotrypsinogen A, B and C, trypsinogen and proelastase; all of these are activated to the respective proteases by proteolytic cleavage [11].

1.6 The mechanism of action of enzymes

The mechanism of action is based on a chemical reaction, in which the enzyme binds to the substrate and finally forms an enzyme–substrate complex. This reaction take place in a relatively small area of the enzyme called the active or catalytic site. In other words, the mechanism of enzyme action is based on the nature of the enzyme–substrate interaction, which accounts for the reaction specificity of the biological catalysts. The active or catalytic site of an enzyme is constituted by several amino acids, located at some distance from each other in the peptide chain. These amino acids are brought close together by the folding resulting from the secondary and tertiary structure of the enzymes. Side chains of amino acid residues at the catalytic site provide groups for binding with specific groups of the substrate. Co-factors assist the catalysis. The substrate forms bonds with amino acid residues in the substrate binding domain of the active site. The binding induces a conformational reaction in the active site. During the reaction, the enzyme forms a transition-state complex. As the products of the reaction disassociate, the enzyme returns to the original state. Two different models postulated for the mechanism of enzyme action are given below.

1.6.1 The Fisher template model (lock and key model)

This is a rigid model of the catalytic site, proposed by Emil Fischer in 1894 [12]. The model explains the interaction between a substrate and an enzyme in terms of a lock and key analogy. In this model, the catalytic site is presumed to be preshaped. The substrate fits as a key fits into a lock. The drawback of this model is the implied rigidity of the catalytic site. The model cannot explain changes in enzyme structure in the presence of allosteric modulators.

1.6.2 Induced fit model

In contrast to the above method, this model suggests a flexible mode for the catalytic site. To overcome the problems of the lock and key model owing to the rigid catalytic site, Koshland [13–15] suggested an induced fit model in 1963. The important feature of this procedure is the flexibility of the active site. In the induced fit model, the substrate induces a conformational change in the active site of the enzyme so that the substrate fits into the active site in the most convenient way so as to promote the chemical reaction. This method suggests competitive inhibition, allosteric modulation and inactivation of enzymes on denaturation.

The Michaelis–Menten theory of enzyme action [16] offers the basis for most current research on the mechanism of enzyme action. This concept of the enzyme–substrate complex scheme assumes the combination of the enzyme and substrate in phase one (occasionally known as the transition phase) of the enzyme activity and liberation of the enzyme and the products of the catalysis in phase two of the reaction.

$$\text{Enzyme} + \text{Substrate} \rightarrow \text{Enzyme -Substrate Complex} \rightarrow \text{Enzyme} + \text{Substrate}$$

1.6.3 Covalent catalysis

Covalent catalysis is evidenced in enzymes capable of forming covalent bonds between the substance and the catalytic group of the active site [17]. A number of enzymes react with their substrates to form very unstable, covalently joined enzyme–substrate complexes, which undergo further reaction to yield products much more readily than in an uncatalyzed reaction. Several of the enzymes that exhibit covalent catalytic behavior are listed in table 1.4.

Table 1.4. Various enzymes exhibiting covalent catalytic behavior.

Enzyme	Reactive group	Typical covalent enzyme–substrate intermediate
Chymotrypsin, trypsin, thrombin, esterase	Serine	Acylserine
	HO–CH_2–CH–	
Phosphoglucomutase, alkaline phosphatase	Serine	Phosphorylserine
	HO–CH_2–CH–	
Glyceraldehyde-3-phosphate dehydrogenase papain	Cysteine	Acylcysteine
	HS–CH_2–CH–	

1.7 Catalysis via chymotrypsin

Hummel and Kalnitzky suggested an enzyme mechanism through the depiction of the sequential transition states experienced by the enzyme–substrate complex during catalysis [18]. Chymotrypsin is a digestive enzyme, responsible for proteolysis (breakdown of proteins and polypeptides) in the duodenum. Chymotrypsin favorably breaks peptide amide bonds (the carboxyl side of the amide bond is a large hydrophobic amino acid). These amino acids contain an aromatic ring in their side chain that fits into a 'hydrophobic pocket' of the enzyme. It is stimulated in the presence of trypsin. Trypsin and chymotrypsin are both serine proteases with high sequence and structural similarities, but with different substrate specificity [19, 20].

1.7.1 Intermediary stages of chymotrypsin

As discussed above, chymotrypsin is a protease enzyme that cuts on the C-terminal phenylalanine, tryptophan and tyrosine on peptide chains [21]. Additionally, it is more specific for aromatic amino acids because of its hydrophobic pocket. Comparable to other serine proteases, chymotrypsin also catalyzes the hydrolysis of certain esters [22]. The molecular events involved in catalysis are called intermediary enzymology. Chymotrypsin, a protease, favorably accelerates breakdown of peptide bonds in which the aromatic amino acid (Phy, Try, or Trp) or bulky nonpolar R group (Met) contribute a carboxyl group. The synthetic substrate p-nitrophenyl acetate allows colorimetric analysis of chymotrypsin activity, as hydrolysis to p-nitrophenol, which is alkali, changes into the chromophore anionic forms.

1.7.2 Kinetic behavior of α-chymotrypsin

The kinetics of chymotrypsin of p-nitrophenyl acetate can be considered in a 'stop-flow' apparatus. This procedure utilizes substrate quantities of enzymes and measures the events in the first few milliseconds [23]. The use of p-nitrophenyl acetate as a substrate offers the prospect of investigating solvent effects on both the acylation of the enzyme and the hydrolysis (deacylation) of the acyl enzyme [23]. The significant features of the slow-flow kinetics of chymotrypsin are:

- Release of p-nitrophenyl anion with chymotrypsin. Hydrolysis of p-nitrophenyl acetate occurs in two different phases:
  - a burst phase featuring rapid liberation of an anion.
  - a subsequent 'steady-state' phase, with slower release of extra anion.
- In catalysis by chymotrypsin, the slow stage is hydrolysis of the chymotrypsin–acetate (CT–Ac) complex. When all the existing chymotrypsin has been converted to CT–Ac, no further release of p-nitrophenyl acetate anion can take place until more free chymotrypsin is released by the slow, hydrolytic elimination of acetate anion from the CT–Ac complex [24, 25]. The free chymotrypsin then is presented for further formation of chymotrypsin–p-nitrophenyl acetate complexes (CT–PNP) and CT–Ac complexes with attendant liberation of PNP. The development and decay of the

enzyme–substrate complex, based on the Michaelis–Menten kinetics can be represented as [24, 25]:

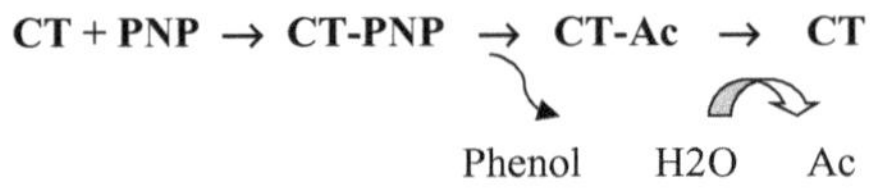

where CT = chymotrypsin, PNP = p-nitrophenyl acetate, CT–PNP = chymotrypsin–p-nitrophenyl acetate complex and CT–Ac = chymotrypsin–acetate complex. In comparison to the hydrolysis of the CT–Ac complex, the formation of CT–PNP and CT–Ac complexs is relatively fast.

- A 'charge relay network' acts as a proton shuttle during catalysis by chymotrypsin. The charge relay network of chymotrypsin encompasses three aminoacyl residues that are far apart in a primary structural sense, but close together in a tertiary structural sense. While most of the charged residues of chymotrypsin are present at the surface of the molecule, those of the charge relay network are hidden in the otherwise nonpolar inner side of the protein. These charges transmit residues which activate sequential proton shifts that shuttle protons in the opposite direction. An equivalent series of proton shifts is assumed to accompany the hydrolysis of the physiologic chymotrypsin substrate, e.g. a peptide.

1.7.3 Selective proteolysis in creation of the catalytic sites of enzymes

Various enzymes, hormones and other physiologically active proteins are produced as inactive precursors (zymogens) that are further transformed to the active form by selective enzymatic cleavage (limited proteolysis) of peptide bonds. The final step to activating enzymatic function is limited proteolysis, either in a single activation step or in a consecutive series (cascade). The specificity of each activation reaction is evaluated by the complementarity of the zymogen substrate and the active site of the attacking protease. The arrangement of successive activation reactions is controlled by the specificity of each enzyme, while the extent of amplification of the initial stimulus is evaluated by the effectiveness of each activating step. Zymogen activation produces a prompt and irreversible response to a physiological stimulus, and is capable of initiating new physiological functions. Classical examples are the processes of hormone production, fibrinolysis, complement activation, blood coagulation, supra-molecular assembly, metamorphosis, fertilization and digestion. The zymogens of the pancreatic serine proteases, in particular, have functioned as models for detailed studies of the nature of the molecular changes that are involved in the intense increase in enzymatic activity that results upon incomplete proteolysis of the zymogen.

Specific proteolysis is a common means of activating enzymes and other proteins in biological systems. A number of proteins are manufactured and released in the form of inactive precursor proteins called proproteins. Various enzymes attain full enzymatic activity as they suddenly fold into their characteristic three-dimensional forms. In contrast, other enzymes are produced as inactive precursors that are

Table 1.5. Gastric and pancreatic zymogens.

Active enzyme	Zymogen	Site of production
Chymotrypsin	Chymotrypsinogen	Pancreas
Trypsin	Trypsinogen	Pancreas
Carboxypeptidase	Procarboxypeptidase	Pancreas
Elastase	Proelastase	Pancreas
Pepsin	Pepsinogen	Stomach

successively activated by breakdown of one or a few specific peptide bonds. The inactive precursor is known as a zymogen (or a pro-enzyme). In other words, when the proteins are enzymes, the proteins are called pro-ezymes or zymogens (table 1.5). An energy source (ATP) is not required for cleavage [11]. Thus, in comparison to reversible regulation by phosphorylation, even proteins sited outside cells can be triggered by this means. An additional noteworthy difference is that proteolytic activation, in comparison with allosteric control and reversible covalent modification, occurs just once in the life of an enzyme molecule. Transformation of a proprotein to the mature protein includes selective proteolysis. This transforms the proproteins by one or more consecutive proteolytic clips to a arrangement in which the individual activity of the mature protein (its enzymatic activity) is expressed, e.g. the hormone insulin (proinsulin), the digestive enzyme chymotrypsin (chymotrypsinogen), a number of factors for blood clotting and for the blood clot dissolution cascades, and the connective tissue protein collagen (procollagen). Chymotrypsinogen consists of 245 amino acid residues, and is practically devoid of enzymatic activity. As the reaction starts, it is converted into a fully active enzyme. This occurs when the peptide bond joining arginine 15 and isoleucine 16 is cleaved by trypsin. The subsequent active enzyme, known as π-chymotrypsin, then acts on other π-chymotrypsin molecules. Two dipeptides are eliminated to form α-chymotrypsin (the stable form of the enzyme) [11]. The three subsequent chains in α-chymotrypsin remain interconnected to each another by two interchain disulfide bonds. The outstanding feature of this process is that cleavage of a single specific peptide bond alters the protein from a catalytically inactive form into one that is fully active. The transformation of prochymotrypsin (Pro-CT), a 2,4,5-aminoacyl residue polypeptide, to the active enzyme α-chymotrypsin includes three proteolytic clips and the formation of an active intermediate called π-chymotrypsin (π-CT) and consequently to the mature catalytically active enzyme α-chymotrypsin (α-CT). Examples of gastric and pancreatic zymogens are listed in table 1.5.

1.7.4 Kinetic models for enzymes

Generally, enzyme kinetics is defined as the study of the rate of reactions, i.e., how the substrate concentration impacts the velocity of the reaction. Enzyme kinetics involves optimization of bio-catalytic reactions to allow process design and scaling up processes to further increase the production and minimize the overall overhead costs of various procedures. Kinetic investigations in the branch of biochemistry concerned with enzymes can be categorized into three types:

- *Transient-state kinetics*: This is the stage of reaction before the steady or rapid-equilibrium state, and involves quick reactions between the enzymes and substrate. These sudden changes in the reaction mixture when the substrate and enzymes are mixed require advance equipment to monitor the reaction before it changes into the steady state. The mechanisms of the reaction are associated with the enzyme structural configuration. Basic steps are involved during an enzyme-catalyzed reaction, which allow the direct study of the intermediates and products formed during a single enzyme cycle, which may further help in direct analysis of individual reaction steps for short times. In this type of reaction a sufficient concentration of enzymes is used to witness the intermediate and product formation.
- *Steady-state kinetics*: This is the phase in which the rate of formation of intermediates and the rate of decomposition remain the same, and thus the concentrations of reactive intermediates remain the same. During this reaction substrate concentration is greater than enzyme concentration. The Michaelis–Menten enzyme kinetic (figure 1.3) can be considered as the most often studied reaction for several enzymes. For example, chymotrypsin (protease) with a high concentration of substrate achieves maximum velocity of the reaction (called the first order of reaction) but at a certain point the substrate occupies all binding sites of the enzyme, after which further addition of substrate does not increase the rate. This is called the zeroth order of reaction (the steady state). It is the phase in which the enzyme and substrate concentrations cannot be determined using the dissociation constant. Thus steady-state enzyme kinetics is based on the theory that a catalytic reaction remains constant if the reaction is not exposed to continuous changes.

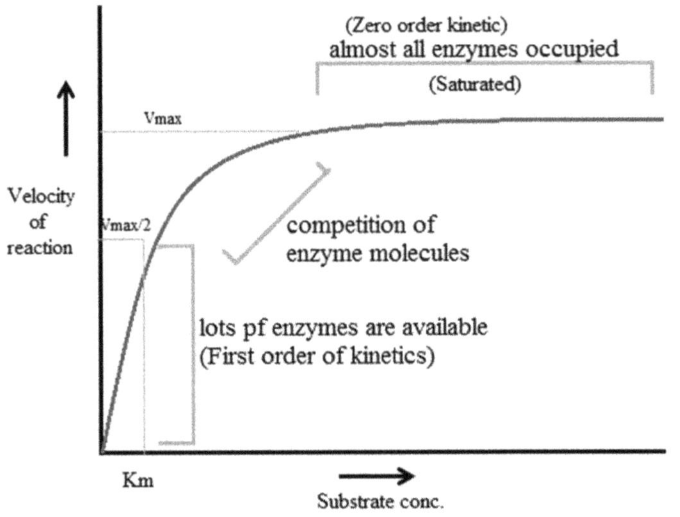

Figure 1.3. The Michaelis–Menten enzyme kinetic.

- *Rapid-equilibrium kinetics*: This the phase in which both the enzyme and substrate concentrations can be determined using the dissociation constant. During this procedure total enzyme concentration remains constant during the reaction and the concentration is very small compared to the amount of substrate. In this reaction, before the rate-determining reaction, the reactions are in equilibrium with their components, thus this stage is called rapid-equilibrium kinetics.

According to reports, factors that affect enzyme-catalyzed reactions also affect the velocity of a reaction. These factors are called modifiers of enzyme-catalyzed reactions. These modifiers can be divided into two classes: inorganic modifiers (enzyme activators) and organic modifiers (enzyme inhibitors). These factors can have different types of effects on the velocity of the reaction; nevertheless the most vital effect is that they offer many pathways to products, e.g. when one modifier is bound to an enzyme, it alters the rate of reaction and thus forms two rate constants. However, when two modifiers participate, there are five self-regulating equilibria, resulting in three paths for making products.

There are two mechanisms, single-substrate and multiple-substrate, that are helpful in studying the different stages of enzymatic reactions. Understanding these stages helps in understanding the properties of enzymes. Certain enzymes have single substrates (a single substrate binding site), e.g. triosephosphate isomerase, whereas certain enzymes have multiple substrates molecules (multiple binding sites), such as dihydrofolate reductase, and bind with multiple substrates. After the exploration of specific RNA sequences required for RNA replication, new biocatalysts in the form of ribozymes have emerged with the potential to catalyze specific biochemical reactions. There is a misconception about biological catalysts that all biological catalysts are made up of proteins, which is not true; some are RNA-based catalysts (ribozymes and ribosomes). Both are important for many cellular functions. A major difference between enzymes and ribozymes is that RNA-based catalysts are restricted to only a few reactions; however, their reaction mechanisms and kinetics can be studied and classified by similar procedures. Enzyme-based mutation, in particular site-directed mutagenesis, is an important approach to alter genes and investigate the functional and structural features of enzymes, e.g. mutation of the enzyme present in *Coprinus cinereus* peroxidase offers an understanding of its increased thermostability. Challenges involved in studying cascades of reactions catalyzed by a multi-enzyme, e.g. proteasome involved in the ubiquitin–proteasome pathway, can be overcome by establishing understanding of the complex structure and the respective biochemical reactions. This understanding allows exploration of active sites, intermediate compounds, final products and their interrelation with complex machinery, as well as biochemical reactions. It has been well understood that enzymes that accelerate complex reactions have numerous substrates and involve complex enzyme kinetic mechanisms. As discussed above, most of the biochemical reactions occurring in the body are multi-substrate reactions. In such reactions two substrates are involved and yield two products (figure 1.4). These types of reactions involve the transfer of a compound from one compoment to another,

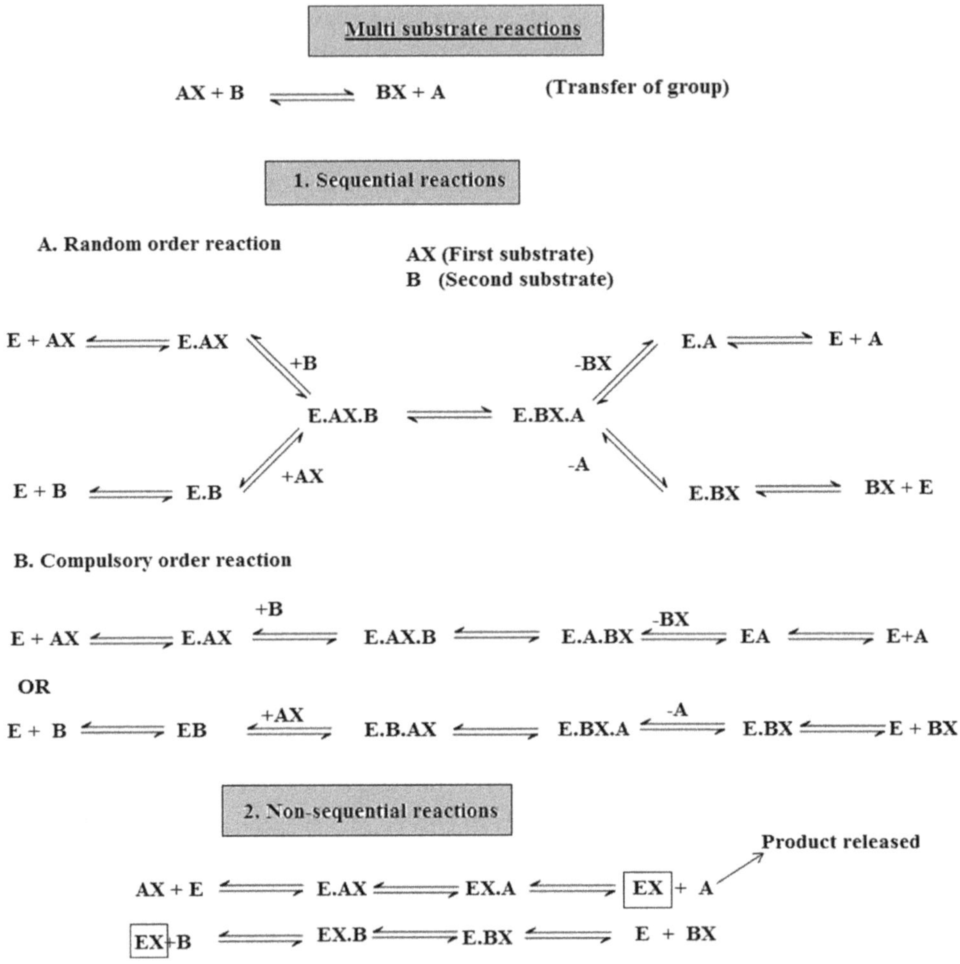

Figure 1.4. Multi-substrate reactions.

e.g. when glucose reacts with ATP in the presence of hexokinase it forms glucose 6-phophaste and ADP. Here, phosphate from ATP is transfered to glucose to form glucose 6 phosphate. The mechanism of catalysis involves two types of reactions: sequential and non-sequential reactions. Sequential reaction results in the formation of a ternary complex. This means that both of the substrates involved in the reaction bind with an enzyme to form the product (figure 1.4). Sequential reaction is further divided into two types: the random and compulsory order mechanisms. As the name suggests, in a 'random' mechanism, either substrate can bind first and any product can leave first. In contrast to the random order mechanism, in the compulsory order mechanism the order of binding of the substrate and order of release of the product is specific; this is also called the Theorell–Chance mechanism (figure 1.4). In a non-sequential reaction, also called the 'ping-pong' mechanism, formation of ternary complex does not take place. In these types of reactions, when the first substrate binds with enzyme its product is released, and then the second substrate binds and its product is released. Such a reaction is called a double placement reaction. Thus only a single substrate binds at a time; this may be due to the presence of a single binding site on the enzyme. Major differences between the sequential and non-sequential reactions are that the formation of a ternary complex takes place only in the sequential reaction, and that in the sequential reaction both substrates bind to the enzyme and release products, while in the non-sequential mechanism the substrates bind and release their products one after the other (figure 1.4).

Another type of sequential mechanism is the systematic mechanism, which involves the addition of substrates and formation of products in a specific order.

1.7.5 Enzyme mediated acid–base (general) catalysis

Several protein enzymes use general acid–base catalysis as a way to increase reaction rates [26]. The amino acid histidine is optimized for this function because it has a pK (a) (where K(a) is the acid dissociation constant) near physiological pH [26].

When the substrate has been bound at the catalytic site, the charged functional groups of the side chains of neighboring aminoacyl residues may contribute in catalysis by behaving as acidic or basic catalysts. There are two extensive groups of acid–base catalysis by enzymes: general and specific (acid or base) catalysis. Specific acid or specific base catalysis are those reactions in which the reaction rates fluctuate under the influence of changes in H^+ or H_3O^+ concentration, but are independent of the concentrations of the other acids or bases present in the solution. In contrast to specific catalysis, general acid or general base catalysis are the reactions whose rates are very reactive to all acids (proton donors) or bases (proton acceptors) present in the solution. To examine whether a given enzyme-catalyzed reaction is a general or specific acid or base catalysis, the rate of reaction is determined under two sets of circumstances:

- at different pH values at a constant buffer concentration, and
- at constant pH values but at different buffer concentrations. Against this background, if the degree of the reaction deviates as a function of pH at a constant buffer concentration, the reaction is specific base/acid catalyzed if the pH is above/below 7.0. If the reaction rate at a constant pH rises as the buffer concentration increases, the reaction is general base/acid catalysis, if the pH is above/below 7.0.

1.7.6 Metallozymes

Almost 25% of all enzymes include tightly bound metal ions or need them for activity. The major role of these metal ions is investigated using techniques such as x-ray crystallography, magnetic resonance imaging (MRI) and electron spin resonance (ESR). A metalloprotein is a protein that contains a metal ion co-factor. Metallozymes contain a certain amount of functional metal ion that is retained during the course of purification [27]. A metal-activated enzyme binds with metals less firmly, but needs to be activated by addition of metals. Four types of complexes are possible for the tertiary complexes of the catalytic site (Enz), a metal ion (M) and substrate (S) that exhibit 1:1:1 stoichiometry:

Enz--S--M

Substrate-bridge complex

M--Enz--S

Enzyme-bridge complex

Enz--M--S

Simple metal-bridge complex

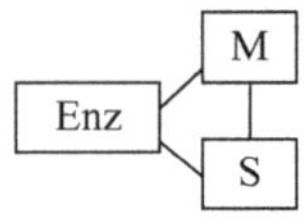

Cyclic metal-bridge complex

All of these complexes are possible for metal-activated enzymes. Metallozymes cannot form the EnzSM complex (substrate–bridge complexes), as the purified enzyme exists as Enz–M. Three generalization can be made:

- The majority of the kinases (ATP: phosphotransferases) form substrate–bridge complexes of the type enzyme–nucleotide–M.
- Phosphotransferases (phosphoenolpyruvate or pyruvate used as the substrate), enzymes catalyzing other reactions of phosphoenolpyruvate and carboxylases, form metal bridge complexes (Enz–M–S).
- A particular enzyme may form one type of bridge complex with one substrate and a different type with another.

The metal ions participate in each of the four mechanisms by which the enzymes are known to accelerate the rates of chemical reaction:

- Approximation of reactants.
- Covalent catalysis.
- General acid–base catalysis.
- Induction of strain in the enzyme or substrate.

Metal ions are electrophiles (attracted to electrons) and share an electron pair forming a sigma bond. They may also be considered as super acids as they exist in neutral solutions, frequently having a positive charge which is greater than their quantity. Mn^{2+}, Ca^{2+} and Mg^{2+} are the metal ions that are most commonly used in enzymatic catalysis. Two metal ions, iron and manganese are used in the form of haemprotein. Metal ions have the potential to accept electrons via sigma or pi bonds to successively activate electrophiles or nucleophiles. By means of donating electrons, metals can activate nucleophiles or act as nucleophiles themselves. The co-ordination sphere of a metal may bring together the enzyme and substrate or form chelate-producing distortion in either the enzyme or substrate [28]. A metal ion may also mask a nucleophile and thus avoid an otherwise probable side reaction. Metals can also function as three-dimensional templates for the co-ordination of basic groups on the enzyme or substrate.

1.8 Enzyme inhibition

Enzyme inhibition decreases the activity of an enzyme without significantly disrupting its three-dimensional macromolecular structure. Inhibition is therefore distinct from denaturation and is the result of a specific action by a reagent directed or transmitted to the active site region. When low molecular weight compounds interfere with the activity of enzymes by partially reducing or completely inhibiting the enzyme activity either reversibly or irreversibly, it is known as enzyme inhibition. The compounds responsible for such inhibition are called enzyme inhibitors. To protect the enzyme catalytic site from any change, a ligand binds with a critical side chain in the enzyme. Chemical modification can be performed to test the inhibitor for any drug value. Studies of enzymes can yield much information about the following:

- A number of drugs useful in medicine, which seem to function because they can inhibit certain enzymes in malfunctioning cells.
- The convenience of elucidating metabolic pathways in cells.
- The mechanism of the catalytic activity.
- The nature of the functional group at the active site.
- The substrate specificity of the enzyme.

The pharmacological action of drugs is mainly based on enzyme inhibition, e.g. sulfonamides and other antibiotics. In the majority of cases the enzyme inhibited is known. The development of nerve gases, insecticides and herbicides is based on enzyme inhibition studies. There are two major types of enzyme inhibition: reversible and irreversible.

Reversible inhibitors efficiently bind to enzymes by forming weak non-covalent interactions, e.g. ionic bonds, hydrophobic interactions and hydrogen bonds. Reversible inhibitors do not form any strong chemical bonds or reactions with the enzyme, they are formed quickly and can easily be removed, in contrast to irreversible inhibitors. Reversible inhibition includes competitive inhibition, uncompetitive inhibition and noncompetitive inhibition. Irreversible inhibition includes group specific inhibition (reacts only to a certain chemical group), reactive substrate analogs (affinity label) and inhibitors that are structurally similar to the substrate and will bind to the active site, and mechanism-based inhibitors (enzymes transform the inhibitor into a reactive form within the active site).

1.9 Pharmaceutical applications

Currently, enzymes are often utilized for a broad range of applications such as: washing powders (e.g. proteases, lipases, amylases); textile manufacture (amylases and catalase to remove the starch); the leather industry (proteases to hydrolyze proteins); the paper industry; improvement of the environment; food production (enzyme-modified cheese/butter), processing (glucose oxidase for dough strengthening) and preservation; and medical applications. According to current reports, several enzymes are produced industrially and there are significant applications in the food industry (45% of use), detergent industry (35%), textiles industry (10%) and leather industry (3%). Details on the applications of individual enzymes are provided in table 1.6.

Table 1.6. Industrially produced enzymes from plant sources and their applications.

Enzyme	Source(s)	Application(s)
β-amylase	Barley, soy bean	Baking, preparation of maltose syrup
Bromelain	Pineapple	Baking
Esterase	Wheat	Ester hydrolysis
Ficin	Fig meat	Tenderizer
Papain	Papaya	Meat tenderizer, tanning, baking
Peroxidase	Horse radish	Diagnostic
Urease	Jack bean	Diagnostic

1.9.1 Diagnostic applications of enzymes

Enzymes have been used widely in diagnostic applications varying from immunoassays to biosensors. Enzyme immunoassay methods hold great promise for application under a wide variety of conditions. Under laboratory conditions they can be as sensitive as radio-immunoassays, but they can also be adapted as simple field screening procedures [29, 30]. The examination of enzyme quantity in the extracellular body fluids (blood plasma and serum, urine, digestive juices, amniotic fluid and cerebrospinal fluid) are vital aids to the clinical diagnosis and management of disease. Most enzyme-catalyzed reactions occur within living cells, however, when an energy imbalance occurs in the cells because of exposure to infective agents, bacterial toxins, etc, enzymes 'leak' through the membranes into the circulatory system. This causes their fluid level to be raised above the normal cell level. Estimation of the type, extent and duration of these raised enzyme activities can then furnish information on the identity of the damaged cell and indicate the extent of injury. Enzyme assays can make an important contribution to the diagnosis of diseases, as a minute change in enzyme concentration can easily be measured. Determination of the changes in enzyme level thus offers a greater degree of organ and disease differentiation in comparison to other possible clinico-chemical parameters, e.g. albumin or gamma globulin. Currently, the diagnostic specificity of enzyme tests is such that they are limited primarily to confirming diagnosis, offering data to be weighed alonside other clinical reports, owing to lack of disease specific enzymes. Table 1.7 includes a number of diagnostically important enzymes which are most often examined in clinic laboratories [29, 30].

1.9.1.1 Enzyme examinations in diseases of the liver and biliary

The diseases of the liver and gastrointestinal tract were among the first to which serum enzyme tests were applied. They have proved to be most effective owing to the large size of the organs and the wide range and abundance of enzymes [32–36]. The liver-based enzymes GOT, GPT and AP are examined to evaluate the site and nature of liver disease. LD, GGT, OCT and CHE are also examined. Several enzymes employed in the diagnosis of liver diseases along with their respective levels are listed in table 1.8.

Table 1.7. Diagnostically significant enzymes.

Enzyme and abbreviation	Tissue source[1]	Reaction
γ-glutamyl transferase GGT	K L	γ-glutamyl peptide to γ-glutamylamino acid
Ornithine carbamoyltransferase	L	Carbamoyl-P to citrulline
OCT triacylglycerol lipase	Pa	Triacylglycerol to diacylglycerol and fatty acid
Lactate dehydrogenase LD	H L M K	Lactate to pyruvate
Isocitrate dehydrogenase ICD	L	Isocitrate to oxoglutarate
Hydroxybutyrate dehydrogenase HBD (LD I)	H	2-hydroxybutyrate to 2-oxybutyrate
Fructose-biphosphate aldolase ALD	M H	Fructose-1,6-biphosphate to triosephosphate
Creatine lipase CPK	M H B	Creatine to creatine phosphate
Chymotrypsin CT	Pa	Proteins to polypeptides
Cholinesterase CHE	L	Acylcholine to fatty acid and choline
Aspartate aminotransferase GOT (AST)	H L M K B	Aspartate to glutamate
Alkaline phosphatase AP	B I L Pl K	Phosphate monoester to alcohol and Pi (pH 8–10)
Alanine aminotransferase GPT (AAT)	L	Alanine to gultamate
Acid phosphatase SP	Pr E	Phosphate monoester to alcohol and Pi (pH 8–10)
Acetylcholinesterase ACHE	B E	Acetylcholine to acetate and choline
α-amylase	Pa S	Starch to maltose
5′-nucleosidase 5.N	Ht Pa	5′-ribonucleotide to ribonucleoside

[1] B, brain; E, erythrocytes; H, heart muscle; Ht, hepatobiliary tract; I, intestinal mucosa; K, kidney; L, M, skeletal muscle; Pa, pancreas; P1, placenta; Pr, prostate gland; S, saliva.

Table 1.8. Liver diseases and enzymes used in diagnosis [32–36].

Disease	Enzyme used	Enzyme level
Solvent poisoning of liver	GOT, GPT and LD	GOT:GPT:LD 6500:3000:10 000 (U ml^{-1})
Hepatobiliary disease (obstructive jaundice)	GOT and GPT	5–10 times normal level
Fatty liver	GPT	2 times normal level
Chronic hepatitis and cirrhosis	All liver transaminases	3–12 times normal level and inflammation of the liver
Acute hepatitis	GOT and GPT	20–50 times normal level

1.9.1.2 Enzyme applications in heart disease

According to previous reports, no single enzyme has yet been reported to cure myocardial damage. The discovery of serum glutamine oxalacetic acid transaminase determination (GOT) in 1954 was considered a significant step forward in the diagnosis of acute myocardial infarction. A mixture of results from assays of CPK (creatine phosphokinase), HBD (α-hydroxybutyrate dehydrogenase) and GOT (glutamine oxalacetic acid transaminase)—each of which has been shown to be elevated in more than 90% of cases—is used for diagnostic purposes [37–39]. The level of CPK starts rising three to four hours after the initial onset of pain, followed in order by GOT and AST (HBD) which appear after approximately eight hours. The maximum levels are reached in the same sequence, CPK after 24 h, LD 1 after 36 h and AST after about two days. The rise in enzyme levels is fairly moderate, AST and CPK increase by four to ten times their respective normal levels and LD 1 is approximately five-fold higher than normal. An enzyme known as hyaluronidase (hyaluronate hydrolysis) has been reported to cure heart attack [38]. The activity of many enzymes including aldolase, malic dehydrogenase, isomerase and ICD may increase following myocardial infarction [38].

1.9.1.3 Diagnosis of muscle disease

Skeletal muscle disorders include diseases of the muscle fibers (myopathies) or of the muscle nerves (neurogenic disorders) [40]. In myopathies CPJ, LD, ALD, GOT and GPT levels are raised. In the case of neurogenic diseases and hereditary diseases, CPK is occasionally raised (2–3 fold) [40]. Damage to the muscle may be due to extensive muscular exercise, drugs, physical trauma, inflammatory diseases, microbial infection or metabolic dysfunction, or it may be genetically predisposed. In muscular disorders the level of CPK is elevated in serum with the highest frequency and is assayed in the diagnosis of these disorders. An additional useful assayed enzyme is acetyl cholinesterase (AChE), which is significant in regulating certain nerve impulses [41]. Various pesticides affect this enzyme, so farm labors are frequently tested to be sure that they have not received accidental exposure to significant agricultural toxins. There are number of enzymes that are characteristically used in the clinical laboratory to diagnose diseases. There are highly specific markers for enzymes active in the pancreas, red blood cells, liver, heart, brain, prostate gland and many of the endocrine glands [41]. From the time when these enzymes became comparatively easy to examine using automated techniques, they have been part of the standard blood tests that veterinarians and medical doctors are likely to need in the diagnosis and treatment/management of diseases.

1.9.2 Enzymes in therapeutics

Enzymes have two significant features that differentiate them from all other types of drugs. First, enzymes frequently bind and act on their targeted sites with high affinity and specificity. Second, enzymes are catalytic and convert numerous target molecules to the desired products. These two important features make enzymes specific and potent drugs that can achieve therapeutic biochemistry in the body that small molecules cannot. These features have resulted in the development of many enzyme-based drugs for a wide range of disorders [42]. Currently, numerous enzymes are used as therapeutic agents, owing to the following features:

- High specificity to their substrates.
- Proficient in producing the desired effect without provoking any side effects.

- Water soluble.
- Extremely effective in a biological environment.

Enzymes as therapeutic agents also have some serious disadvantages which restrict their application. Their bulky structure, due to their large molecular weight, excludes them from the intracellular domain. Owing to their high proteinaceous nature they are highly antigenic and are rapidly cleared from blood plasma. Extensive purification from pyrogens and toxins is essential for parenteral enzymes, which increases the cost. Table 1.9 lists some therapeutically important enzymes.

1.9.2.1 Enzyme therapy of cancer

In traditional medicine, proteolytic enzymes derived from plant extracts have been used for a long time In addition to proteolytic enzymes from natural resources such as plants, 'modern' enzyme therapy includes pancreatic enzymes. Therapeutically,

Table 1.9. Therapeutically important enzymes.

Enzyme preparation	Source	Therapeutic application
Aspargenase	*Escherichia coli*, guinea pig serum	Cytotoxic agents
Bromelain	*Ananas comosus*	Inflammation, edema
Chymotrypsin	Bovine pancreas	Inflammation edema ophthalmology and upper respiratory tract diseases
Deoxyribonuclease (DNA hydrolysis)	Bovine pancreas	Reduces viscosity of pulmonary secretions
Dextranase (dextran hydrolysis)	*Penicillium funiculosum*	Dental plaque restriction
Diastase (starch hydrolysis)	Malt	Amylaceous dyspepsia
Galactosidase (lactose hydrolysis)	*Aspergillus niger*	Inherited β-galactosidase deficiency
Hyaluronidase (mucopolysaccharide hydrolysis)	Animal testes	Increase absorption rate, increase effectiveness of local anesthetics
Pancreatin	Animal pancreas	Pancreatitis
Papain (protease)	*Carica papaya*	Dyspepsia and gastritis
Penicillinase	*Bacillus cereus*	Penicillin allergy
Plasmin (protease)	Plasminogen	Thrombotic disorders anticoagulation
Streptodornase (DNAase)	Streptococci	Depolymerization of DNA in purulent exudates
Streptokinase (protease)	Streptococci	Thromboemolic diseases
Tissue plasminogen activator (protease)	Recombinant DNA technology	Thromboeniolic diseases
Trypsin (protease)	Animal pancreas	Cleaning necrotic tissue
Urokinase (protease)	Human urine	Thromboemolic diseases

the use of proteolytic enzymes is partly based on scientific reports and is partly empirical [43]. Clinical evidence of the use of proteolytic enzymes in cancer studies has typically been obtained with an enzyme preparation comprising a combination of papain, trypsin and chymotrypsin. Earlier reports proved that enzyme therapy can reduce the adverse effects caused by radiotherapy and chemotherapy. There is also a report available that, in some types of tumors, survival may be sustained. The positive effects of systemic enzyme therapy appear to be based on its anti-inflammatory potential. Nevertheless, the exact mechanism of action of systemic enzyme therapy remains unsolved. The proportion of proteinases to antiproteinases, which is regularly used as a prognostic marker in cancer studies, is likely to be influenced by the oral administration of proteolytic enzymes, most likely via induction of the synthesis of antiproteinases. In addition, there are many alterations of cytokine composition during treatment with orally administered enzymes, which might be a sign of the efficacy of enzyme therapy [44].

Proteases and their inhibitors have long been studied in several tumor systems. However, out of numerous promising serine and metalloproteinase inhibitors, not a single one is included in oncology at present. The present exploration for active antiproteolytic agents is in contrast to the traditional approach, as evidenced by John Beard, who proposed the management of advanced cancer using fresh pancreatic extracts whose antitumor activity was based on their proteolytic potential.

The enzymatic treatment of tumors is based on the idea of denying the abnormal cells their essential metabolic precursors such as amino acids, nucleic acids and folates. A number of enzymes have been examined and evidenced as antitumor agents. L-serine dehydratase, L-arginase, carboxypeptidase G (folate depletion), L-asparaginase, L-methioninase, L-phenylalanine ammonia lyase, L-glutaminase, L-tyrosinase and xanthine oxidase have been studied for their anticancer activity. Enzyme preparations such as asparaginase (amidase), bromelain (protease) and chymotrypsin (protease) have also been studied as cancer treatments (table 1.9).

L-asparaginase is the most widely investigated enzyme. It has been reported in treatment against three neoplastic diseases, acute lymphoblastic leukemia, leukemic lymphosarcoma and myeloblastic leukemia. It deprives the cancerous cells of their nutritional asparagine supply. Asparagine is essential for protein synthesis, which takes place inside the cell, and decreased protein synthesis perhaps accounts for the immunosuppression and toxic effects of asparaginase-based treatment.

$$\text{L-Asparagine} + \text{PPi} + \text{AMP} \xrightleftharpoons{\text{Aspargenase}} \text{L-Aspartate} + \text{NH3} + \text{ATP}$$

The prospects of enzyme-based treatment against cancer are very bright, but the difficulties of antigenicity and short circulation time remain to be overcome.

1.9.2.2 Enzymes in thrombolytic treatment

Activation of the blood clotting mechanism during inflammation is part of the body's defense mechanism which requires therapeutic intervention. Under normal physiological conditions there is an equilibrium between blood coagulation (clotting) and fibrinolysis (the process of dissolving the clotted blood) [47]. Biocatalysts such as enzymes, ribozymes, pro-enzymes, activators and pro-activators are responsible for maintaining equilibrium between clot formation and fibrinolysis. Imbalances in the concentration of these bio-activators may disturb physiology. In the biological process of fibrogenesis, clot formation takes place due to the plasma protein (soluble fibrinogen), which is ultimately converted to insoluble fibrin by the enzyme thrombin. This process is dependent on the conversion of thrombin from prothrombin. This bio-conversion takes place after the cascade of enzymatic reactions which involved certain key biological compounds called clotting factors. A blood clot dissolving enzyme known as plasmin is present in the blood as the pro-enzyme plasminogen. During clot dissolution activators convert the plasminogen to plasmin. This biological process is well regulated by certain process such as vasoconstriction, formation of a fibrin and clot platelet aggregation [46].

As the body utilizes enzymes in conserving this key balance of homeostasis, in a similar way we can utilize enzymes to repair or restore the homeostatic balance once it is lost. Several reports have shown that one of the best approaches for treating such clinical conditions is the administration of enzymes capable of converting plasminogen to plasmin (the enzyme which dissolves the clot) via intraveneous injection. This type of treatment is called therapeutic thrombolysis or thrombolytic therapy. In this treatment, pharmacological agents are used to medically induce clot breakdown [47]. Various novel thrombolytic agents have been derived from different sources for therapeutic use, such as from bacteria (streptokinase), the venom of the Malayan pit viper (Arvin), a filamentous fungus Koji mold *Aspergillus oryzae* (brinase), a South American snake (reptilase) and human urine (urokinase) [47].

Current advancements in thrombolytic therapy are more focused on the treatment of occlusions (blockages) of blood vessels. These types of therapy can be considered as life-saving and emergency medicine for life-threatening conditions such as myocardial infarction and massive pulmonary embolism, which are the most common reasons for cardiac arrest. This life-saving treatment is more reliable in preventing the blockages of vessels in the lungs and heart. Artery blockage conditions such as pulmonary embolism in the lungs by the formation of a clot creates tension on the right side of the heart, resulting in shortness of breath and chest pain mainly upon breathing in. Enzyme-based thrombolysis for treating massive pulmonary embolism has been considered as an effective approach to dissolving clots in these large vessels. Since surgical removal raises the chances of new blood clot formation that can cause another pulmonary embolism at the same or a different site, it is considered a dangerous practice and thrombolytic therapy is considered the more effective treatment [47]. Nevertheless, reoccurrence of clot formation or clot re-formation is very common in patients who have undergone enzyme-based thrombolytic treatment. Researchers from various organizations (1971) determined the effectiveness of streptokinase over heparin in reducing the chances of death in acute myocardial infarction patients. Significant results were obtained during this experiment. As discussed above, re-formation of the clot is one of the major concerns in fibrinolytic therapy. Most clinicians start treatment with a high dose of fibrinolytic agents, which is reduced later on. This approach may reduce disease progression for some time, but often increases the chances of clot re-formation. Even after the dissolution of the clot it is very difficult to maintain the same physiologically balanced environment (homeostasis) at the site of damaged tissues and the chance of new clot formation at that particular location is very high. Therefore, fibrinolytic based treatment is always accompanied by anticoagulants, such as heparin [46].

Major concerns associated with streptokinase therapy are fever, a tendency for bleeding, antigenicity (as with any foreign protein) and the difficulty of determining the proper dose [47]. Post-enzymatic treatment bleeding is one of the major concerns and it is also a concern when anticoagulants are used alone. According to current research, urokinase (produced in the kidneys and obtained from human urine) is considered safer than streptokinase. For the production of urokinase, 2300 l of urine is required to yield only 29 mg of purified urokinase, thus considering the expense involved in its manufacture, its clinical utilization has been restricted. Other examples are Arvin and reptilase. Utilization of these has been restricted for several reasons, but they are still considered as potential replacements for heparin as anticoagulants. Some researchers have noticed that optimum dose plays an important role and is one of the key factors in determining re-clot formation. Thorough investigation is required to overcome any shortcomings and increase the acceptance of these enzymes in therapeutic use [47].

1.9.2.3 The role of enzymes in digestive disorders and inflammations

Enzymes play an essential role in the management of various digestive disorders, such as exocrine pancreatic insufficiency [48]. Supplementation with enzymes may also be advantageous for other conditions associated with poor digestion, such as lactose intolerance. Generally, pancreatic enzymes such as porcine and bovine have been the preferred form of supplementation for exocrine pancreatic insufficiency [48]. Utilization of microbe-derived lipase has presented promise with reports showing benefits alike to pancreatic enzymes, but with a lower dosage concentration and a broader pH range. The safety and efficacy of enzymes derived from microbial species in the treatment of conditions such as malabsorption and lactose intolerance is promising. Plant-derived enzymes, e.g. bromelain from pineapple, serve as active digestive aids in the breakdown of proteins. Synergistic properties have also been reported using a combination of animal-based enzymes and microbe-derived enzymes or bromelain. Buccal administration of pancreatin (derived from an alcoholic extract of animal pancreas) enhances the enzymatic digestion of starch and proteins in patients with pancreatic cysts and pancreatitis. Pancreatin in combination with lipase is used to treat patients with fatty stools. Hydrolytic enzymes such as papain and fungal extracts (*Aspergillus niger* and *Aspergillus otyzae*) are used to enhance absorption from the small intestine [49]. These fungal extracts comprise amylases and proteases along with cellulases, which support the breakdown of the otherwise indigestible fibers of cabbages, etc, and thus reduce dyspepsia and flatulence [50]. Currently, micro-organisms are used at a large scale for the production of therapeutic enzymes. Among various micro-organisms *Saccharomyces cerevisiae, Saccharomyces fragilis, Bacillus subtilis* and two *Aspergillus* species are considered safe by the FDA (USA) for obtaining oral β-galactosidase (from *A. oryzae*) which is often used by patients suffering from inherited intestinal disease lactose deficiency [51]. Children with this genetic disorder children are incapable of digesting milk lactose. Enzymatic preparations such as β-galactosidase catalyze the conversion of lactose to glucose and galactose, which are quickly absorbed by the intestine. Other enzymatic preparations, e.g. penicillinase (from *B. subtilis*) are often used to treat hypersensitivity reactions caused by the antibiotic penicillin [52]. This enzyme catalyzes the conversion of penicillin to penicillanic acid, which is non-immunogenic. In addition, microbial and plant hydrolases are also used to decrease inflammation and edema [53]. Thrombin, trypsin, chymotrypsin, papain, streptokinase, streptodornase and sempeptidase are under clinical trial investigation. These enzymatic preparations are administered orally and have considerable proteolytic activity in the serum. Streptodornase has also displayed pain-relieving action on systemic injection [54]. Preparations have also been used to clean dirty wounds and necrotic tissue and to remove debris from second and third degree burns.

1.10 Plants and algae enzyme systems

Plant based foods are usually consumed in their raw form [68]. This eases the main concern with animal-based enzymes by preserving the integrity of the enzymes themselves. Moreover, plant-based digestive enzymes are effective over a broad scope of pH levels. This range is usually between 3.0 and 9.0, which is highly well-matched with the human gastrointestinal environment [55–72]. Thus plant-based enzymes are compatible for supporting comprehensive digestive health. Protease, amylase, lipase and cellulose are the important enzymes and are present in plants. Protease breaks down protein that can be present in meat, fish, poultry, eggs, cheese and nuts. Amylase assists your body with the breakdown and subsequent absorption of carbohydrates and starches. Lipase aids the digestion of fat. When your diet includes lipase-rich foods, it eases the production burden on the gall bladder, liver and pancreas. *Cellulase* is present in many fruits and vegetables, and it breaks down food fibers, which increases their nutritional value to our bodies. The presence of *cellulase* in plant-based sources is important, because it is not naturally present in the human body. Fruits and vegetables are an ideal source for enzymes. They are enzyme-rich and easily consumed without needing to be cooked or processed, ultimately preserving the full functionality of the enzymes. By using plant biotechnology several enzymes can be produced from plants as well algal resources [56–72].

During algal photosynthesis various proteins and enzymes are produced which can be utilized in economic development and environment management, such as in wastewater treatment, production of fine chemicals, and biodiesel production [56–72]. Due to their potential to capture and fix carbon dioxide using solar energy, photosynthetic marine algae are considered as potential models for the production of proteins. It has been recently observed that algal chloroplasts can be transformed for the production recombinant proteins [55]. Five different classes of recombinant enzymes; *xylanase, α-galactosidase, phytase, phosphate anhydrolase,* and β-

mannanase, D. tertiolecta or C. reinhardtii were in the plastids of *D. tertiolecta* or *C. reinhardtii.* Similar strategies should allow for recombinant protein production in many species of marine algae [55].

References

[1] Franzén R and Tois J 2003 Purine and sugar chemistry on solid phase—100 years after the Emil Fischer's chemistry Nobel Prize 1902 *Comb. Chem. High Throughput Screen.* **6** 433–44

[2] Simoni R D, Hill R H and Vaughan M 2002 Urease, the first crystalline enzyme and the proof that enzymes are proteins: the work of James B Sumner *J. Biol. Chem.* **277** 23e

[3] Northrop J H 1930 Crystalline pepsin: isolation and tests of purity *J. Gen. Physiol.* **13** 739–66

[4] Berg J M, Tymoczko J L and Stryer L 2002 *Biochemistry* 5th edn (New York: W H Freeman)

[5] Percudani R and Peracchi A 2003 A genomic overview of pyridoxal-phosphate-dependent enzymes *EMBO Rep.* **4** 850–4

[6] Fischer J D, Holliday G L and Janet M 2010 Thornton. The cofactor database: organic cofactors in enzyme catalysis *Bioinformatics* **26** 2496–7

[7] Agarwal P K 2006 Enzymes: an integrated view of structure, dynamics and function *Microb. Cell Fact.* **5** 2

[8] Lee H C 2006 Structure and enzymatic functions of human CD38 *Mol. Med.* **12** 317–23

[9] Rodriguez R, Menendez-Arias L, Gonzalez de Buitrago G and Gavilanes J G 1985 Amino acid sequence of pigeon egg-white lysozyme *Biochem. Int.* **11** 841–3

[10] Canfield R E 1963 The amino acid sequence of egg white lysozyme *J. Biol. Chem.* **238** 2698–707

[11] Berg J M, Tymoczko J L and Stryer L 2002 Many enzymes are activated by specific proteolytic cleavage *Biochemistry* 5th edn (New York: W H Freeman) section 10.5

[12] Lemieux R U and Spohr U 1994 How Emil Fischer was led to the lock and key concept for enzyme specificity *Adv. Carbohydr. Chem. Biochem.* **50** 1–20

[13] Csermely P, Palotai R and Nussinov R 2010 Induced fit, conformational selection and independent dynamic segments: an extended view of binding events *Trends Biochem. Sci.* **35** 539–46

[14] Koshland D E 1995 The key–lock theory and the induced fit theory *Angew. Chem. Int. Ed. Engl.* **33** 2375–8

[15] Boehr D D, Nussinov R and Wright P E 2009 The role of dynamic conformational ensembles in biomolecular recognition *Nat. Chem. Biol.* **5** 789–96

[16] Johnson K A and Goody R S 1913 The original Michaelis constant: translation of the Michaelis–Menten paper *Biochemistry* **50** 8264–9

[17] Cooper G M 2000 The central role of enzymes as biological catalysts *The Cell: A Molecular Approach* 2nd edn (Sunderland, MA: Sinauer Associates)

[18] West E S, Todd W R, Mason H S and Van Bruggen J T 1966 *Textbook of Biochemistry* 4th edn (Delhi: Oxford and IBH), 477

[19] Gutfreund H and Sturtevant J M 1956 The mechanism of chymotrypsin-catalyzed reactions *Proc. Natl Acad. Sci. USA* **42** 719–28

[20] Ma W, Tang C and Lai L 2005 Specificity of trypsin and chymotrypsin: loop-motion-controlled dynamic correlation as a determinant *Biophys. J.* **89** 1183–93

[21] Fink A L 1976 Cryoenzymology of chymotrypsin: the detection of intermediates in the catalysis of a specific anilide substrate *Biochemistry* **15** 1580–6

[22] Brandt K G, Himoe A and Hess G P 1967 Investigations of the chymotrypsin-catalyzed hydrolysis of specific substrates. III. Determination of individual rate constants and enzyme-substrate binding constants for specific amide and ester substrates *J. Biol. Chem.* **242** 3973–82

[23] Kezdy F J and Bender M L 1962 The kinetics of the α-chymotrypsin-catalyzed hydrolysis of p-nitrophenyl acetate *Biochemistry* **1** 1097–106

[24] Neurath H and Walsh K A 1976 Role of proteolytic enzymes in biological regulation (a review) *Proc. Natl Acad. Sci. USA* **73** 3825–32

[25] Ulusu N N 2015 Evolution of enzyme kinetic mechanisms *J. Mol. Evol.* **80** 251–7

[26] Nakano S, Chadalavada D M and Bevilacqua P C 2000 General acid–base catalysis in the mechanism of a hepatitis delta virus ribozyme *Science* **287** 1493–7

[27] Vallee B L and Williams R J 1968 Metalloenzymes: the entatic nature of their active sites *Proc. Natl Acad. Sci. USA* **59** 498–505

[28] Shook R L and Borovik A S 2010 The role of the secondary coordination sphere in metal-mediated dioxygen activation *Inorg. Chem.* **49** 3646–60

[29] Bodansky O 1959 Diagnostic applications of enzymes in medicine. General enzymological aspects *Am. J. Med.* **27** 861–74

[30] Voller A, Bidwell D E and Bartlett A 1976 Enzyme immunoassays in diagnostic medicine: theory and practice *Bull. World Health Organ.* **53** 55–65

[31] Robinson P K 2015 Enzymes: principles and biotechnological applications *Essays Biochem.* **59** 1–41

[32] Wolf P L 1999 Biochemical diagnosis of liver disease *Indian J. Clin. Biochem.* **14** 59–90

[33] Ellis T G, Goldberg D M, Spooner R J and Ward M 1978 Serum enzyme tests in diseases of the liver and biliary *Am. Soc. Clin. Pathol.* **70** 248–58

[34] Giannini E G, Testa R and Savarino V 2005 Liver enzyme alteration: a guide for clinicians *CMAJ* **172** 367–79

[35] Wieme R J and Demeulenaere L 1970 Enzyme assays in liver disease *J. Clin. Pathol. Suppl. (Assoc. Clin. Pathol.)* **4** 51–9

[36] Schmidt F W 1967 Enzymes in diagnosis of liver disease *Proc. R. Soc. Med.* **60** 1247

[37] Agress C M 1965 Ischemic heart disease: use of serum enzyme determinations in diagnosis *Calif. Med.* **103** 21–6

[38] Chowdhury B, Xiang B, Liu M, Hemming R, Dolinsky V W and Triggs-Raine B 2017 Hyaluronidase 2 deficiency causes increased Mesenchymal cells, congenital heart defects, and heart failure *Circ. Cardiovasc. Genet.* **10** e001598

[39] Nissen N I, Ranl P and Weis-fogh J 1965 Evaluation of four different serum enzymes in the diagnosis of acute myocardial infarction *Br. Heart J.* **27** 520

[40] Sobel B E and Shell W E 1972 Serum enzyme determinations in the diagnosis and assessment of myocardial infarction *Circulation* **45** 471–82

[41] Plebani M 2001 Enzymes and muscle diseases *Reumatismo* **53** 158–65

[42] Vellard M 2003 The enzyme as drug: application of enzymes as pharmaceuticals *Curr. Opin. Biotechnol.* **14** 444–50

[43] Gurung N, Ray S, Bose S and Rai V 2013 A broader view: microbial enzymes and their relevance in industries, medicine, and beyond *Biomed. Res. Int.* **2013** 329121

[44] Leipner J and Saller R 2000 Systemic enzyme therapy in oncology: effect and mode of action *Drugs* **59** 769–80

[45] Novak J F and Trnka F 2005 Proenzyme therapy of cancer *Anticancer Res.* **25** 1157–77

[46] Collen D and Lijnen H R 2005 Thrombolytic agents *Thromb. Haemost.* **93** 627–30

[47] Cichoke A 1990 Enzymes in thrombolytic therapy *Dynamic Chiropractic* **08** 17

[48] Roxas M 2008 The role of enzyme supplementation in digestive disorders *Altern. Med. Rev.* **13** 307–14

[49] Rao M B, Tanksale A M, Ghatge M S and Deshpande V V 1998 Molecular and biotechnological aspects of microbial proteases *Microbiol. Mol. Biol. Rev.* **62** 597–635

[50] Fieker A, Philpott J and Armand M 2011 Enzyme replacement therapy for pancreatic insufficiency: present and future *Clin. Exp. Gastroenterol.* **4** 55–73

[51] Montalto M *et al* 2006 Management and treatment of lactose malabsorption *World J. Gastroenterol.* **12** 187–91

[52] Blumberg P M and Strominger J L 1974 Interaction of penicillin with the bacterial cell: penicillin-binding proteins and penicillin-sensitive enzymes *Bacteriol. Rev.* **38** 291–335

[53] Emanuela R and FitzGerald G A 2011 Prostaglandins and inflammation *Arterioscler Thromb. Vasc. Biol.* **31** 986–1000

[54] Miller R R, De Young D V and Paxinos J 1970 Enzymes for trauma *Postgrad. Med. J.* **46** 154–5

[55] Georgianna D R *et al* 2013 Production of recombinant enzymes in the marine alga Dunaliella tertiolecta *Algal Res.* **2** 2–9

[56] Bhatia S, Rathee P, Sharma K, Chaugule B B, Kar N and Bera T 2013 Immuno-modulation effect of sulphated polysaccharide (porphyran) from *Porphyra vietnamensis Int. J. Biol. Macromol.* **57** 50–6

[57] Bhatia S, Kumar V, Sharma K, Nagpal K and Bera T 2014 Significance of algal polymer in designing Amphotericin B nanoparticles *Sci. World J.* **2014** 1

[58] Bhatia S, Sharma K, Nagpal K and Bera T 2015 Investigation of the factors influencing the molecular weight of porphyran and its associated antifungal activity *Bioact. Carb. Diet. Fiber* **5** 153–68

[59] Bhatia S, Sharma K, Nagpal K, Sharma S and Bera T 2015 Evaluation of pharmacognostical, phytochemical and anti-microbial properties of *Porphyra vietnamensis Int. J. Green Pharm.* **2** 1–5

[60] Bhatia S, Sharma K and Bera T 2015 Structural characterization and pharmaceutical properties of porphyran *Asian J. Pharm.* **9** 1–3

[61] Bhatia S, Sharma K, Sharma A, Nagpal K and Bera T 2015 Anti-inflammatory, analgesic and antiulcer properties of *Porphyra vietnamensis Avicenna J. Phytomed.* **5** 1–7

[62] Bhatia S *et al* 2009 Novel algal polysaccharides from marine source: Porphyran *Pharmacogn. Rev.* **2** 271–6

[63] Bhatia S *et al* 2011 Mycosporine and mycosporine-like amino acids: a paramount tool against ultra violet irradiation *Pharmacogn. Rev.* **5** 138–46

[64] Bhatia S *et al* 2010 Broad-spectrum sun-protective action of Porphyra-334 derived from *Porphyra vietnamensis. Pharmacogn. Res.* **2** 45–9

[65] Bhatia S, Namdeo A G and Nanda S 2015 Factors affecting the gelling and emulsifying property of natural polymer *Sys. Pharm. Rev.* **9** 93–101

[66] Bhatia S, Sharma K, Sharma A, Namdeo A G and Chaugule B B 2011 Anti-oxidant potential of Indian porphyra *Pharmacologyonline* **1** 248–57

[67] Bhatia S *et al* 2015 *Modern Applications of Plant Biotechnology in Pharmaceutical Sciences* (Amsterdam: Elsevier), pp 277–87
[68] Bhatia S 2016 *Nanotechnology in Drug Delivery: Fundamentals, Design, and Applications* (Boca Raton, FL: CRC press), pp 121–7
[69] Bhatia S and Goli D 2016 *Leishmaniasis: Biology, Control and New Approaches for Its Treatment* (Boca Raton, FL: CRC press), pp 177–9
[70] Bhatia S 2016 *Natural Polymer Drug Delivery Systems: Nanoparticles, Plants, and Algae* (Berlin: Springer Nature), pp 117–27
[71] Bhatia S 2016 *Systems for Drug Delivery: Safety, Animal, and Microbial Polysaccharides* (Berlin: Springer Nature), pp 122–7
[72] Bhatia S 2018 *Introduction to Pharmaceutical Biotechnology* Volume 1 (Bristol: IOP Publishing), pp 178–87

IOP Publishing

Introduction to Pharmaceutical Biotechnology, Volume 2
Enzymes, proteins and bioinformatics
Saurabh Bhatia

Chapter 2

Technologies and procedures involved in enzyme production

2.1 Introduction

Enzyme production has become a significant area in the modern biotechnology industry. With advancements in traditional enzyme production methodologies, different industrial sectors can now easily exploit new applications offering novel activities and/or improved stability. Enzymes have been used by human beings since ancient times, either in the form of vegetables rich in enzymes, or in the form of microorganisms. These microorganisms have been used for many purposes, such as in brewing, and the baking industry. In the late nineteenth century numerous biological conversions were attributed to the action of enzymes. Isolation was first introduced by Payen and Persoz in 1833 [1]. They isolated an enzyme complex from malt, which they named 'diastase', that converts gelatinized starch into sugars, primarily maltose. In 1874, the history of modern enzyme production really began when Christian Hansen first synthesized rennet by isolating it from dried calves' stomachs with saline solution [2]. This was the first enzymatic preparation of relatively high purity used for commercial purposes. The Christian Hansen Company is still developing rennet today, in an almost identical way. In 1896, the utilization of the mold fungus koji in the production of certain foodstuffs and flavor additives based on soya protein and fermented beverages was the basis on which the Japanese researcher Takamine established a fermentation process for the industrial production of fungal amylase.

This process involves culturing of *Aspergillus oryzae* on moist rice or wheat bran, and the product is called 'Takadiastase' which is currently used as a pharmaceutical agent to cure digestive disorders [3]. The market value of the industrial enzymes market was estimated as approximately $2 billion, and has increased at an average annual rate of 3–5% in the last decade. Several enzyme production based companies have emerged, such as Novozymes dominating with 45% sales, followed by Danisco which holds a 20% share of the market. Bulk enzymes, such as proteases, amylases,

doi:10.1088/978-0-7503-1302-5ch2

lipases, etc, which are required in large quantities have, however, an intrinsically low unit value, so they demand significantly lower manufacturing costs. In contrast, the therapeutics sector includes enzymes such as urokinase, which are produced in lower quantities and at greater manufacturing cost. The technologies involved in the production of enzymes unite the disciplines of microbiology, genetics, biochemistry and engineering. Requirements for new enzymes can be achieved by the development of new procedures or the development of new procedures to address the unsatisfactory performance of known enzymes. Recent developments in gene technology have had a large impact on the enzyme industry. Genetic engineering tools have allowed enzyme manufacturers to synthesize adequate amounts of any enzyme irrespective of source, whereas protein engineering facilitates alteration of the enzymes before production. This chapter offers an outline of enzyme production procedures beginning from raw materials to the finished product, and also offers an understanding of the different alternative technologies existing for different phases of production.

2.1.1 Sources of enzymes

These enzymes are used for applications in industries on commercial scales [4]. There are various applications of these enzymes, for example, softening hides by using the feces of dogs and pigeons before tanning. A German researcher named Otto Rohm reported in 1905 that extracts from animal organs (pig and cow pancrease) could be used as the source of the enzyme protase, for leather softening [4]. The utilization of enzymes (mainly proteases) for cleaning laundry began in 1915. However, this was not initially successful due to allergic reactions to impurities in the enzymes. Currently, advance techniques are available for purification and the use of enzymes in washing powders (without allergic reactions) is common. A real boom in the large-scale industrial production of enzymes from micro-organisms occurred after the 1950s. From the industrial point of view, enzymes can be derived from different sources, however, microbes act as an active source for unique biological enzymes and thus most enzymes are derived from microbial sources [4]. Enzymes can be derived from a wide range of living organisms, as shown in figure 2.1.

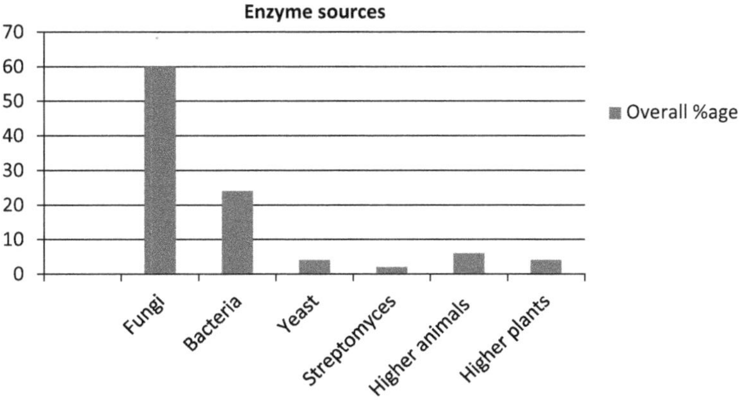

Figure 2.1. Enzyme sources.

2.1.1.1 Enzymes from animal and plant sources

A few decades ago, plants and animals were considered the major sources for enzymes. Even today, they are still major sources of certain enzymes [4], such as lipases, esterases and protease. Hen's eggs and human milk contain high amounts of lysozyme. This lysozyme with bovine lactoferrin offers innate immunity to an infant by killing gram-negative bacteria. Certain plants such as papaya (papain) and pineapple (bromelain) are excellent sources of enzymes. The following lists a few recent breakthroughs and discoveries:

- Enzymes present in herbivore gut fungi and their potential applications in fuels and chemical industry [5].
- Bacterial enzymes responsible for human infection [6].
- Gro3P phosphatase responsible for the conversion of glucose and fats into other compounds (i.e., regulate how your body converts sugar and fats) [7].
- A class of DNA repair enzymes (DNA glycosylase) [8].
- An enzyme (reductive aminase) that will make a drug used to treat Parkinson's disease [9].

Enzyme production involves many critical steps which present major challenges in the production of enzymes from different sources. Lack of suitable isolation and purification procedures, low yield and high cost of overall process are the major challenges facing enzyme-based industries. Certain enzymes are more closely associated with cell walls and face certain problems such as low concentration, insolubility and maintaining biological activity. Offering an optimal pH and isoelectric point during extraction is an important factor in determining the biological activity of these enzymes. Thus selection of the purification procedure and the correct physical and chemical conditions also helps to retain the biological activity of an enzyme. Certain enzymes are very sensitive to variation in environmental factors, and can immediately lose activity, so their catalytic activity needs to be maintained by using a continuous process to accelerate the purification process. One of the recent developments for purification is a miniaturization automated approach that not only accelerates but also optimizes and statistically analyzes the enzymes during purification. The utilization of chromatographic techniques (aided by the development of new resins) and ultrafiltration are the main approaches to enzyme purification. The contamination and toxicity of enzymes are again major challenges that can be minimized during the production procedure, e.g. enzymes derived from bovine sources involve a high risk of contamination with bovine spongiform encephalopathy (an illness caused by the ingestion of abnormal proteins called prions). These prions can be carefully heat-inactivated at a certain temperature which will not affect the enzyme properties. Due to these serious concerns, microbial sources are generally preferred over plant- and animal-based enzymes [10, 11].

2.1.1.2 Enzymes derived from mammalian cell cultures

Mammalian cell cultures offer a continuous supply of enzymes that are commercially important (table 2.1), however, a major constraint is the cost, which can be extremely high [12]. With more advancements in tissue culture, in particular in the

Table 2.1. Commercially produced enzymes from animal sources and their applications.

Enzyme(s)	Source(s)	Application(s)
Amylase, esterase	Lamb, calf	Digestive aids, preparation of cheese
Pepsin, trypsin	Bovine	
Lipase, rennin (chymosin) phospholipase, phytase	Porcine	
Lysozyme	Hen eggs	Cell wall breakage in bacteria
Human urine	Urokinase	For dissolution of blood clots

mass cultivation of cells and the availability of cheaper culture media, it is now possible to affordably produce many therapeutic enzymes such as urokinase. Further advancements in mammalian cell culture are exploring the application of cell immobilization, capillary culture, large-scale suspension culture and perfusion techniques. By using such approaches it is now possible to produce therapeutic enzymes, such as tissue plasminogen activator, at a large scale.

2.1.1.3 Microbial enzymes

Enzymes derived from microbes are referred to as microbial enzymes. Microbial enzymes were exploited for many decades without a complete understanding of how they work. Microbes are considered as the most reliable source for the production of commercial enzymes. In fact, many such enzymes are already being used on a commercial scale (table 2.2).

Since the rate of cell growth and proliferation in microbes is very high, a large amount of enzymes can be produced by culturing micro-organisms under controlled *in vitro* environments. The media and other facilities involved in culturing microbes are economical in comparison to using plants and animals [4]. Moreover, plant and animal cells are more susceptible to contamination than micro-organisms. Recombinant DNA technology can be utilized to increase the production of a desired protein in micro-organisms, which is more challenging in plant and animal cells because of their high genetic complexity as eukaryotes compared to prokaryotes. This notion is based on the sexual mode of reproduction in eukaryotes which offers more genetic variation then prokaryotes (asexual reproduction).

To satisfy industrial demands, microbial sources must allow the proper recovery, isolation and purification of enzymes. The production procedure for enzymes is described in figure 2.2. The initial step in isolation is to select the best source that allows easy production of the enzyme within the shortest amount of time. Since microbial strains can grow rapidly and can be manipulated with the help of genetic engineering, i.e., rDNA and gene editing tools, they are considered as the best sources for the production of a large variety of enzymes. Several therapeutic enzymes have been successfully produced from different microbial sources, as shown in table 2.2.

Table 2.2. Selected list of industrially (microbially) produced enzymes, their sources and applications.

Enzyme	Source(s)	Application(s)
α-amylase	*Aspergillus oryzae*, *Aspergillus niger*, *Bacillus subtilis*, *Bacillus licheniforms*	Production of beer and alcohol; preparation of glucose syrups; as a digestive aid; removal of starch sizes
Amyloglucosidase	*A. niger*, *Rhizopus niveus*	Starch hydrolysis
Cellulase	*A. niger*, *Tricoderma koningi*	Alcohol and glucose production
Glucoamylase	*A. niger*, *Bacillus amyloliquefaciens*	Production of beer and alcohol; starch hydrolysis
Glucose isomerase	*Arthrobacter* spp., *Bacillus* spp.	Manufacture of high fructose syrups
Glucose oxidase	*A. niger*	Antioxidant in prepared foods
Invertase	*Saccharomyces cerevisiae*	Sucrose inversion; preparation of artificial honey; confectionaries
Keratinase	*Streptomyces fradiae*	Removal of hair from hides
Lactase	*Kluyveromyus* spp., *Saccharomyces fragilis*	Lactose hydrolysis; removal of lactose from whey
Lipase	*Candida lipolytica*, *A. niger*	Preparation of cheese; flavor production
Pectinase	*Aspergillus* spp., *Sclerotina medina*	Clarification of fruit juices and wines; alcohol production; coffee concentration
Penicillin acylase	*Escherichia coli*	Production of 6-aminopenicillanic acid
Penicillanase	*B. subtilis*	Removal of penicillin
Protease, acid	*A. niger*	Digestive aid; substitute for calf rennet
Protease, neutral	*B. amyloliquefaciens*	Fish and meat tenderizer
Protease, alkaline	*A. oryzae*, *Streptomyces griseus*, *Bacillus* spp.	Meat tenderizer; detergent additive; beer stabilizer
Pollulanase	*Klebsiella aerogens*	Hydrolysis of starch
Takadiastase	*A. oryzae*	Bread supplement; digestive aid

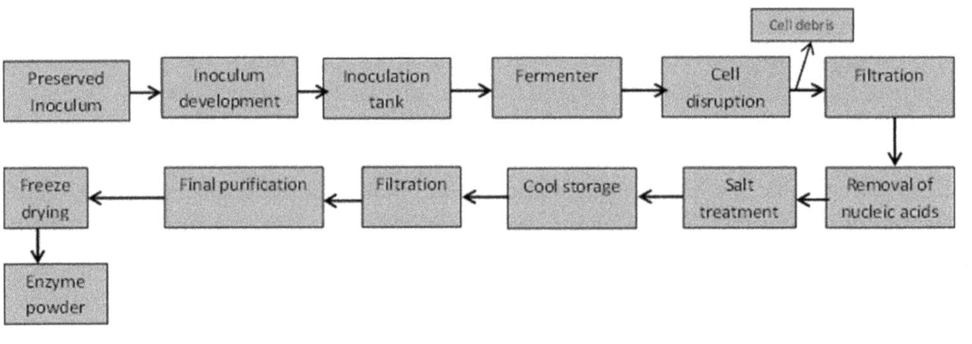

Figure 2.2. Schematic representation of the production of enzymes by micro-organisms.

2.1.1.3.1 Aspergillus niger for the production of bulk enzymes

Enzymes derived from fungal sources such as *Aspergillus*, *Rhizopus* and *Penicillium* are considered as safe. These sources all produce extracellular enzymes which can be conveniently recovered. Although various micro-organisms produce extracellular enzymes, the most frequently utilized in industrial applications is *Aspergillus niger*. *A. niger,* a filamentous fungi, produces significant amounts of extracellular enzymes and citric acid, which are widely used for industrial purposes. So far, more than 50 enzymes have been isolated from this source, including various commercial enzymes that are conveniently produced by *A. niger.* Some of the key enzymes are protease, insulinase phytase, catalase α-amylase, lipase, pectinase and cellulase [4]. This acid tolerant micro-organism has significant hydrolytic potential in the production of α-amylase, and it also prevents bacteria-mediated contamination [13].

2.2 Enzyme production technology

Generally, the procedures used for microbial production of enzymes are equivalent to the methods used for the production of other industrial products. The significant features are, briefly [4]:

- Selection of organisms.
- Formulation of medium.
- Production process.
- Recovery and purification of enzymes.

A flow chart for enzyme production by micro-organisms is depicted in figure 2.3.

2.2.1 Selection of micro-organisms

Appropriate selection of micro-organisms allows a high rate of production of enzymes. Selection can be achieved using different approaches, such as mutagens and radiation, in particular UV rays. The selection procedure for micro-organisms is described in figure 2.4. The most significant factor in selecting micro-organisms is that the organism should produce a high amount of the desired enzyme in a very short time, while the quantities of other metabolites produced should be low. After selecting the organism, strain improvement to optimize enzyme production can be achieved using suitable methods (as above, using mutagens or UV rays). After selection of the micro-organisms, inoculum can be prepared in a liquid medium.

In enzyme research, the identification of enzymes is an important goal. The classical selective microbial screening method helps in discovering enzymes in populations of micro-organisms. During this procedure organisms are selected and then individually isolated to develop pure cultures. A pure culture can be propagated to a higher level, however, only few colonies are obtained from this pure culture. Due to syntrophic benefits (interspecies interaction) it is again challenging to isolate pure colonies under standard conditions. Thus only 1% of pure culture can be derived from this classical selection method [14].

However, if pure cultures are only considered for enzyme isolation, then the maximum potential of the micro-organisms, which are capable of producing other

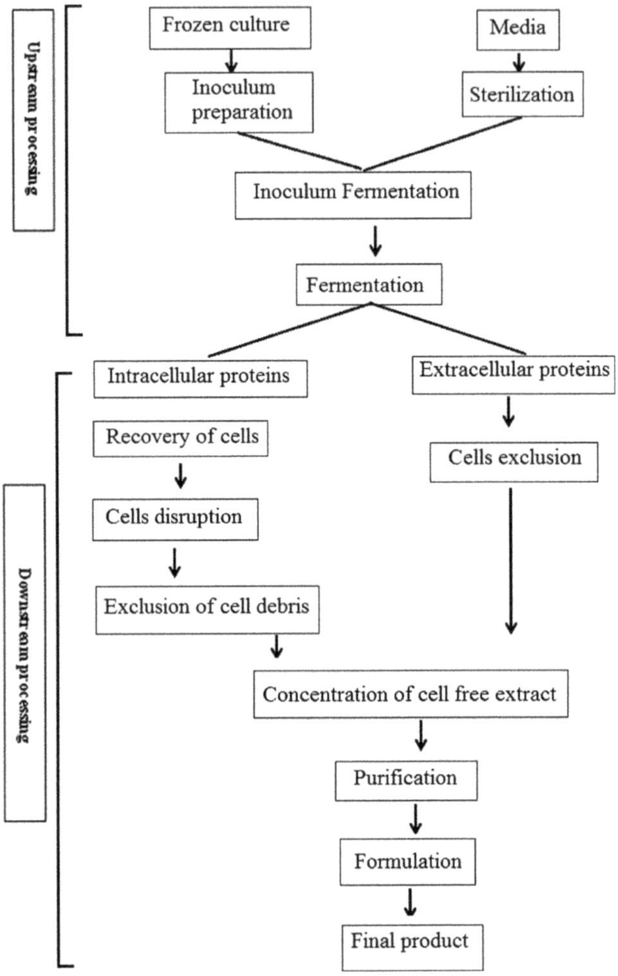

Figure 2.3. Enzyme production at an industrial scale.

novel enzymes, will be ignored. There are vast numbers of micro-organisms present in nature and it is challenging to cultivate all of them using classical microbiological techniques. The arrival of metagenomics allowed the retrieval of genetic material from the microbial environment without using any cultivation methods. At present, metagenomics are extensively used as a technique to isolate and identify enzymes with unique properties from the uncultivable part of microbial populations. These techniques allow researchers to establish the relationship between the genetic material (the genes responsible for enzyme production), the enzyme utilized (or whether an enzyme is utilized or not) and their properties. Recent developments in the field of enzyme technology allow the exploration of new approaches, such as metaproteomics. Most importantly, using metagenomics it is possible to discover the genetic potential required by a cell to produce enzymes, i.e., identification of those particular genes that are responsible for the synthesis of the desired enzyme. After

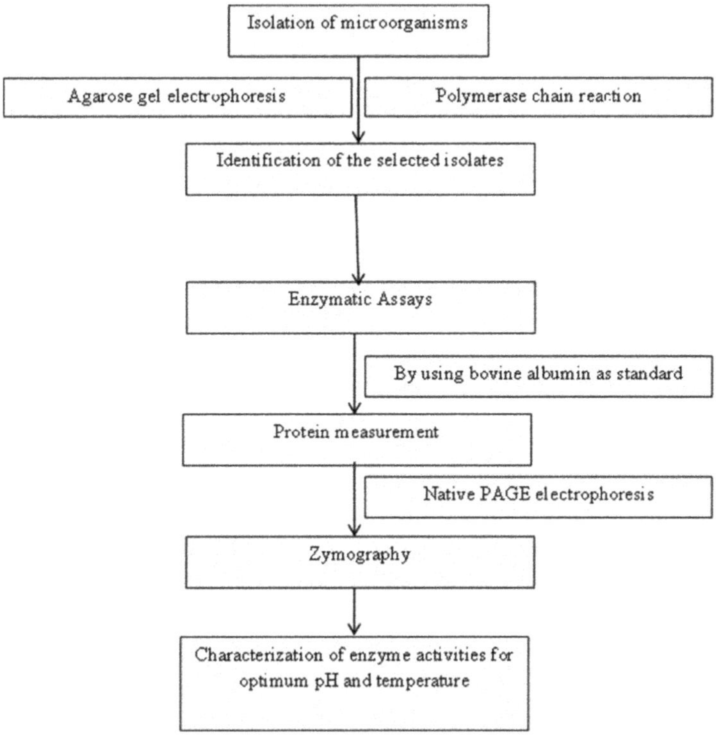

Figure 2.4. Micro-organism selection for enzyme production by the classical method.

identification of the gene, there remains the challenge to discover whether the enzyme is synthesized under a particular environment and also whether the enzyme produced will carry all the desired properties. To further improve the selection process, the study of symbiotic environments (consortiums) was suggested, by integrating various omics-techniques, such as metagenomics (shotgun sequencing methods), metatranscriptomics and metaproteomics (extensive characterization of the whole protein complement of one microbiota condition at a particular time) [15, 16]. 'At a particular time' is significant, as the protein expression at different times helps in understanding protein dynamics. Consequently, metaproteomics allows temporal interpretation studies of protein expression, in comparison to the observational studies of the static metagenome.

2.2.2 Medium selection

In the classical approach, a suitable culture medium with defined composition is selected that contains all the vital components to encourage the growth of only those micro-organisms that are capable of producing relatively large amounts of enzymes. The nutritional component of the culture medium should be available at a low price and should be safe. For the growth of anchorage-dependent micro-organisms, a number of substrates in the medium have been explored, such as yeast extract, whey, starch hydrolysate, corn steep liquor, molasses and soybean meal. In traditional approaches of fermentation, cereals (wheat) and pulses (peanut) have been utilized as substrates. Optimal pH is required for significant microbial growth and also plays an important role in enzyme production.

2.2.3 Production process

The production of enzymes is often performed at larger scales using fermentation techniques, in particular submerged fermentation (the development of micro-organisms in a liquid broth) and solid-substrate fermentation (the development of micro-organisms on a solid substrate, e.g. rice bran or wheat bran). In submerged fermentation a liquid broth is utilized to offer nutrients that result in the production of industrial enzymes, antibiotics or other products. At the industrial scale, submerged liquid fermentation procedures are classically employed for the production of enzymes from microbial sources, whereas solid-substrate fermentation is rarely used for this purpose. This is because submerged liquid fermentation offers a number of advantages, such as ease of handling and greater control of environmental factors (such as temperature and pH). However, the solid-substrate fermentation method can also be utilized to improve the yield and decrease the cost of enzyme production [17, 18].

Solid-substrate fermentation is an alternative approach to submerged fermentation which is often employed for the production of fungal enzymes such as cellulases, amylases, proteases and pectioases [19]. During this process micro-organisms are developed under a controlled environment to produce enzymes, fuel and nutrients. Sterilization is not essentially required for solid-substrate fermentation, as the fermentation substrate starts sterilization and the microbes inhibit the growth of micro-flora. It has been reported that both approaches (submerged and solid-substrate fermentation) are considered as effective techniques to produce several valuable products. Due to advancements in fermentation technology, particularly in solid-substrate fermentation, bioreactors of different sizes have now been designed. The design of bioreactors is based on the principles of biochemical engineering, including mathematical modeling, which is important in defining the cellular development of micro-organisms. Several problems, such as heat and mass transfer, can be overcome using such bioreactors.

For fermentation (if required), batch and continuous sterilization techniques can be employed to sterilize a suitable medium. Afterwards the medium is inoculated with a suitable micro-organism to initiate the fermentation process. The culture's environmental conditions, such as pH, temperature, O_2 and nutrient supplementation are initially optimized and later maintained to achieve the desired growth of micro-organisms. The addition of an antifoaming agent is required to inhibit surface tension. Production also involves fermentation kinetics, the mathematical analysis of certain factors that helps in designing or improving the batch and continuous process. It has been observed that continuous fermentation processes give a lower enzyme yield, thus enzyme production is often conducted using batch fermentation, and only rarely using the continuous process. The vessel should be sterile during the fermentation process. Aseptic conditions can be maintained by effective management of vessel design, operating procedures, continuous monitoring and maintenance, and skilled technicians. Usually, the duration of the fermentation process varies from 2–7 days. Certain additional metabolites are also produced in the medium which can interfere with recovery of the desired enzyme, however, only desired enzyme(s) should be recovered and purified.

2.2.4 Recovery and purification of enzymes

After developing the selected micro-organisms or establishing a pure culture, the recovery of suitable extracellular enzymes is again a challenge, since the culture medium can be full of unwanted end-products released by the stationary phase of the micro-organisms. These unwanted substances can interfere with the recovery of suitable enzymes. Since the recovery of extracellular enzymes is easier than that of intracellular enzymes, based on the recovery process used, commercial enzyme preparations are available in crude or highly purified form. However, other factors such as the form of medium (solid or liquid), viscosity, the nature of the enzyme and the degree of purity desired, etc, affect the recovery of enzymes. As discussed, downstream processing, which involves recovery and purification steps for extracellular enzymes, is easier compared to intracellular enzymes which require more advance techniques for cell lysis. The selection of a suitable hydrolyzing agent for intracellular enzyme recovery is a crucial step. It requires a basic understanding of cell membrane physiology and should be carried out such that the enzyme properties are not affected. Different physical, chemical and enzymatic hydrolyzing agents have been reported to break the cell wall and release their intracellular material [20]. However, the recovery of enzymes closely associated with the membrane is challenging. Cell lysis is always followed by differential centrifugation which allows the separation of cytoplasmic particles on the basis of size and specific gravity. For recovery from microbial sources, the cell membrane can be lysed by physical means, such as sonication, high pressure and glass beads. Lysozyme is often used as a hydrolyzing agent for the lysis of the bacterial cell wall. For cell wall breakdown in yeasts, the carbohydrate enzyme β-glucanase is used to break down the glycosidic bonds within beta-glucans, which allows the release and solubilization of these polymers from the cell walls. However, enzymatic procedures have the disadvantage of their high cost. Cellulose-degrading enzyme methods are often employed for intracellular enzyme recovery. The methodologies utilized for both intracellular and extracellular enzyme extraction always remain same, however, the hydrolyzing agents often differ. The objective of both procedures is to decrease the loss of the anticipated enzyme and its activity [20].

2.2.5 Cell debris removal

Procedures such as crossflow microfiltration with rapid back pulsing and differential centrifugation can be employed to remove cell debris. For thermostable enzymes, a more convenient method is to denature the unwanted proteins by heating, then cooling the solution, and finally unwanted proteins can be removed by centrifugation.

2.2.6 Nucleic acid removal

Non-proteinaceous contamination, such as nucleic acids, restricts recovery and purification of enzymes. These contaminants can be removed by precipitation processes using polyamines, streptomycin and polyethyleneimine [21]. The presence of DNA/RNA contaminants can be examined by blotting techniques and simple wavelength scan assessment of DNA/RNA.

2.2.7 Precipitation of enzymes

In a solution, salt concentration generally determines the protein solubility. During the process called salting-in, the concentration of salts is low, which can stabilize several charged groups over the protein molecule, and which can ultimately increase the solubility of the protein. In contrast, an increase in salt concentration increases the solubility to a particular point, afterwards a further increase in the salt concentration increases precipitation of proteins as water molecules are now replaced with salt concentration, so insufficient water molecules for solubilization of the protein are available. Precipitation of protein in the presence of a surplus amount of salt is called salting-out. Various salts can be utilized for this phenomenon. Ammonium sulfate is often used for this purpose as it has high solubility and is comparatively inexpensive. Ammonium sulfate reduces the solubility of proteins and enhances the precipitation, which may further increase the stability of the native conformation. Precipitation is very beneficial in downstream processing, as the precipitated enzyme can be solubilized in a minimal volume to concentrate the enzyme [22].

2.2.8 Liquid–liquid partition

The desired enzymes can be obtained by liquid–liquid chromatography using polyethylene glycol or polyamines. Liquid–liquid partition chromatography can also be utilized to identify conformational changes in well-characterized enzymes [23]. Various advanced liquid–liquid fractionation techniques have emerged, such as the aqueous two-phase system, which was initially accidently utilized by Martinus Willem Beijerinck (1896) by mixing an aqueous solution of starch and gelatin [26, 27]. The influence of organic solvents on the partitioning of enzymes in aqueous two-phase systems was studied in 1987 [25]. As per this report, partition coefficients for alkaline phosphatase decrease with ethylene glycol, glycerol, sucrose and urea. In 1996, the effects of low temperature (−18 °C) on glycolytic enzymes (e.g. phosphofructokinase) were investigated in terms of the stability and partitioning within an aqueous two-phase system [24]. It was found that the presence of ethylene glycol, phase polymers and low temperature stabilized enzyme activities.

This technique has now gained much attention because of its great potential for the extraction, separation, purification and enrichment of proteins, in particular enzymes, nucleic acids and other biomolecules. This complex partition behavior is formed by mixing a variety of components in water. As water is the main component of both phases in aqueous two-phase systems, it offers a protected environment to biomolecules for their separation and polymers to stabilize their structure and biological activities. In contrast, other liquid–liquid extraction methods containing organic solvents can cause severe damage to biological products [28].

One of the more reliable systems for enzymes is the PEG–salt aqueous two-phase systems. This system is an excellent and useful liquid–liquid extraction approach for the downstream processing of proteins and enzymes.

Several enzymes have been purified using liquid–liquid extraction. For purification using liquid–liquid partition, a novel, efficient and economical extraction method with a high purification factor, composed of surfactant and acetonitrile, has recently been developed to purify polygalacturonase from *Durio zibethinus* [29].

2.2.9 Chromatographic separation

Various chromatographic procedures for separation and purification of enzymes are available such as:

- *Ion-exchange chromatography*: A pH-dependent process which is governed by the enzyme structure and isoelectric point. The enzymes' charge becomes positive (they bind to cation exchangers) when present in a solution with pH less than the isoelectric point and becomes negative when present in a solution of pH greater than the isoelectric point (they bind to anion exchangers).
- *Size exclusion chromatography*: In this process enzymes in a solution are separated by their size, and in some cases molecular weight.
- *Hydrophobic interaction*: Proteins are not denatured in hydrophobic interaction chromatography, and the enzymes eluted from the column retain their enzymatic activity. Hydrophobic interaction is performed to separate the amylase, lipase and trypsinogen present in dog pancreatic juice through high-performance liquid chromatography [30].
- *Dye ligand chromatography*: A unique and selective purification technique in which ligands may act as substrate analogs, offering affinity interactions with their corresponding enzymes, e.g. the interaction of Cibacron blue F3GA with proteins and enzymes. Another example of synthetic affinity ligands is chlorotriazine dyes, often used as affinity-based protein purification methods for a number of reasons, such as their low cost, easy chemical immobilization of the dyes to the matrix and the fact that the final product dye-adsorbents are resistant to chemical and biological degradation.

Among all of these, ion-exchange chromatography is the most frequently used for enzyme purification. Many fractionation issues have been faced in the separation of certain enzymes, such as cellulolytic enzymes. For separation of such enzymes ion-exchange chromatography and isoelectric focusing (chromatofocusing) are considered beneficial [31].

2.2.10 Drying and packing

Drying is the most suitable method to obtain a concentrated form of an enzyme. This step is crucial in preserving the original properties and function of an enzymatic preparation. Drying can be achieved using several evaporators or dryers, e.g. film evaporators, freeze dryers (lyophilizes) and spray dryers. The dried enzyme can be packed and marketed. For some enzymes, stability can be attained by keeping them in ammonium sulfate suspensions.

Most of the enzymes used in foods or medical treatments should be of high-grade purity, and must meet the required specifications set by the regulatory bodies. These enzymes must be completely free from toxic materials, harmful micro-organisms and should not cause any allergic reactions.

Based on a recent breakthrough, the innate bacterial system for the formation of outer membrane vesicles can be used to protect the enzyme function. This type of packing prevents the degradation of the enzyme and protects its natural integrity. Thus outer membrane vesicle packing provide the enzyme with increased stability across a extensive range of storage conditions [32].

The drying of enzymes has usually been achieved by spray-drying and freeze-drying. In a recent study, two traditional methods of α-amylase drying were studied. The product derived from both procedures showed high enzymatic activity, however, spray-drying can be considered as more economical, because in freeze-drying the process duration can be considered as a limiting factor [33].

2.2.11 Regulation of microbial enzyme production

Fermentation conditions such as nutrients, pH, O_2, temperature, etc, can be optimized to obtain the maximum production of therapeutic proteins such as microbial enzymes. To achieve maximum production, a deep understanding of the genetic elements responsible for the regulation of enzyme synthesis is required. Enzyme synthesis can be repressed by regulatory proteins. These proteins actively bind with DNA and increase or inhibit the function of RNA polymerase. Thus these proteins can act as repressors or activators. The regulation of microbial enzyme synthesis and the genetic elements involved are briefly discussed here.

2.2.12 Induction

Enzyme induction is defined as an increase in the production of enzymes as a result of any stimulus, whereas repression can be defined as the decrease in enzyme production after a stimulus [34]. The process of induction is often present in bacteria and other microbes, whereas it is more rarely observed in animal metabolism, e.g. cholesterol synthesis and regulation of gluconeogenesis provide examples of enzyme induction and repression [34].

Several inducers should be explored to make the enzyme more efficiently. These inducers are available in different forms: substrate, product and intermediate. Sucrose, starch and galactosides are examples of substrates, whereas fatty acid, phenylacetate and xylobiose are products or intermediates. These inducers selectively and efficiently allow the expression of gene coding for a particular enzyme and hence exert the opposite action to an enzyme repressor. Some examples of enzymes and their respective inducers are listed in table 2.3. The high cost of these inducers and their treatment (sterilization, addition at specific time) has restricted their utilization. Thus recently several efforts have been made

Table 2.3. Selected examples of inducible enzymes along with the inducers.

Enzyme	Inducer
β-galactosidase	Galactosides
Amylase	Starch
Invertase	Sucrose
Lipase	Fatty acids
Penicillin G amidase	Phenylacetate
Xylanase	Xylobiose

to produce mutants of micro-organisms in which inducer dependence is eliminated. There are certain proteins that are responsible for mutation, and reports suggest that it is the proteins responsible for induction of the stress response pathways that allow mutation [35, 36].

2.2.13 Feedback repression

During feedback repression, repression of the synthesis of an enzyme (required at an initial stage of the pathway) takes place, e.g. the supplementation of a higher concentration of addition of such an end-product (usually a small molecule) to the culture medium leads to the inhibition of the enzymes synthesis of the specific pathway. This type of end-product addition significantly affects the enzyme production. In this method the desired organisms are cultured to achieve large-scale production of desired enzymes. Feedback inhibition takes place once the reaction end-product inhibits the enzyme that helped in their production. Thus feedback regulation by the end-product significantly affects enzyme synthesis. This takes place when the end-product accumulates in large quantities. The synthesis of feedback regulated enzymes in the laboratory is somewhat challenging. However, mutants that do not carrying feedback repression can be produced to prevent this issue. Mitchell *et al* reported feedback repression of ornithine decarboxylase synthesis mediated by an antizyme [37].

2.2.14 Nutrient repression

Overproduction of microbial metabolites is linked to the growth phases of micro-organisms. Different types of overproduction are caused by inducers, effectors, inhibitors and various signal molecules. In microbial cells, the biosynthesis of enzymes catalyzing metabolic reactions is regulated by distinct positive and negative mechanisms, such as induction, nutritional regulation (carbon or nitrogen source regulation), feedback regulation, etc [38]. The native metabolism of the micro-organism is so developed that no production of unnecessary enzymes occurs. In other words, the micro-organisms do not produce enzymes that are not required by them, as this is a wasteful exercise. The inhibition of surplus enzyme production is achieved by nutrient repression. The nutrients can be carbon, nitrogen, phosphate or sulfate suppliers in the growth medium [38]. For large-scale synthesis of enzymes, nutrient repression must be overcome. In using carbon sources and energy from artificial media, certain micro-organisms, in particular heterotrophic bacteria, can utilize an extensive range of sugars, organic acids and other organic compounds. Usually, glucose is considered as the ideal carbon source for *B. subtilis*. The presence of a large amount of glucose supports the strong repression of genes encoding the enzymes for the utilization of alternative carbon sources in the presence of glucose [39, 40]. Some other carbohydrates such as malate also act as a second preferred carbon source for *B. subtilis*. Malate is also responsible for causing a strong catabolite repression of transporters for alternative carbon sources [41]. This strong catabolite repression allows the cells to select suitable carbon sources. Generally, it involves the regulation of gene expression to prevent transcription of catabolic genes, which permits the cells to select among several available carbon sources. Regulation of gene expression allows the regulation of protein activity to avert the formation of specific inducers.

Glucose repression is a type of nutrient-based repression which prevents the production of certain enzymes that are required for metabolism. To prevent glucose repression carbohydrate can be supplemented to the fermentation medium in such a manner that the amount of glucose becomes almost zero at any given time. To date, many efforts have been made to explore mutants that are resistant to catabolize repression by glucose, e.g. development of mutants of *Trichoderma reesei* that are resistant to catabolite repression [42].

Moreover, many other carbon sources such as succinate, citrate, lactate and pyruvate act as catabolite repressors for some micro-organisms. Several examples of nitrogen sources as repressors are also present in micro-organisms. It is already known that not all nitrogen sources support growth equally. Thus yeast chooses its nitrogen sources to allow the best growth through a mechanism referred to as nitrogen catabolite repression. As per the literature, two nitrogen sources, ammonia and glutamine (due to presence of ammonium ions or amino acids), are considered to elicit nitrogen catabolite repression. Thus inexpensive ammonium salts are often utilized as nitrogen sources. By developing mutants resistant to this nitrogen source, repression by ammonium salts can be prevented [43].

2.3 Procedures involved in enzyme production

2.3.1 Source and location of enzymes

Every cell synthesizes enzymes, thus they can be derived from plant tissues, animal tissues and micro-organisms. The amount of enzymes produced on an industrial scale from plants and animal sources is significant, nevertheless microbial enzymes have gradually overtaken them for both technical and commercial reasons. Plant-based enzymes need a large amount of plant material and the amount of enzyme recovered is very small. Animal-based enzymes are end-products of the meat industry. The only animal enzyme to be produced in quantities greater than 2 ton/year is rennet or chymosin, obtained from the calf stomach. Most plant-based enzymes are hydrolytic enzymes, e.g. α- and β-amylases, β-glucanase, and papain. Most of these enzymes are used by the food industry. Therefore, initially, plant and animal enzymes were preferred over microbial enzymes mainly for considerations of safety and the fear of contamination through micro-organisms, toxins, etc. However micro-organisms have received significant attention during the last three decades. According to the literature, microbial enzymes are not subjected to any of the production and supply limitations of plant and animal enzymes. The production size of microbial enzymes may be expanded and the types of enzymes accessible from micro-organisms are almost limitless. Microbes offer the following advantages over other sources of enzymes:

- Micro-organism growth is very fast and they can be cultured on a medium comprising a cheap raw material.
- Genetic engineering and manipulation of microbial cells can be achieved in the research laboratory to enhance the final yield of enzymes.
- Enormous quantities of enzymes can be produced from microbes.
- Animal sacrifice can be stopped.

In a living cell (plant or animal), the mitochondria have a highly organized structure and contain a large number of enzymes, such as glutamate dehydrogenase. Cellular organelles, e.g. granular microsomes, lysosomes and ribosomes, also contain key enzymes; ribosomes are the site of protein biosynthesis and lysosomes contain many hydrolases. The soluble portion of the cytoplasm includes enzymes responsible for glycolysis. In the bacterial cell, certain structures of plant and animal cells are not present. In the cytoplasm, outside the nucleus, the cell is filled with a granular cytoplasm. Enzymes are present in the granules, the soluble cytoplasm and adsorbed on the cell membrane.

Enzymes produced by hyperthermophiles (bacteria and archaea with optimal growth temperatures of >80 °C), also known as hyperthermophilic enzymes, are characteristically thermostable (i.e. resistant to irreversible inactivation at high temperatures) and are optimally active at high temperatures. These enzymes share the same catalytic mechanisms with their mesophilic counterparts. These thermophilic and hyperthermophilic enzymes can be potentially utilized as research reagents and as catalysts for industrial processes [44].

2.3.2 The variety of micro-organisms

Selection of a micro-organism for enzyme production is performed on the following basis [4]:

- Fermentation duration should be low.
- The micro-organisms must adapt themselves to the physical and chemical properties of the culture medium, e.g. temperature, pH, the availability of substrates, etc. Currently, two genera, *Aspergillus* and *Bacillus*, are used for the large-scale production of enzymes.
- The micro-organisms must be nonpathogenic.
- The micro-organisms must cultivate on an economical raw material.
- The micro-organisms must offer high yields of enzymes.

- The micro-organisms should synthesize extracellular enzymes, as their isolation and separation is simple and economic, however, only hydrolases have been found extracellularly.
- The strain must not synthesize end-products which hinder the growth of micro-organisms.

2.3.3 Media for fermentation

The enzyme production medium should include sources of carbon, nitrogen, energy, minerals, macro-nutrients and micro-nutrients. Growth factors are required in the case of auxotrophic micro-organisms. If an inducible enzyme is to be synthesized, the inducer should be supplemented to the medium. At times, co-enzymes act as an inducer or sometimes constituents of the medium may have an induction effect. Enzyme-catalyzed reaction end-products may also act as inductors. Catabolic enzyme fermentation is inhibited by the direct or indirect effect of products on their activity, such as the production of proteases in *Bacillus* species by amino acids. Different sources of carbon, nitrogen and other growth substances are mentioned in table 2.4.

2.3.4 Fermentation

There are several measures available for fermentation; however, for the synthesis of enzymes only three methods are used [45]:

- Submerged culture.
- Solid-substrate culture.
- Deep-bed cultivation.

2.3.4.1 Submerged culture

Submerged fermentation is the cultivation of micro-organisms in liquid nutrient broth. In submerged cultures, the synthesis usually takes place in mechanically stirred bioreactors with capacities ranging from 20 000–100 000 l batch fermentation [45]. Usually the main fermentation persists for 50–200 h depending upon the enzyme and micro-organism used. Continuous fermentation has some limitations due to the difficulty in sterilization of the nutrient media, and instability of highly mutated production strains. On a large scale, continuous fermentation has been utilized for the

Table 2.4. Fermentation medium constituents and their sources.

Constituents	Various sources
Source of carbon and energy	Cereal meal, soybean meal, potato starch, wheat or rice bran, molasses
Source of nitrogen	Fish meal, gelatin, casein, soybean meal, bran, distillers' solubles' peptones
Source substances and trace elements	Yeast extract, corn steep liquor, plant oil, meal of oil-bearing seed bran

production of glucose isomerase. The overall yield of enzyme depends upon the degree of enzyme synthesis during different growth phases. Maximum enzyme production is frequently derived in the stationary phase in an inducible system with catabolic repression, when the micro-organism has fallen to zero. Currently, the process is carried out in a two-stage cascade, in which a high quantity of cell mass is first synthesized and then the synthesis of the enzyme takes place in a second reactor under different conditions. Additionally, factors such as nutrient medium, pH, temperature, partial pressure of oxygen and aeration must be taken into consideration. The final yield of extracellular enzymes can be improved by the supplementation of surfactants. Once fermentation is completed, the ferment is cooled and the cell mass is separated. Extracellular enzymes are present in the culture filtrate and therefore the biomass is discarded, however, for intracellular enzymes the biomass contains the enzymes and thus the culture filtrate is discarded [45].

While isolating enzymes from the fermentation medium one must eliminate insoluble products, e.g. microbial cells. This is usually achieved by centrifugation. As most industrial enzymes are extracellular (secreted by cells into the external environment), they remain in the fermented broth after the biomass has been removed. The biomass can be reused as a fertilizer, but it must be treated with lime to inactivate the micro-organisms and stabilize it during storage. The enzymes in the rest broth are then concentrated by evaporation, membrane filtration or crystallization, depending on their intended application. If pure enzyme preparations are required, they are usually isolated by gel or ion-exchange chromatography. Some applications require solid enzyme products, so the crude powder enzymes are converted into granules to make them more suitable to use. Occasionally liquid formulations are selected as they are easier to handle and dose along with other liquid ingredients. Enzymes employed during starch conversion to convert glucose into fructose are immobilized, usually on the surfaces of inert granules held in reaction columns or towers. This is done to extend their working life as these enzymes usually go on working for over a year.

2.3.4.2 Solid-substrate culture

Solid-substrate fermentation is currently used in a variety of applications, e.g. antibiotic and enzyme production, recently developed products such as bioactive compounds and organic acids, new trends regarding bioethanol and biodiesel as sources of alternative energy, and biosurfactant molecules with environmental purposes of utilizing unexploited biomass [46]. Solid-substrate fermentation is primarily used for traditional food processing and also for fungal enzyme production. During this process, the micro-organism is propagated on a solid substrate supplemented with a high concentration of nutrients, micro-nutrients and minerals and with a large surface area, e.g. cereal meal, wheat bran and/or rice bran. This process is an alternative to the production of enzymes in liquid by submerged fermentation. This process is suitable for the extraction of enzymes from fungi, such as *Penicillium, Aspergillus,* etc [46]. The moisture content of the medium is low which hinders the growth of bacteria. There are two procedures available for growths of fungi: the drum process and tray process. In the drum process horizontally rotating drums are utilized for the propagation of fungi. In the tray process, the fungi are propagated in trays of size 2 × 40 cm. The substrate is spread in a thin layer (thickness 1–10 cm). The micro-organisms, in the form of spores, are inoculated and then the trays are incubated in an air-conditioned room. A method called the high-heap process has also been used, in which a constant stream of air is enforced through the nutrient substrate. This ensures a supply of oxygen as well as elimination of the heat of reaction. The growth phase continues for 1–7 days. After completion of growth and fermentation, the fungi are homogenized and dried (moisture content 10–15%). For this process, homogenized powder can be used directly or the fungal mycelium is extracted with water. There are numerous substrates that can be utilized for the production of enzymes by solid-substrate fermentation such as wheat bran, rice bran, sugar beet pulp, and wheat and corn flour. The selection of the substrate depends on several factors, chiefly related to cost and the availability. Additional factors such as particle size and the level of moisture also play an important role. Smaller substrate particles have a larger surface area for the proliferation of the micro-organisms, however, if the surface is too small the efficiency of respiration will be obstructed and poor development and hence poor synthesis of enzymes will result. Larger particles offer more effective aeration and respiration, although there is a reduction in the surface area. Solid-substrate fermentation needs moisture to be present on the substrate for the micro-organisms to synthesize enzymes. Therefore, the water content of the substrate must also be optimized, as a higher or lower presence of water may badly affect the microbial activity. Water also has effects on the physico-chemical properties of the solid substrate. Some of the enzymes of industrial importance which have been produced by solid-substrate fermentation are proteases, pectinases, glucoamylases and cellulases. Revankar *et al* reported solid-substrate fermentation for enhanced production of laccase using indigenously isolated *Ganoderma* spp. [47].

2.3.4.3 Deep-bed cultivation

The deep-bed or pile cultivator method process was initially developed to meet the huge demand for enzymes for soya fermentation [48]. During this procedure the micro-organisms are allowed to cultivate in rectangular vessels of dimensions 18 × 200 inches. The nutrient medium is added into the vessels up to a height of 2 feet. In deep-bed cultivation, cereal meal, wheat bran, rice bran, soybean, potato flakes, etc are used as the medium. The medium is sterilized and then inoculated with the organism. The nature of the culture vessel is maintained to increase the growth of the organism. At desirable growth, the microbes are used in the extraction of enzymes. This process solves most of the problems related to the traditional process. It uses substrate layers up to 2–6 feet. The deep-bed process is a modification of solid-substrate fermentation and it is fully automated.

2.3.5 Enzyme extraction

The extraction of enzymes can be defined as the liberation of enzymes from cells or cellular constituents. Extraction is done by mechanical, physical, chemical, or a combination of these procedures to disrupt the cell wall or membrane. To enable extraction of either intra- or extracellular enzymes, it may be essential to modify the nature of the liquid medium to complete the dissociation. Down-streaming of enzyme fermentation through drying of the substrate is simple, although it can be complicated depending on whether a crude or high-grade enzyme is to be produced, or whether the enzyme is extracellular or intracellular [48]. For fungal enzymes, centrifugation or filtration are the two procedures which are most often used to separate enzymes, however, bacterial enzymes are difficult to concentrate and purify [48]. Certain enzymes necessitate the presence of a co-factor, lipid or carbohydrates to preserve their activity during extraction.

2.3.5.1 Animal and plant tissue breakdown

Most animal enzymes are confined in a particular organ or in muscles. In this procedure, these organs are initially minced in a vertical cutter mixer after removal of fat [49]. Freezing the animal tissue frequently supports grinding and prevents blockage by wet tissue. For this purpose frozen meat grinders are used. The pulverized tissues are then allowed to pass through a colloid which produces maximum cell disintegration. Grinding of certain plant tissues is a major concern, e.g. seeds which contain more enzymes than green tissues. Grinding of green tissues can be achieved by a process called maceration, in which the plant material is macerated by grinding in a hammer mill or some other chopper mill, and the pulp is pressed. Lytic enzymes can also be used to disrupt the cell wall [50].

There are several ways of extracting enzymes from animal tissue. Initially, aqueous phase separation was proposed for extracting and purifying proteins from animal tissue [51]. Boland *et al* reported the purification of enzymes from animal tissue using aqueous two-phase systems in pilot scale studies.

2.3.5.2 Disruption of microbial cells

Microbial cells synthesize both extracellular and intracellular enzymes. Extracellular enzymes do not require cell disruption, but the release of intracellular enzymes from micro-organisms requires a more vigorous method of cell breakage [52–55]. The different methods of microbial cell disruption are illustrated in figure 2.5.

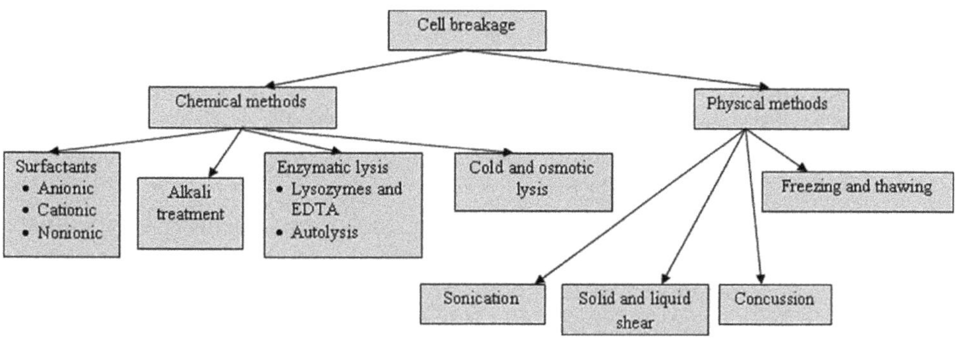

Figure 2.5. Various methods of microbial cell disruption.

For industrial scale cell disruption, the mechanical procedures of disruption seem to be the most popular [52–55]. Of these, the high-pressure homogenizer and the bead mill are the most frequently used. The high-pressure homogenizer is best suited to some bacteria and yeasts, whereas the bead mill is more appropriate for the disintegration of yeast cells and for mycelial organisms and algae. There is sufficient evidence to demonstrate that mechanical cell disruption methods do not harm most intracellular enzymes and proteins; membrane-associated enzymes and multi-enzyme complexes may be the exceptions. Additional techniques which may have potential large-scale applications are ultrasonication, freeze-pressing and enzymatic lysis. An arrangement of two or more disruption techniques for disruption of more resistant organisms may have economic advantages [52–55].

2.3.5.2.1 Disruption by chemical methods

Surfactants. Among surfactants, ionic (cationic and anionic) and nonionic surfactants are used to lyse microbial cells. Surfactants have the potential to solubilize the microbial cell wall. Frequently used surfactants are sodium dodecyl sulfate, cetyl triethyl ammonium bromide, triton X-100, various tweens, etc. Selective extraction of cholesterol oxidase from *Nocardia rhedocrous* by cell permeabilization, using the surfactant Triton X-100, has been evidenced [53]. This surfactant is quite costly, however, and industrial scale use may not be possible. Additionally, contamination of the product with the surfactant is another disadvantage.

Enzymatic lysis. During large-scale operations autolysis is the most frequently used method for the extraction of enzymes. A suspension of cells is maintained at high temperature (23–37 °C) for several hours. After cooling, the cell extract is harvested by centrifugation [54]. This procedure is used in the extraction of transaldolase from frozen *Candida utilis*, invertase from baker's yeast, and glucose-6-phosphate dehydrogenase from yeast. Autolysis is also used for the extraction of intracellular enzymes. The procedure has a risk of thermal denaturation of enzymes or their damage by cellular proteases. Microbial cell lysis can also be achieved by using lysozymes, e.g. egg white lysozyme. Lysozyme hydrolyzes the glycosidic bonds in the glycopeptide component of the bacterial cell wall, releasing intracellular enzymes [54].

Enzyme cell lysis is an attractive tool in terms of its delicacy and specificity to just the cell wall structure, but is restricted by the high cost of the enzyme, which is usually lost into the extract [54]. This can be prevented by the use of an immobilization technique in which soluble immobilized enzymes in ultrafilter reactors are used. Reduction in the cost of cell lytic enzymes may also be possible by increasing the scale of production [52, 54–56]. The sensitivity of micro-organisms against various lytic enzymes changes significantly with the development phase and fermentation conditions. In certain cases, autolysis of microbial cells without any foreign enzyme may be possible.

Cold and osmotic shock. In this procedure, the normal growth temperature is reduced to 0 °C, resulting in loss of viability of the micro-organisms. This technique is not applicable to large-scale production, as cell suspensions with a density greater than 10^8 ml^{-1} show no effect of osmotic shock and bacteria are more vulnerable to cold shock.

Some degradative enzymes such as alkaline phosphatase, cyclic phosphodiesterase, 5′-nucleotidase, acid phosphatase, and ribonucleic acid inhibited endonuclease are explicitly released from exponentially growing *E. coli* by osmotic shock [57]. Cell disruption by osmotic shock involves the following steps:

- Bacteria washing to remove growth medium by a buffer solution.
- Resuspending the washed cells in a 20% buffered sucrose solution.
- Removal of cell by centrifugation. After following these steps, the final paste is then suspended in water. The rapid increase in osmotic pressure inside the cells causes the release of cell constituents.

Alkali treatment. A rapid increase in pH will cause the lysis of the bacterial cell wall. This method is used in the large-scale extraction of L-asparaginase from *Erwinica chrysantheme* [58]. The effectiveness of this method depends upon the pH stability of the desired enzyme. If the desired enzyme is inactivated by alkaline pH, then this method is not applicable [58, 59].

2.3.5.2.2 Disruption by physical methods

Sonication. Sonication of proteins results in the formation of aggregates that resemble amyloids. In addition, sonication is often used to break aggregates into smaller pieces for seeding new batches. At the pilot scale, ultrasonic waves are often used for cell lysis. The ultrasonic or sonic waves passing via a liquid consist of alternate compression and rarefaction. If the wave amplitude is high, then a cavity is produced, which leads to the formation and breaking of bubbles. The formed bubbles or cavities create many cycles, generating high local pressures of about 20 000 atmospheres. The mechanical shocks are sensed at a distance of a few microns. Ultrasonic waves have been used successfully in a number of extraction methods [60], in particular in the extraction of enzymes and proteins at small scales.

Freezing and thawing. The reason for the disruption of cells caused by freezing and thawing is still unclear, although the plasma membrane is often considered the primary site for freezing injury [61]. Tansey (2006) reported freeze–thaw lysis for extraction of proteins from mammalian cells [62]. This efficient method for mild solubilization of inclusion bodies uses a freeze–thaw process in the presence of a low concentration of urea. The extraordinary effects of freezing and thawing are very similar to those reported during cold and osmotic shock [63]. Despite the differences in osmotic pressure between intra- and extracellular fluids, intra- and extracellular crystal formation further damages the cell wall. For this procedure 10% of available soluble protein release has been reported.

Solid and liquid shear. These types of mechanical procedures of lysing do not involve adding chemicals or enzymes to the system [64]. However, the energies required when using these 'harsh' protocols can be high and destroy the very proteins being extracted [64]. The cell membrane destruction is influenced by exposing the cells to the following:

- shear through liquid flow;
- exploding through pressure differences between the inside and outside of the cell;

- collision forces through the effects of beads or paddles; or
- a combination of these forces.

All these energies should be optimized properly to expose the cell to the optimum level of energy [64].

During solid shear, frozen (−20 °C) microbial paste is placed in a cylindrical hole (0.1–1 mm in diameter) in a metal block. Then a tight fitting plunger is inserted to introduce 1000–4000 atmospheric pressure. This level of pressure is applied from the top of this plunger. The majority of the cells are ruptured by this method. In liquid shear, the microbial suspension is permitted to pass through a narrow orifice at pressure up to 40 000 psi. Occasionally a French press is also used for this purpose. Cells experience disruption owing to the extreme level of shear during passage from the orifice. Nearly 90% of cell breakage was reported with baker's yeast after a single passage at 20 000 psi.

Concussion. During concussion, microbial cells can be disrupted by bombardment with hard particles such as glass beads. Numerous types of mills are used for this purpose, such as agitator bead mills, ball mills, colloidal mills, etc [65].

2.3.5.3 Aqueous and solvent extraction

The growing interest in industrial enzymes emphasizes the need for the development of new downstream approaches to make best use of enzyme recovery [66]. Considerable efforts have been focused on the development of newly adapted technologies to purify enzymes in their catalytically active form. Recently, an aqueous two-phase system was developed as a promising tool for effective extraction and purification of enzymes due to the versatility, lower cost, process integration capability and easy scale-up of this system [66]. Three-phase partitioning has also been championed for enzyme enrichment as a promising technique for efficiently integrated bioseparation [66]. Currently, CLEAs technology and organic–inorganic nanoflower preparation are also utilized as novel approaches for simultaneous extraction, purification and immobilization of enzymes [66]. Microbial cells lysis does not necessarily result in breakdown of the molecular complexes of other cell components, e.g. lipids, nucleic acids and carbohydrates. The liquid extraction of the enzyme from insoluble particles can offer a technique for fractionation, in which extractants of increasing eluting power are employed. Amid *et al* reported a novel liquid–liquid extraction process composed of surfactant and acetonitrile for purification of the enzyme polygalacturonase from *Durio zibethinus.* Glyk *et al* reported polyethylene glycol (PEG)–salt aqueous two-phase systems as a promising, efficient liquid–liquid extraction technology for the downstream processing of various biomolecules, such as proteins and enzymes [29, 67]. There are two general procedures for extraction of enzymes from broken cells:

- Using salt solution as a solvent (in which most of the enzymes are stable), extract all possible materials under mild conditions.

- Selective extraction of enzymes utilizing the different solubility of enzymes in different solvents. In this procedure the initial extraction is achieved by using a solvent in which only a few enzymes are soluble and then repeating the same process with other solvents in which other enzymes are soluble. In stirred jacketed vessels, microbial cells are extracted batchwise. In this procedure, particle size governs not only the extent of extraction, but also the amount of extract which gets adsorbed on the particle surface. With a known quantity of extracting fluid, the use of many fractions of the liquid results in a more nearly complete removal of the desired solute than does the use of the entire liquid in a single extraction. Further improvements in extraction efficiency may be obtained through arranging the staged extraction in a countercurrent continuous flow. After extraction of enzymes, the spent solids are removed by filtration, centrifugation or sedimentation.

Aqueous two-phase systems (ATPSs) are one of the most common tools for the separation of proteins. They involve mixing a polymer (usually polyethylene glycol, PEG) and a salt (e.g. phosphate, sulfate or citrate), or two polymers and water, to successfully separate and purify the proteins [68]. The surface properties of the proteins and the properties of the two-phase system determine the partitioning between both phases. The complex mechanism of partitioning is somewhat dependent on the level of protein hydrophobicity. Hydrophobicity is the main factor in the partitioning of proteins and can be determined in many different ways. The procedures that are more advantageous, depending on the ATPS used, are those that consider the three-dimensional structure and the degree of hydrophobicity of aminoacids on the surface and that based on precipitation with ammonium sulfate [68]. The influence of charge has a comparatively small effect on the partitioning of proteins in PEG/salt systems, but is more important in PEG/dextran systems. Protein concentration plays an important role in the partitioning of proteins in ATPSs [68], where it produced higher levels of solubility of the protein in both phases, and thus the partitioning reported at low protein concentrations can produce very different results to that reported at high concentrations. This protein precipitate is in equilibrium with the protein solubilized in each of the phases. The practial application of ATPSs has been shown in numerous cases, including a number of industrial applications with excellent levels of purity and yield [68].

2.3.6 Purification of enzymes

Batchwise adsorption and elution from ion-exchange celluloses is now often used and has largely replaced traditional practices such as precipitation techniques for large-scale purification [69]. Procedures such as concentration, dilution and reconcentration with the use of hollow-fiber ultrafiltration equipment has replaced dialysis in the effective purification of enzymes [69]. Additionally, antiphonally direct scaling-up of column chromatographic procedures can also be utilized for purification purposes [69]. Alterations to column geometry to enhance flow rates are often desirable, however, the purification factors and recovery are comparable to those derived at the laboratory scale and can be achieved comparatively easily. Conventional affinity chromatographic tools have not been considered for large-scale purification, primarily because of their cost and the rather short life of the matrices. However, quasi-affinity chromatography performed using triazine dye conjugates has proved to be of great benefit. The materials are cheap to prepare [69]. The coupling methods are both easy and quick and do not contain the use of poisonous chemicals such as cyanogen bromide [69]. In addition, the triazine linkage is more stable under different conditions than the isourea formed in cyanogen bromide coupling [69].

2.3.6.1 Lipid removal

Lipid removal is usually done as early as possible as lipids obstruct in the separation of proteins. The traditional procedure of lipid removal is to make an acetone powder of the ground cells. In this process, the ground cells are mixed with cold acetone (−10 °C). This is repeated with acetone at 0 °C and finally with acetone at 20 °C. The method predominantly eliminates water and some lipid. The lipid is removed later with petroleum ether or ether at room temperature.

Triton X-100 (using a nickel column) is frequently used to aid lysis and solubilize lipids and membranes [70]. Removing lipids from membrane proteins can be involved, but most protein purifications remove lipids as a matter of course. Hydrophobic interaction chromatography can remove even minute quantities of lipopolysaccharides and lipids, and BioBeads can be used in spin preparations [70]. Separation is again dependent on factors such as the structural configuration of the protein, its solubility and whether it is free or membrane bound. Simply washing inclusion bodies with levels of detergent and/or urea can also be done until the sample is lipid free and quite pure. 1 M urea and 1% Triton X-100 are regularly used, although both concentrations should be optimized to prevent protein solubilization [70]. Different enzymes and their respective solvents for extraction are listed in table 2.5.

Additional methods of lipid removal are the separation of lipids from solutions of enzymes in 20–50% v/v n-butanol at 0 °C, and lipid extraction using different gases under supercritical conditions. In the latter procedure, liquid CO_2 is most frequently used for the removal of lipids from biological materials.

Table 2.5. Various enzymes and their respective solvent for extraction.

Enzyme	Source	Solvent
Cholesterol esterase	Pancreas	Dil. H_2SO_4, in the presence of protease inhibitor
Fungal lipase	Fungi	Water
Glycerol phosphate dehydrogenase (GPDH)	Rabbit muscle	1 mM disodium EDTA containing 2 mM 2-mercaptoethanol

2.3.6.2 Nucleic acid removal

Several methods have been developed to remove nucleic acid contaminants, given the general concern of accidentally introducing recombinant genes into humans. Due to this concern, FDA has recommended an upper limit for nucleic acid of 10 pg/dose [71]. A number of methods are available for removing nucleic acid contaminants during protein isolation, e.g. treatment with cation exchangers or precipitation by salts (e.g. MnCl, NaCl), acids (e.g. acetic, trifluoroacetic), streptomycin sulfate, polycations (e.g. protamine, clupeine, polyethyleneimine), and heat [71]. The selection of the method to be used in any specific protein purification scheme depends mainly on the nature of the protein of interest, for example strongly basic protein products can form ionic complexes with nucleic acids leading to co-purification of nucleic acid contaminants with the product [71].

Purification of intracellular enzymes is a challenge as most of the intracellular enzyme preparations contain nucleic acids which can increase the viscosity, which may further interfere with enzyme purification procedures, in particular ultra-filtration. This type of purification is complicated by the presence of nucleic acid. Nucleic acid can be precipitated by high molecular weight polyvalent cations such as protamine sulfate, cetyl triethyl ammonium bromide, streptomycin sulfate and polyethyleneimine [71]. These polyvalent cations form a complex between negatively charged phosphate residues of the nucleic acid molecules and positively charged groups of precipitants. The resulting complex is then separated by centrifugation [71]. Several micro-organisms contain satisfactory nuclease enzyme to circumvent this problem, however, the nucleic acids must be eliminated by precipitation or degraded by the supplementation of exogenous nucleases. Ammonium sulfate precipitation can be very effective in eliminating nucleic acids but also eliminates some protein at the same time. Many positively charged material precipitants such as polyethyleneimine, cetyltrimethyl ammonium bromide, streptomycin sulfate and protamine sulfate have been used. Generally, these can easily form complexes with the negatively charged phosphate residues of the nucleic acids. All of these precipitants are expensive and sometimes toxic (mainly streptomycin sulfate). Moreover, they form complexes with certain enzymes.

2.3.6.3 Purification by precipitation and differential solubility

Differential protein precipitation is a rapid and less expensive stage in protein purification and is based on using the inherent physio-chemical properties of the polypeptide [72]. Precipitation of proteins, lysed from the host cell, is generally used to concentrate the protein of choice before further refining procedures with more selective purification columns (e.g. His-tag, size exclusion, etc) [72]. Proteins can also precipitate naturally as inclusion bodies owing to various influences during over-expression in the host cell. While this phenomenon allows convenient initial separation from native proteins, these inclusion bodies must be carefully differentially solubilised so as to achieve functional, correctly folded proteins. Solubilization is the most frequently used method for enzyme purification. It is based on precipitation of the active enzyme or other proteins and soluble substances [72].

Salting-out The solubility of globular proteins increases upon addition of salt (<0.15 M), which is called salting-in [73]. At higher salt concentrations protein solubility usually declines, resulting in the precipitation of protein, which is called salting-out [73]. A salt that decreases the solubility of proteins also tends to increase the stability of the native conformation. In comparison, salting-in ions are usually denaturants. The salting-out mechanism is based on superior solvation due to elimination of the cosolvent (salt) from the layer of water closely associated with the surface of the protein (the hydration layer) [73]. The hydration layer, typically 0.3–0.4 g water per gram protein, plays a significant role in conserving solubility and the correctly folded native conformation. During this procedure, the precipitation of proteins is achieved by a high concentration of neutral salts. The most often used salts are ammonium sulfate and sodium sulfate. Ammonium sulfate is inexpensive and is highly soluble (767 g l^{-1}) which permits salting-out of practically all proteins [73].

Temperature and pH Biological macromolecules have been developed to execute their function in a particular cellular environment (subcellular compartments or tissues); thus, they should be adjusted to the biophysical features of the corresponding environment. Several macromolecular features are pH-dependent, e.g. activity and stability [74]. However, only activity is biologically significant, while stability may not be essential for the corresponding reaction [74]. Most proteins exhibit increased solubility with temperature elevation [74]. At high temperatures, the differential stability of enzymes is quite significant and selective heat denaturation is frequently used at the industrial scale. Thus, inert protein materials may be denatured by controlled heating of extracts. This process is used in the purification of α-amylase, ribonuclease, adenyl kinase, cholesterol esterase and glutamate dehydrogenase [74]. Alteration of pH is also used as a method of fractional precipitation. The different proportions of basic and acidic groups of different enzymes lead to a wide range of pH values at which enzymes exhibit isoelectric zero net charge characteristics. The principal difficulty in using differential pH precipitation is the pH stability of the desired enzymes, which is limited. Almost every biological process is pH-dependent, which indicates the significance of the local pH on all processes in the cell [74]. Therefore, various biologically relevant effects are significantly influenced by the pH of the water phase including:

- functional pH dependence;
- structural conformations associated with the function; and
- subcellular translocation.

Additionally, the role of the concentration of H^+ ions on protein stability is confirmed by acidic/basic unfolding and protein pH-dependent stability. Variation in pH is proven to activate the development of amyloid fibers and aggregation.

Organic solvents Organic solvents have not been commonly used for the separation of enzymes, as they denature proteins at room temperature. During World War II, an ethanol precipitation method was developed. At low temperatures and low ionic strengths, various protein fractions were isolated from blood plasma. Ethanol fractionation has several benefits over the salting-out method. At low ionic strengths, the interaction between proteins and salts depends on specific features of protein molecules, e.g. the distribution of electric charges [75]. Small differences in dielectric constant and ionic strength will produce large and specific changes in the solubility of proteins. As discussed above, organic solvents are frequently used in fractional precipitation of proteins based on their dielectric constants [75]. Once an organic solvent is mixed with the aqueous solution of an enzyme, the dielectric constant of the solution is decreased, thus the solubility of the enzyme is reduced, however, enzymes should retain their internal hydrophobic amino acid residues and remain comparatively loosely folded [75]. Sometumes adding an organic solvent causes the molecule to refold into a new and inactive form with the hydrophobic residues exposed on the surface. The risk of refolding and subsequently denaturation is larger as temperature increases. This results in a requirement for low temperature, often below 0 °C, and fractionation with organic precipitants. Some organic solvents used in protein purification are methanol, ethanol, isopropyl alcohol and acetone [75]. Ethanol is the most frequently and extensively used organic solvent owing to its suitability for use in the food and pharmaceutical industry [75].

Nonionic polymers and multiphase systems Biological macromolecules with high molecular weight polymers, e.g. dextrans and polyethylene glycols (PEG), have been used extensively in the isolation of biological material in three main ways:

- For the concentration of biological components including viruses by dialysis and by liquid–liquid extraction in aqueous two-phase mixtures.
- As a stabilizer on a particular component in a solution.
- As a precipitant which causes the formation of a solid protein phase.

Two-phase systems can be used to separate enzymes from cell homogenates and at the same time achieve a certain degree of purification. Dextran–PEG two-phase systems have been used in the production of commercial intracellular food enzymes. Based on a report from 1981, PEG helps in the precipitation of various proteins ranging in size between 14 000–670 000 daltons (Da). This was analyzed by excluded volume [76–78].

The nondenaturing, water-soluble, high molecular weight polymer referred to as PEG can precipitate protein from aqueous solutions, which can be qualitatively examined in terms of an excluded volume mechanism. The increase in PEG amount required to decrease solubility is distinctive for a given protein–polymer pair. This distinctiveness is not sensitive to solution conditions and is mainly reliant on the size of the protein and polymer [77]. The solubility of specific proteins can be selectively manipulated by controlling their state of association or ligand environment. PEGs for protein precipitation are typically classified by molecular weight. The higher molecular weight PEGs can efficiently precipitate but this may result in higher viscosity [79]. It has now been realized that the hydrodynamic radius of the PEG plays an important role in selecting suitable PEG precipitants. Moreover, it also offers more insight into understanding precipitation mechanisms [79].

Crystallization Crystallization is compulsory to derive the three-dimensional structure of proteins and nucleic acids; it frequently signifies the bottleneck in structure determination [80, 81]. The crystallization of biological macromolecules (proteins, nucleic acids and viruses), depends on the formation of a solution that is supersaturated in the macromolecule, but also displays conditions that do not considerably perturb its natural state [80, 81]. Supersaturation is produced by the supplementation of mild precipitating agents, e.g. neutral salts or polymers, and by the manipulation of different parameters, which include temperature, ionic strength and pH. Factors that can disturb the structural state of the macromolecule, such as metal ions, inhibitors, co-factors or other conventional small molecules, play an important role in the crystallization process [80, 81]. A number of approaches have been established that combine a range of factors that affect and promote crystallization. Among the most extensively used are vapor diffusion, dialysis, batch and liquid–liquid diffusion. Crystallization is often performed after extraction and purification of the enzyme. The first crystals of the enzyme may contain a number of other proteins. Crystallization and recrystallization are effective procedures in enzyme purification. The most extensively used procedure of crystallization is via ammonium sulfate solutions. In this procedure the salt is supplemented to a concentrated enzyme solution until a slight turbidity appears. It is then permitted to stand, however, the salt concentration is enhanced by:

- Supplementing a strong solution of the salt dropwise at long intervals.
- Through a fine capillary.
- Through a dialysis membrane.
- The solution may simply be permitted to evaporate slowly.

Crystallization may be introduced by altering pH or temperature at a constant salt concentration. Glycerol phosphate dehydrogenase, pyruvate kinase, lactate dehydrogenase, urease and glutamate dehydrogenase are the most prominent examples of enzymes that have been purified using crystallization procedures [80, 81].

2.3.6.4 Protein purification using chromatographic methods

To attain a high level of purity in the purification of recombinant proteins for therapeutic or analytical applications, it is essential to use numerous chromatographic steps [82], which are based on the differential migration of enzyme. A variety of techniques is available, e.g. adsorption, anion and cation exchange (which can be performed at different pHs), hydrophobic interaction chromatography, gel filtration and affinity chromatography [82].

Adsorption chromatography Adsorption in stable expanded beds allows proteins to be recovered directly from particulate-containing feedstocks, e.g. fermentation broths and preparations of disrupted cells, without the requirement for prior elimination of the suspended solids, which would usually lead to the blockage of packed beds [83]. The implementation of this method significantly reduces the complexity of downstream processing by removing certain filtration, centrifugation and concentration steps [83]. Aspects that are vital for the success of the procedure include the right choice of adsorbent, together with careful design of the apparatus in which the separation is performed. The design, optimization and scale-up of appropriate operating protocols for expanded-bed procedures are very similar to those used for the operation of packed beds [83].

The first chromatographic purifications were performed using inorganic compounds that were adsorbed by the biochemicals via van der Waal's forces and steric interaction [83]. The most extensively used adsorbent in enzyme purification is calcium phosphate gel, in particular its crystalline form hydroxyapatite (HA). The calcium and phosphate ions on the surface of HA crystals form bonds with charged groups of proteins. Acidic and neutral proteins form bonds with the calcium sites on the HA. The elution of acidic and neutral proteins is usually achieved with low concentrations of phosphate buffers of pH 6.8. Basic proteins associate and form bonds with the phosphate group on the HA crystals. Bacteria-derived alkaline phosphatase was purified using HA. *E. coli* derived amino transferases have also been purified, with a six-fold increase in specific activity.

Ion-exchange chromatography The principle of ion-exchange chromatography is based on the separation of proteins, and hence has numerous applications in protein science, such as the discovery of proteins, high-resolution purification and the industrial production of proteins. Ion-exchange chromatography includes the separation of ionizable molecules based on their total charge, often used for the separation and purification of biomolecules such as proteins, polypeptides and nucleic acids [84–86]. Virtually all enzymes are polar in nature and can be charged. Ion-exchange chromatography is usually used for enzyme purification. Separation is achieved with the help of derivatives of cellulose, agarose, dextrans or resins. Agarose is the most commonly used cross-linked matrix for the purification of bioactive enzymes and pharmaceuticals because of the following properties:

- stable bed volume;
- high capacity;
- high flow rate and resolving power;
- good chemical stability between pH 3–10; and
- good thermal stability up to 70 °C, which makes it suitable for autoclaving.

The anion exchangers most frequently used are diethylaminoethyl cellulose, triethylamino cellulose and triethanolamine coupled to cellulose through glycerol and polyglycerol chain mixed groups (ECTEOLA cellulose) [84–86]. Carboxymethyl cellulose, phospho cellulose, sulfoethyl cellulose, acrylic acid resins are used as cation exchangers. Ion-exchange chromatography based enzyme fractionation can be achieved in a column or by a batch method. Enzymes, whose isoelectric points are well removed from neutral pH, are preferably separated by ion-exchange chromatography. For L-asparaginase (with an isoelectric point of 6.8) a 100-fold purification is possible using a carboxymethyl cellulose batch column. The batch method is also beneficial in eliminating nucleic acids from enzymes [84–86].

Gel filtration chromatography Gel filtration (GF) chromatography is termed as separation of proteins exclusively on the basis of molecular size. This is done by using a porous matrix to which the molecules, for steric reasons, have different degrees of access, i.e., smaller molecules have greater access and larger molecules are excluded from the matrix. Therefore, proteins are eluted from the gel filtration column in decreasing order of size [87]. The sample is applied on the surface of a column of appropriate porous beads of hydrated gel and solvent is percolated through the column [87]. The molecules that are too large in size to penetrate porous structure of the beads are excluded and allowed to pass through in the void volume of the column. Molecules which are smaller in size can enter porous beads and move more slowly via the column, as they spend a proportion of their time in the beads, thus molecules are eluted in order of decreasing molecular size. This technique is also known as permeation, molecular sieve, molecular exclusion, restricted diffusional or steric chromatography. For enzyme purification there are three available gel filtration media:

- Partially cross-linked dextrans with a fractionation range up to 250 000 Da. Dextran is a polysaccharide made up of glucose residues. Dextran gels (Sephadex) are the first for which a close association between molecular size and elution behavior was established.
- Cross-linked granulated polyacrylamide gels with a fractionation range up to 400 000 Da. Cross-linked polyacrylamides form gels with water, which are used in various biochemical separations.
- Granulated agarose gel (Sepharose) with a fractionation range up to 50 000–40 000 000 Da.

Allyl dextran cross-linked with *N*,*N*-methylene bis acrylamide (Sepharoyl) is preferred for large-scale production. By using a 25 $\times$ 80 cm column of Ultrogel Ac and Sepharcyl S-200, two enzymes, alkaline phosphatase and restriction endonucleases, have been purified.

Affinity chromatography Affinity chromatography is among the best chromatographic methods for the purification of a specific molecule or group of molecules from complex mixtures [88]. The principle is based on highly specific biological interactions between two molecules, such as interactions between an enzyme and substrate, receptor and ligand, or antibody and antigen [88]. These reversible interactions are used for purification by placing one of the interacting molecules, called the affinity ligand, onto a solid matrix to form a stationary phase, whereas the target molecule is in the mobile phase. Effective affinity purification needs a certain degree of knowledge of the nature of interactions between the target molecule and the ligand to help determine the selection of an appropriate affinity ligand and purification procedure [88]. Recently, matrices with distinctive features which bypass the limitations of more traditional materials have been established. Affinity purification can offer significant time savings and higher purification, however, the effectiveness of this method depends on the type of procedure used [88]. Therefore, it is important to optimize the purification protocol to attain efficient capture and maximum recovery of the target.

The entire procedure depends upon exact interaction between the analyte and the opposite group which is covalently bound to the column packing. A column comprising an insoluble polymer or gel to which a specific competitive inhibitor or other ligand has been covalently bound is used for this purpose. The solution containing macromolecules to be purified is allowed to pass through this column. Those proteins that have high affinity with the ligand will be retained, whereas those with less affinity will pass unretarded through the column. Using certain procedures, such as washing the ligand–protein complex with a solution of displacing agent such as an inhibitor, or by changing the pH or ionic strength of the elution solvent to favor dissociation, exactly retained proteins can then be eluted. Various insoluble supports are used for enzyme purification by affinity chromatography, such as hydrophilic cellulose derivatives, polystyrene gels, cross-linked dextrans, beaded agarose, glass beads and polyacrylamide gels. Some chemical derivatives of agarose and polyacrylamide have also been used for enzyme purification. The affinity matrices used for enzyme purification can be divided into two categories:

- Those which are specific for the desired enzyme by the specificity of the ligand such as the substrate, substrate analogs, inhibitors or antibody of the enzyme.
- Those which will interact with a related group of enzymes because of an immobilized general ligand such as a co-factor (5′-AMP, 2′,5′-ADP, NAD′ and others), which is specific for a class of enzymes and hydrocarbon ligands or dyes. This will interact with a large number of different enzymes.

The total expense of enzyme isolation can be minimized by the incorporation of an affinity step. One of the most common examples is purification of tissue plasminogen activator (t-PA) from cultures of human kidney cells, which is purified using α-benzylsulfonyl-p-amino-Sepharose [89].

Electrophoresis and ultracentrifugation During electrophoresis, a molecule with an overall charge will move in an electric field. Electrophoresis offers a potential means of separating proteins and other macromolecules, such as DNA and RNA. The velocity of migration (v) of a protein (or any molecule) in an electric field depends on the electric field strength (E), the net charge on the protein (z) and the frictional coefficient (f). Considering all these, one can see that electrophoretic purification is based on the movement of proteins when placed in an electrical field. This process is capable of the highest resolution of enzymes employing physico-chemical separation. This procedure is now mainly used as a technique for separation of isoenzymes and other enzymes linked for diagnostic purposes. The procedures often used at the laboratory scale for enzyme purification and characterization are: free-boundary electrophoresis; zone electrophoresis with the help of supports such as paper, starch, and cellulose powder; electrophoresis in agarose and acrylamide gels; and isoelectric focusing. Recently, a continuous electrophoretic separator has been developed for large-scale separation of proteins. The separator can draw as many as 29 separate fractions. This application can be utilized for the isolation and purification of high value pharmaceutical enzymes and other proteins.

Electrophoresis-based separations are always performed in gels as the gel assists as a molecular sieve that enhances separation. Molecules that are small in comparison with the pores in the gel easily move from the gel, while molecules with a larger size than the pores are almost immobile. Intermediate-sized molecules move from the gel with different degrees of ease. Electrophoresis is performed in a thin, vertical slab of polyacrylamide and the direction of flow is from top to bottom [90]. One of the most suitable supporting media for electrophoresis is polyacrylamide gels, formed by the polymerization of acrylamide and cross-linked by methylenebisacrylamide, as they are chemically inert and are readily formed. Electrophoresis is the reverse of gel filtration, in that all of the molecules, irrespective of size, are forced to move through the same matrix [90]. In a solution, proteins have the tendency to sediment at high centrifugal force, so it is tricky to separate a protein from a mixture by centrifugation. Enhanced separation of proteins has been accomplished by density gradient centrifugation. In this process, sucrose or glycerol is used as density gradient [90]. With more sophisticated continuous flow ultracentrifugation, whole cells, viruses and subcellular components can be separated. With ultracentrifugation, we can determine parameters such as mass and density, study something about the shape of a molecule and examine the interactions between molecules [90].

2.3.7 Finishing operations

Its well-known that all enzymes are antigenic. After issues occurred in the late 1960s when workers manufacturing enzymes suffered from severe allergic responses after breathing enzyme dusts, methods have now been applied to reduce dust formation [3]. These include delivering enzymes as liquids wherever possible, or enhancing the particle size of dry powders from 10 μm to 200–500 μm by either prilling (mixing the enzyme with polyethylene glycol and preparing small spheres by atomization) or marumerizing (mixing the enzyme with a binder and water, extruding long filaments, converting them into spheres in a marumerizer, drying them and covering them with a waxy coating) [3]. Finishing operations such as enzyme desalting, concentration, purity control and storage are of greater importance [3].

2.3.7.1 Desalting

Ammonium sulfate (AS) fractionation is frequently used in protein and enzyme purification; however, the final protein pellets have a high salt content and desalting by dialysis is required prior to subsequent analysis [91, 92]. Unnecessary amounts of inorganic salts must be eliminated from enzyme preparations. The salts existing in one enzyme are perhaps inhibitory to other enzymes present in clinical diagnostic reagents. Also, desalting may be crucial at intermediate stages of enzyme isolation, e.g. before ion-exchange adsorption. The simplest and oldest method of salt removal is dialysis [91, 92]. The main function of dialysis is to eliminate small molecules such as salts from larger ones, in particular the enzymes to be purified. Mixed bed ion exchange, the conventional method of removing small ions, has been applied to some of the more robust enzymes. Mixed bed cation and anion exchange resins are used to eliminate both positive and negative ions. Wang *et al* developed a phenol-based method for rapid desalting and concentration of proteins after ammonium sulfate fractionation of complex olive leaf protein extract [91, 92]. After re-dissolving, ammonium sulfate precipitates were desalted with phenol extraction and a lot of β-glucosidase was observed in each fraction using a specific antibody [91, 92].

2.3.7.2 Concentration of enzymes

Once extraction is performed, both intracellular and extracellular enzymes are received as a dilute solution and must be further concentrated prior to further use. On the pilot scale, precipitation and adsorption are employed to concentrate the enzymes. On the large scale ultrafiltration has been used.

Ultrafiltration A semipermeable membrane allows the separation of solvent molecules from larger enzyme molecules, as only the smaller molecules can enter the membrane when the osmotic pressure is exceeded. Ultrafiltration is performed to separate macromolecules and colloidal particles by means of a membrane. Hydraulic pressure is employed as the active force for the solvent molecules to pass through the membrane, whereas the microporous membrane does not allow the passage of large solute molecules. Membranes with exclusion limits of 2000–300 000 Da are available for ultrafiltration. Ultrafiltration has a tendency to remove small molecules. The enzyme solution is fed into a cell fitted with a membrane which retains the selected protein while being permeable to the solvent and small molecules. Negative pressure is applied to the collecting chamber and positive pressure to the solution, which acts as the driving force that causes the flow of solvent and small solute molecules across the membrane. Large volumes can be reduced to a few milliliters in only one or two hours. Diafiltration is a very useful procedure for eliminating salts and other low molecular weight substances in the form of contaminants from dilute enzyme solution [93]. During this procedure the water is fed to the ultrafilter cell so that molecules other than water are successfully eliminated from the system. Cellulose acetate and organic polymers, e.g. polysulfone, have proven to be useful as membrane materials. With the exception of cellulose acetate, these membranes can easily be cleaned with alkali or acid and stream sterilized [93].

Drying Drying procedures have long been established for food and pharmaceuticals and have now been applied successfully to enzymes. The extracellular enzymes in the food industry are often used as dried free-flowing powders. The drying of enzymes mainly depends upon the thermal stability of enzymes. The drying of enzymes has commonly been carried out by spray-drying and freeze-drying [33]. Spray-drying is a common method for producing powder from a liquid, whereas lyophilization, or freeze-drying, is a procedure for preserving food and biological materials (enzymes, proteins, vitamins, etc), floral products, drugs and whole animals. Robust enzymes may be vacuum dried or spray-dried, while the delicate ones are freeze-dried or spray-dried. Spray-drying is extensively used for extracellular bacterial enzymes, while most intracellular enzymes are dried using freeze-drying processes [33].

Vacuum evaporation Enzymes are labile molecules, so evaporation must be conducted under rather high vacuum conditions to prevent denaturation. During the production of extracellular enzymes, the diluted enzyme solution is reduced under vacuum pressure at temperatures below 40 °C. At the industrial scale, enzymes are concentrated in a similar manner, with the same equipment as is used for the concentration of fruit juices at a reduced temperature [94]. In a characteristic application, an enzyme solution was reduced in two stages from 8% or 12% to 35% solids, then to 65% solids. Vacuum-based evaporation of enzymatic solution is always achieved by the addition of sugar alcohol such as sorbitol or sugars such as lactose in higher concentrations [94].

2.3.7.3 Storage

Good storage conditions are of paramount importance in the industrial production of enzymes. One of the major concerns is that the loss of activity during storage will result in the loss of product in its purest and most expensive form. Most enzymes have a restricted shelf-life owing to inherent active-conformation liability in the minority of molecules present in high energy states, at any temperature at any particular time. This loss can be suspended by cross-linking, immobilization or by using enzyme stabilizers, e.g. sugars, certain ions and co-factors. Most enzyme products are marketed as dry powders; a few enzymes are available in a liquid or suspension form. These products usually contain compounds which decrease the loss of enzyme activity during storage and check microbial growth, i.e., enzyme stabilizers and co-factors. Proteins (in particular enzymes) must be stored at in a suitable temperature and pH range and frequently in the presence of concentrated (approximately 1 M) glycerol, sucrose, or a similar substance, for the proteins to retain activity and prevent aggregation [95]. The compounds generally used for these purposes are glycols, propylene glycol, ethylene glycol, sorbitol, mannitol, thiol, reducing agents, sodium chloride, salts of organic hydroxy acids and other salts of buffers, sodium benzoate, and esters of parahydroxy benzoic acid. Recently, compounds such as gelatin, dextrans, partially hydrolyzed collagen, gum arabic, albumin, polyamines, poly-L-lysine and glycerol monoethers have been used as stabilizers. Enzyme containers should be kept sealed to prevent the escape of the stabilizing atmosphere and also to maintain the low moisture content (below 50%) of the solid product. The storage temperature should be low but not freezing, as freezing may cause denaturation of certain enzymes [95].

2.4 Recombinant proteins from algae

Plants produce many types of proteins, including enzymes [96], hormones [97], functional vaccines [98, 99], antibodies [100, 101], and a variety of other therapeutic proteins [102]. Examples of pharmaceutical proteins that are being expressed in plants include human growth hormone, human serum albumin, interferons, and

Table 2.6. Recombinant proteins from algae.

Algae name	Products	Expression location
Amphidinium sp	β-glucoronidase	Nucleus
C. reinhardtii	Anti-HSV glycoprotein D Isc, Anti-CD22-gelonin sc, Anti-CD22-ETA sc, VP1-CTB, 14FN3, Metallothionein-2, TRAIL, Allophycocyanin, Anti-PA 83 anthrax IgG1, E7GGG, VEGF, HMGB1	Chloroplast
C. reinhardtii	GBSS-PfMSP$_{1-19}$ GBSS-PbAMA$_{1\text{-}C}$	Nucleus

erythropoietin [103]. Several recent reviews describe the significant advances in this field [104–111]. The first Food and Drug Administration-approved plant-made biopharmaceutical is glucocerebrosidase; this injectable product produced in carrot cells is used to treat a genetic/metabolic disorder [115]. Another major advancement is the application of chloroplast-generated biopharmaceuticals for toxic antibody suppression [112]. Chan and Daniell noted the challenges of plant cells-produced vaccine antigens toward clinical development, though these oral vaccines may be quite attractive for both human and veterinary purposes [113, 114].

As photosynthetic organisms, microalgae can efficiently convert solar energy into biomass. Microalgae are currently used as an important source of valuable natural biologically active molecules, such as carotenoids, chlorophyll, long-chain poly-unsaturated fatty acids, phycobiliproteins, carotenoids and enzymes. Significant advances have been achieved in microalgae biotechnology over the last decade, and the use of microalgae as bioreactors for expressing recombinant proteins is receiving increased interest. Compared with the bioreactor systems that are currently in use, microalgae may be an attractive alternative for the production of pharmaceuticals, recombinant proteins and other valuable products (table 2.6) Products synthesized via the genetic engineering of microalgae include vaccines, antibodies, enzymes, blood-clotting factors, immune regulators, growth factors, hormones, and other valuable products, such as the anticancer agent Taxol. Here, we briefly compare the currently used bioreactor systems, summarize the progress in genetic engineering of microalgae, and discuss the potential for microalgae as bioreactors to produce pharmaceuticals.

References

[1] Payen A and Persoz J 1833 Mémoire sur la Diastase, les principaux Produits de ses Réactions, et leurs applications aux arts industriels *Ann. Chim. Phys.* **53** 73–92

[2] Garg S K and Johri B N 1994 Rennet: current trends and future research *Food Rev. Int.* **10** 3

[3] Robinson P K 2015 Enzymes: principles and biotechnological applications *Essays Biochem.* **59** 1–41

[4] Gurung N, Ray S, Bose S and Rai V 2013 A broader view: microbial enzymes and their relevance in industries, medicine, and beyond *Biomed. Res. Int.* **2013** 329121

[5] Haitjema C H *et al* 2017 A parts list for fungal cellulosomes revealed by comparative genomics *Nat. Microbiol.* **2** 17087

[6] Eckhard U *et al* 2017 Discovery of a proteolytic flagellin family in diverse bacterial phyla that assembles enzymatically active flagella *Nat. Commun.* **8** 521

[7] Mugabo Y 2016 A mammalian glycerol-3-phosphate phosphatase *Proc. Natl Acad. Sci.* **113** E430–9

[8] Jacobs A L and Schär P 2012 DNA glycosylases: in DNA repair and beyond *Chromosoma* **121** 1–20

[9] Aleku G A 2017 Turner a reductive aminase from *Aspergillus oryzae Nat. Chem.* **9** 961–9

[10] Anbu P, Gopinath S C B, Cihan A C and Chaulagain B P 2013 Microbial enzymes and their applications in industries and medicine *Biomed. Res. Int.* **2013** 204014

[11] Murdoch B M and Murdoch G K 2015 Genetics of prion disease in cattle *Bioinform. Biol. Insights* **9** 1–10

[12] Lieberman I and Gingold J L 1962 Techniques for study of mammalian cell cultures: extraction of enzymes *Methods Enzymol.* **5** 119–22

[13] Djekrif-Dakhmouche S, Gheribi-Aoulmi Z, Meraihi Z and Bennamoun L 2006 Application of a statistical design to the optimization of culture medium for α-amylase production by *Aspergillus niger* ATCC 16404 grown on orange waste powder *J. Food Eng.* **73** 190–7

[14] Amann R I, Ludwig W and Schleifer K H 1995 Phylogenetic identification and *in situ* detection of individual microbial cells without cultivation *Microbiol. Rev.* **59** 143–69

[15] Ferrer M, Martínez-Martínez M, Bargiela R, Streit W R, Golyshina O V and Golyshin P N 2016 Estimating the success of enzyme bioprospecting through metagenomics: current status and future trends *Microb. Biotechnol.* **9** 22–34

[16] Narayanasamy S, Muller E E, Sheik A R and Wilmes P 2015 Integrated omics for the identification of key functionalities in biological wastewater treatment microbial communities *Microb. Biotechnol.* **8** 363–8

[17] Ghildyal N P, Lonsane B K, Sreekantiah K R and Sreenivasa M V 1985 Economics of submerged and solid state fermentation for the production of amyloglucosidase *J. Food Sci. Technol.* **22** 171–6

[18] Lin H, Cheng W, Ding H T, Chen X J, Zhou Q F and Zhao Y H 2010 Direct microbial conversion of wheat straw into lipid by a cellulolytic fungus of *Aspergillus oryzae* A-4 in solid-state fermentation *Bioresour. Technol.* **101** 7556–62

[19] Souza P M *et al* 2015 A biotechnology perspective of fungal proteases *Braz. J. Microbiol.* **46** 337–46

[20] Liu H F, Ma J, Winter C and Bayer R 2010 Recovery and purification process development for monoclonal antibody production *MAbs.* **2** 480–99

[21] Liang J, Niu Q, Xu X, Luo Y, Zhou X, Deng Z and Wang Z 2009 Effective elimination of nucleic acids from bacterial protein samples for optimized blue native polyacrylamide gel electrophoresis *Electrophoresis* **30** 2454–9

[22] Wingfield P T 2001 Protein precipitation using ammonium sulfate *Curr. Protoc. Protein Sci.* APPENDIX 3: Appendix–3F

[23] Wingren C and Hansson U B 1996 Enzyme conformational alterations detected by partition column chromatography *J. Chromatogr. B Biomed. Appl.* **680** 113–22

[24] Grimonprez B and Johansson G 1996 Liquid–liquid partitioning of some enzymes, especially phosphofructokinase, from *Saccharomyces cerevisiae* at sub-zero temperature *J. Chromatogr.* B **680** 55–63

[25] Johansson G and Kopperschläger G 1987 Effects of organic solvents on the partitioning of enzymes in aqueous two-phase systems *J. Chromatogr.* **388** 295–305

[26] Grilo A L, Raquel Aires-Barros M and Azevedo A M 2016 Partitioning in aqueous two-phase systems: fundamentals, applications and trends *Sep. Purif. Rev.* **45** 68–80

[27] Albertsson P Å 1986 *Partition of Cell Particles and Macromolecules* 3rd edn (Hoboken, NJ: Wiley Intersciences)

[28] Hatti-Kaul R 2001 Aqueous two-phase systems. A general overview *Mol. Biotechnol.* **19** 269–77

[29] Amid M, Manap Y, Azmira F, Hussin M and Sarker Z I 2015 A novel liquid/liquid extraction process composed of surfactant and acetonitrile for purification of polygalacturonase enzyme from *Durio zibethinus J. Chromatogr.* B **993–4** 1-8

[30] Yamadera K and DiMagno E P 1994 Separation of canine pancreatic juice proteins by hydrophobic interaction chromatography preserves enzyme activity *Pancreas* **9** 32–6

[31] Tomme P, McCrae S, Wood T M and Claeyssens M 1988 Chromatographic separation of cellulolytic enzymes *Methods Enzymol.* **160** 187–93

[32] Alves N J, Turner K B, Medintz I L and Walper S A 2016 Protecting enzymatic function through directed packaging into bacterial outer membrane vesicles *Sci. Rep.* **6** 24866

[33] Jesus S S and Filho M R 2014 Drying of α-amylase by spray drying and freeze-drying—a comparative study *Braz. J. Chem. Eng.* **31**

[34] Roskoski R 2007 *Modulation of Enzyme Activity. xPharm: The Comprehensive Pharmacology Reference* pp 1–11

[35] Cirz R T, Chin J K, Andes D R, de Crécy-Lagard V, Craig W A and Romesberg F E 2005 Inhibition of mutation and combating the evolution of antibiotic resistance *PLoS Biol.* **3** e176

[36] Fijalkowska I J, Dunn R L and Schaaper R M 1993 Mutants of *Escherichia coli* with increased fidelity of DNA replication *Genetics* **134** 1023–30

[37] Mitchell J L, Choe C Y and Judd G G 1996 Feedback repression of ornithine decarboxylase synthesis mediated by antizyme *Biochem. J.* **320** 755–60

[38] Sanchez S and Demain A L 2008 Metabolic regulation and overproduction of primary metabolites *Microb. Biotechnol.* **1** 283–319

[39] Fujita Y 2009 Carbon catabolite control of the metabolic network in *Bacillus subtilis Biosci. Biotechnol. Biochem.* **73** 245–59

[40] Stülke J and Hillen W 2000 Regulation of carbon catabolism in *Bacillus* species *Annu. Rev. Microbiol.* **54** 849–80

[41] Kleijn R J, Buescher J M, Le Chat L, Jules M, Aymerich S and Sauer U 2010 Metabolic fluxes during strong carbon catabolite repression by malate in *Bacillus subtilis J. Biol. Chem.* **285** 1587–96

[42] Iwakuma H, Koyama Y, Miyachi A, Nasukawa M, Matsumoto H, Yano S, Ogihara J and Kasumi T 2016 Generation of a glucose de-repressed mutant of *Trichoderma reesei* using disparity mutagenesis *Biosci. Biotechnol. Biochem.* **80** 486–92

[43] ter Schure E G, van Riel N A W and Verrips C T 2000 The role of ammonia metabolism in nitrogen catabolite repression in *Saccharomyces cerevisiae FEMS Microbiol. Rev.* **24** 67–83

[44] Vieille C and Zeikus G J 2001 Hyperthermophilic enzymes: sources, uses, and molecular mechanisms for thermostability *Microbiol. Mol. Biol. Rev.* **65** 1–43

[45] Renge V C, Khedkar S V and Nandurkar N R 2012 Enzyme synthesis by fermentation method: a review *Sci. Rev. Chem. Commun.* **2** 585–90

[46] Lizardi-Jiménez M A and Hernández-Martínez R 2017 Solid state fermentation (SSF): diversity of applications to valorize waste and biomass *Biotech* **7** 44

[47] Revankar M S, Desai K M and Lele S 2007 Solid-state fermentation for enhanced production of laccase using indigenously isolated *Ganoderma* sp *Appl. Biochem. Biotechnol.* **143** 16–26

[48] Gruyter H A B 2015 *Process Technology: An Introduction* (Berlin: Walter de Gruyter), pp 78–87

[49] Owusu-Apenten R 2004 *Introduction of Food Chemistry* (Boca Raton, FL: CRC Press), p 196

[50] Manen J F, Sinitsyna O, Aeschbach L, Markov A V and Sinitsyn A 2005 A fully automatable enzymatic method for DNA extraction from plant tissues *BMC Plant Biol.* **5** 23

[51] Boland M J 1992 Extraction and purification of proteins from animal tissue using aqueous phase systems *Australas Biotechnol.* **2** 47-52

[52] Geciova J, Bury D and Jelen P 2002 Methods for disruption of microbial cells for potential use in the dairy industry—a review *Int. Dairy* **12** 541–53

[53] Chisti Y and Moo-Young M 1986 Disruption of microbial cells for intracellular products *Enzyme Microb. Technol.* **8** 194–204

[54] Crapisi A, Lante A, Pasini G and Spettoli P 1993 Enhanced microbial cell lysis by the use of lysozyme immobilised on different carriers *Process Biochem.* **28** 17–21

[55] Harrison S T L 2011 Cell distruption *Comprehensive Biotechnology* 2nd edn ed M Moo-Young (Oxford: Elsevier), vol 2, 619–39

[56] Middelberg A 1995 Process-scale disruption of microorganisms *Biotechnol. Adv.* **13** 491–551

[57] Nossal N G and Heppel L A 1966 The release of enzymes by osmotic shock from *Escherichia coli* in exponential phase *J. Biol. Chem.* **241** 3055–62

[58] Feliciello I and Chinali G 1993 A modified alkaline lysis method for the preparation of highly purified plasmid DNA from *Escherichia coli Anal. Biochem.* **212** 394–401

[59] Chaiyarit S and Thongboonkerd V 2009 Comparative analyses of cell disruption methods for mitochondrial isolation in high-throughput proteomics study *Anal. Biochem.* **394** 249–58

[60] Stathopulos P B, Scholz G A, Hwang Y M, Rumfeldt J A O, Lepock J R and Meiering E M 2004 Sonication of proteins causes formation of aggregates that resemble amyloid *Protein Sci.* **13** 3017–27

[61] McGann L E, Yang H Y and Walterson M 1988 Manifestations of cell damage after freezing and thawing *Cryobiology* **25** 178–85

[62] Tansey W P 2006 Freeze-thaw lysis for extraction of proteins from mammalian cells *CSH Protoc.* **2006(7)**

[63] Qi X, Sun Y and Xiong S 2015 A single freeze-thawing cycle for highly efficient solubilization of inclusion body proteins and its refolding into bioactive form *Microb. Cell Fact.* **14** 24

[64] Goldberg S 2008 Mechanical/physical methods of cell disruption and tissue homogenization *Methods Mol. Biol.* **424** 3–22

[65] Giza C C and Hovda D A 2001 The neurometabolic cascade of concussion *J. Athl. Train.* **36** 228–35

[66] Nadar S S, Pawar R G and Rathod V K 2017 Recent advances in enzyme extraction strategies: a comprehensive review *Int. J. Biol. Macromol.* **101** 931–57

[67] Glyk A, Scheper T and Beutel S 2015 PEG-salt aqueous two-phase systems: an attractive and versatile liquid-liquid extraction technology for the downstream processing of proteins and enzymes *Appl. Microbiol. Biotechnol.* **99** 6599–616

[68] Asenjo J A and Andrews B A 2011 Aqueous two-phase systems for protein separation: a perspective *J. Chromatogr.* A **1218** 8826–35

[69] Bruton C J 1983 Large-scale purification of enzymes *Philos. Trans. R. Soc. Lond.* B **300** 249–61

[70] Palmer I and Wingfield P T 2004 Preparation and extraction of insoluble (inclusion-body) proteins from *Escherichia coli Curr. Protoc. Protein Sci.* ch 6

[71] Zabriskie D W and DiPaolo M 1987 Removal of nucleic acid contaminants using nuclease enzymes during protein isolation *Biotechnol. Bioeng.* **32** 100–4

[72] Ryan B J 2011 Differential precipitation and solubilization of proteins *Methods Mol. Biol.* **681** 203–13

[73] Duong-Ly K C and Gabelli S B 2014 Salting out of proteins using ammonium sulfate precipitation *Methods Enzymol.* **541** 85–94

[74] Talley K and Alexov E 2010 On the pH-optimum of activity and stability of proteins *Proteins* **78** 2699–706

[75] Askonas B A 1951 The use of organic solvents at low temperature for the separation of enzymes. Application to aqueous rabbit muscle extract *Biochem. J.* **48** 42–8

[76] Mahadevan H and Hall C K 1990 Statistical-mechanical model of protein precipitation by nonionic polymer *AIChE* **36** 1517–28

[77] Ingham K C 1984 Protein precipitation with polyethylene glycol *Methods Enzymol.* **104** 351–6

[78] Atha D H and Ingham K C 1981 Mechanism of precipitation of proteins by polyethylene glycols. Analysis in terms of excluded volume *J. Biol. Chem.* **256** 12108–17

[79] Sim S L, He T, Tscheliessnig A, Mueller M, Tan R B and Jungbauer A 2012 Protein precipitation by polyethylene glycol: a generalized model based on hydrodynamic radius *J Biotechnol.* **157** 315–9

[80] Durbin S D and Feher G 1996 Protein crystallization *Annu. Rev. Phys. Chem.* **47** 171–204

[81] McPhersona A and Gavirab J A 2014 Introduction to protein crystallization *Acta Crystallogr.* F **70** 2–20

[82] Asenjo J A and Andrews B A 2009 Protein purification using chromatography: selection of type, modelling and optimization of operating conditions *J. Mol. Recognit.* **22** 65–76

[83] Chase H A 1994 Purification of proteins by adsorption chromatography in expanded beds *Trends Biotechnol.* **12** 296–303

[84] Yigzaw Y, Hinckley P, Hewig A and Vedantham G 2009 Ion exchange chromatography of proteins and clearance of aggregates *Curr. Pharm. Biotechnol.* **10** 421–6

[85] Jungbauer A and Hahn R 2009 Ion-exchange chromatography *Methods Enzymol.* **463** 349–71

[86] Rahimzadeh M, Poodat M, Javadpour S, Qeshmi F I and Shamsipour F 2016 Purification, characterization and comparison between two new L-asparaginases from bacillus PG03 and bacillus PG04 *Open Biochem J* **10** 35–45

[87] Hagel L 2001 Gel-filtration chromatography *Curr Protoc Mol Biol.* ch 10

[88] Urh M, Simpson D and Zhao K 2009 Affinity chromatography: general methods *Methods Enzymol.* **463** 417–38

[89] Bernik M B and Kwaan H C 1969 Plasminogen activator activity in cultures from human tissues. An immunological and histochemical study *J. Clin. Invest.* **48** 1740–53

[90] Berg J M, Tymoczko J L and Stryer L 2002 The purification of proteins is an essential first step in understanding their function *Biochemistry* 5th edn (New York: Freeman) section 4.1

[91] Smarason S V and Smith A V 2003 Method for desalting nucleic acids *United State patent US 2003/0186247 A1*, deCODE genetics ehf

[92] Wang W, Liu Q J and Cui H 2007 Rapid desalting and protein recovery with phenol after ammonium sulfate fractionation *Electrophoresis* **28** 2358–60

[93] Aehle W 2008 *Enzymes in Industry* 3rd edn (New York: Wiley), p 36

[94] Fazaeli M, Hojjatpanah G and Emam-Djomeh Z 2013 Effects of heating method and conditions on the evaporation rate and quality attributes of black mulberry (*Morus nigra*) juice concentrate *J. Food Sci. Technol.* **50** 35–43

[95] Simpson R J 2010 Stabilization of proteins for storage *CSH Protoc.* **5** pdb.top79

[96] Georgianna D R *et al* 2013 Production of recombinant enzymes in the marine alga *Dunaliella tertiolecta Algal. Research* **2** 2–9

[97] Yan N, Fan C, Chen Y and Hu Z 2016 The potential for microalgae as bioreactors to produce pharmaceuticals *Int. J. Mol. Sci.* **17** 962

[98] Bhatia S, Rathee P, Sharma K, Chaugule B B, Kar N and Bera T 2013 Immunomodulation effect of sulphated polysaccharide (porphyran) from *Porphyra vietnamensis Int. J. Biolog. Macromol.* **57** 50–6

[99] Bhatia S, Kumar V, Sharma K, Nagpal K and Bera T 2014 Significance of algal polymer in designing Amphotericin B nanoparticles *Sci. World J.* **2014** 1

[100] Bhatia S, Sharma K, Nagpal K and Bera T 2015 Investigation of the factors influencing the molecular weight of porphyran and its associated antifungal activity *Bioact Carbohydr. Diet. Fiber* **5** 153–68

[101] Bhatia S, Sharma K, Nagpal K, Sharma S and Bera T 2015 Evaluation of pharmacognostical, phytochemical and anti-microbial properties of *Porphyra vietnamensis Int. J. Green Pharm.* **2** 1–5

[102] Bhatia S, Sharma K and Bera T 2015 Structural characterization and pharmaceutical properties of porphyran *Asian J. Pharm.* **9** 1–3

[103] Bhatia S, Sharma K, Sharma A, Nagpal K and Bera T 2015 Anti-inflammatory, analgesic and antiulcer properties of *Porphyra vietnamensis Avicenna J. Phytomed.* **5** 1–7

[104] Bhatia S *et al* 2009 Novel algal polysaccharides from marine source: porphyran *Pharmacog. Rev.* **2** 271–6

[105] Bhatia S *et al* 2011 Mycosporine and mycosporine-like amino acids: A paramount tool against ultra violet irradiation *Pharmacogn Rev.* **5** 138–46

[106] Bhatia S *et al* 2010 Broad-spectrum sun-protective action of Porphyra-334 derived from *Porphyra vietnamensis. Pharmacog. Res.* **2** 45–9

[107] Bhatia S, Namdeo A G and Nanda S 2015 Factors affecting the gelling and emulsifying property of natural polymer *Sys. Pharm. Rev.* **9** 93–101

[108] Bhatia S, Sharma K, Sharma A, Namdeo A G and Chaugule B B 2011 Anti-oxidant potential of Indian Porphyra *Pharmacologyonline* **1** 248–57

[109] Bhatia S *et al* 2015 *Modern Applications of Plant Biotechnology in Pharmaceutical Sciences* (Amsterdam: Elsevier), pp 277–87

[110] Bhatia S 2016 *Nanotechnology in Drug Delivery: Fundamentals, Design, and Applications* (Boca Raton, FL: CRC Press), pp 121–7
[111] Bhatia S and Goli D 2016 *Leishmaniasis: Biology, Control and New Approaches for Its Treatment* (Boca Raton, FL: CRC Press), pp 177–9
[112] Bhatia S 2016 *Natural Polymer Drug Delivery Systems: Nanoparticles, Plants, and Algae* (Berlin: Springer), pp 117–27
[113] Bhatia S 2016 *Systems for Drug Delivery: Safety, Animal, and Microbial Polysaccharides* (Berlin: Springer), pp 122–7
[114] Bhatia S 2018 *Introduction to Pharmaceutical Biotechnology* Volume 1 (Bristol: IOP Publishing), pp 178–87

IOP Publishing

Introduction to Pharmaceutical Biotechnology, Volume 2
Enzymes, proteins and bioinformatics
Saurabh Bhatia

Chapter 3

Industrial enzymes and their applications

3.1 Industrial enzymes

Microbial enzyme production is an essential industrial process, owing to the extraordinary performance and multitude of applications of enzymes from various microbes. These microbes are active under a varied range of physical and chemical conditions [1]. Microbial enzymes are a potential source of replacement in the absence or deficiency of human enzymes. In addition, microbial enzymes are the favored source of industrial enzymes as they can be produced in large quantities in a short time. Microbes have short generation times, and genetic manipulation can be performed more easily on bacterial cells to enhance enzyme production. The industry is still seeking novel microbial strains to furnish different enzymes to meet requirements [1]. Of the 3000 enzymes reported to date, only a few are industrially used, such as hydrolytic enzymes which degrade naturally occurring polymers such as starch, proteins, pectins and cellulose. Progress in biochemistry leading to the isolation and characterization of many enzymes made it essential to standardize the enzyme nomenclature. Based on the type of reaction catalyzed, enzymes are classified into six main classes, as listed in table 3.1.

Broad applications of enzymes in different bioprocesses can be employed to deliver a range of products in different industries. Table 3.2 summarizes a number of applications used to deliver a range of products. The industrial production of microbial enzymes is presented in figure 3.1.

3.2 Bacterial α-amylases

Amylases are one of the key enzymes used in industry. These enzymes breakdown starch molecules into polymers composed of glucose units [2, 3]. Amylases have wide applications at the industrial level, e.g. in the food, fermentation and pharmaceutical industries. Based on the source, α-amylases are classified into various categories such as α-amylases from bacteria, fungi, plants, animals and microorganisms, whereas enzymes from fungal and bacterial sources have a

doi:10.1088/978-0-7503-1302-5ch3

Table 3.1. Enzyme classification.

S.no.	Enzymes	Class	Reactions
1	Dehydrogenases, oxidases, oxygenases, peroxidases	Oxidoreductases	Transfer of hydrogen or oxygen or electrons between molecules.
2	Fructosyltransferases, transketolases, acyltransferases, transaminases	Transferases	Transfer of groups of atoms from one molecule to another.
3	Isomerases, epimerases, racemases	Isomerases	Transfer of a group from one position to another within one molecule.
4	Pectate lyases, hydratases, dehydratases, decarboxylases, fumarase, argino succinase	Lyases	Non-hydrolytic cleavage by elimination or addition reactions.
5	Proteases, amylases, acylases, lipases, phosphatases, cutinases	Hydrolases	Hydrolytic cleavage of bonds.
6	Synthetases, ligases	Ligases	Covalent joining of two molecules coupled with the hydrolysis of an energy rich bond in ATP or similar triphosphates.

considerable number of applications in industry. α-amylase production is necessary for the conversion of starches into oligosaccharides. Starch, a major storage product of many economically important crops, is a vital constituent of the human diet. Starch-converting enzymes are used in the production of maltodextrin, modified starches, and glucose and fructose syrups. A large number of microbial α-amylases have applications in different industrial sectors such as the food, textiles, paper and detergent industries [2, 3]. α-amylase production is usually performed using submerged fermentation, although solid-substrate fermentation systems also provide reliable technology. The thermostability, pH profile, pH stability, Ca independence and other properties of each α-amylase play an important role in the development of fermentation processes [2, 3].

α-amylases (both bacterial and fungal) are enzymes which have been produced industrially on a large scale. The general procedure for amylase production is depicted in figure 3.2. The bacterial α-amylase endoenzyme breaks down α-1,4 bonds in amylose and amylopectin, resulting in a sudden fall in the viscosity of gelatinized starch solutions (endohydrolysis of starch, also known as starch liquefaction) [2, 3]. After the action of α-amylase, the final product recovered is dextrans, together with small quantities of glucose and maltose. Before treatment with amylases, the native starch is exposed to hydration (gelatinization of starch). According to the literature, α-amylase derived from *Bacillus amyloliquefaciens* is

Table 3.2. Enzymes and their applications.

Enzyme	Microorganism(s)	Function
Acid proteinase	*Aspergillus* spp.	Milk coagulation
Acyltransferase	*Bacillus* sp. APB-6	Synthesis of hydroxamic acids
Alkaline protease	*Alcaligenes faecalis*	Dehairing, bating
Amidase	*Rhodococcus erythropolis*	Degradation of nitriles containing wastes
Aminopeptidase	*Lactobacillus* spp.	Faster cheese ripening
Aminopeptidases	*Lactobacillus brevis*, *Lactobacillus plantarum*	Protein breakdown during mashing
Amylase	*Aspergillus* spp., *Bacillus* spp.	Flour adjustment, bread softness
Amylase	*Bacillus licheniformis*	De-inking, drainage improvement
Amylase	*Aspergillus* spp., *Bacillus subtilis*	Carbohydrate stain removal
Amylase	*Aspergillus* spp., *B. subtilis*	Fiber splitting
Amylase	*B. licheniformis*, *Aspergillus* spp.	Bioremediation of vegetable waste
Amyloglucosidase	*Aspergillus niger*	Starch hydrolysis for bioremediation
Catalase	*A. niger*	Cheese processing
Cellulase	*A. niger*, *Trichoderma atroviride*	Fruit liquefaction
Cellulase	*Bacillus* spp., *A. niger*	De-inking, drainage improvement
Cellulase	*A. niger*, *Bacillus* spp.	Color clarification
Cutinase	*Fusarium solani f. pisi*	Triglyceride removal
Cutinase	*F. solani f. pisi*	Degradation of plastics, polycaprolactone
Endoglycosidase	*Mucor hiemalis*	Teeth and gum tissue care
Glucose isomerase	*Corynebacterium* spp., *Streptomyces murinus*	Production of high-fructose corn syrup
Glucose oxidase	*A. niger*, *Penicillium chrysogenum*	Dough strengthening
Glucose oxidase	*A. niger*	Oxygen removal from beer
Glucose oxidase	*A. niger*, *P. chrysogenum*	Polymerization of anilines
Glycosyl tranferase	*Bacillus* spp.	Synthesis of oligosaccharides
Laccase	*B. subtilis*	Non-chlorine bleaching, delignification
Laccase	*Trametes hirsute*	Polymerization of bisphenol A
Laccase	*B. subtilis*, *Trametes versicolor*	Hair dye
Laccase	*T. versicolor*, *B. subtilis*	Production of textile dyes, cosmetic pigments, flavor agents and pesticides
Laccase	*T. versicolor*	Degradation of waste containing olefin unit, polyurethane and phenolic compounds
Lactase (β-galactosidase)	*Escherichia coli*, *Kluyveromyces* spp.	Lactose reduced milk and whey products

(*Continued*)

Lignin peroxidase	*Phanerochaete chrysosporium*, *Coprinus cinereus*	Degradation of phenolic compounds
Limoninase	*A. niger*, *Aspergillus oryzae*	Debittering
Lipase	*A. niger*, *A. oryzae*	Faster cheese ripening, flavor customized cheese
Lipase	*A. niger*	Dough stability and conditioning
Lipase	*Candida antarctica*	Pitch control
Lipase	*C. antarctica*	Polycondensation, ring-opening polymerization of lactones, carbonates
Lipase	*A. oryzae*, *Aspergillus flavus*	Fat stain elimination
Lipase	*A. oryzae*, *A. flavus*,	Degreasing
Lipase	*A. oryzae*, *A. flavus*	Skin care
Lipase	*A. oryzae*, *A. flavus*	Synthesis of pharmaceuticals, polymers, biodiesels, biosurfactants
Lipase	*A. oryzae*, *Candida tropicalis*	Degradation of crude oil hydrocarbons
Maltogenic α-amylase	*Bacillus stearothermophilus*	Enhancing the shelf life of breads
Manganese peroxidase	*P. chrysosporium*, *C. cinereus*	Degradation of phenolic compounds
Mannanase	*Bacillus* spp.	Mannan spot removal
Naringinase	*A. niger*	Debittering
Neutral protease	*A. niger*, *A. flavus*, *B. subtilis*	Dehairing, soaking
Neutral proteinase	*B. subtilis*, *A. oryzae*	Faster cheese ripening, debittering
Nitrile hydratase	*Rhodococcus rhodochrous* PA-34, *Bacillus* spp. APB-6	Synthesis of acrylamide, butyramide and nicotinamide
Nitrile hydratase	*Rhodococcus* spp.	Degradation of nitriles containing wastes
Oxygenase	*Pseudomonas* spp., *Rhodococcus* spp.	Degradation of halogenated contaminants
Pectinase	*A. oryzae*, *Penicillium funiculosum*	Depectinization
Phytase	*A. niger*	Hydrolyze phytic acid to release phosphorous
Protease	*A. niger*	Restrict haze formation
Protease	*B. subtilis*	Biofilm removal
Protease	*A. oryzae*, *B. subtilis*	Protein stain removal
Protease	*A. niger*, *A. flavus*, *B. subtilis*	Removal of dead skin
Protease	*Chrysosporium keratinophilum*	Bioremediation of keratinic wastes
Pullulanase	*Bacillus* spp., *Klebsiella* spp.	Starch saccharification
Superoxide dismutase	*Corynebacterium glutamicum*, *Lactobacillus plantarum*	Free radical scavenging, skin care

Transglutaminase	*Streptomyces* spp.	Protein crosslinking
Transglutaminase	*Streptoverticillium* spp., *Streptomyces* spp.	Laminated dough strength
Transglutaminase	*Streptomyces mobaraensis*	Protein crosslinking
Tyrosinase	*Trichoderma reesei*	Polymerization of lignin and chitosan
Xylanase	*A. niger*	Dough conditioning
Xylanase	*Aspergillus* spp., *Bacillus* spp.	Enhanced digestibility of starch
Xylanase	*Trichoderma reesei*, *Thermomyces lanuginosus*, *Aureobasidium pullulans*	Bleach boosting
α-amylase	*Bacillus* spp., *Aspergillus* spp.	Starch hydrolysis
β-amylase	*Bacillus* spp., *Streptomyces* spp., *Rhizopus* spp.	Starch hydrolysis
β-glucanase	*B. subtilis*, *Aspergillus* spp.	Restrict haze formation
β-glucanase	*A. niger*	Digestive aid

active up to 90 °C and this high temperature stability (for gelatinization purpose) has been used at the industrial scale. Another strain belonging to the same genus, *Bacillus licheniformis*, has also been used industrially for the production of α-amylases. α-amylases are metallo-proteins, containing at least one mol of calcium ions (they are stabilized by Ca^{++} and inhibited by chelate forming agents) [2, 3]. The activity and stability of α-amylase obtained from *B. licheniformis* is not dependent on the calcium content of the solution, in contrast to α-amylase from *B. amylolequefaciens*. pH 6.5–7 is the optimum range of pH, at which these enzymes show maximum activity. At neutral pH and temperature (30–40 °C), fermentation of bacterial amylases is performed in submerged culture. Cereal meal and starch rich medium are usually used along with an organic source of nitrogen. After an interval of 10–20 h, α-amylase formation starts and lasts for another 100 h [2, 3]. The pH of the medium must be less than 6 during fermentation to avoid the denaturation of α-amylase. The α-amylase is accompanied by other extracellular enzymes. *B. licheniformis* synthesizes a serine protease, whereas *B. amyloliquefaciens* synthesizes a neutral protease along with hemicellulase and β-glucanase. Starch liquefaction using α-amylase is performed in a continuous or batch reactor. The degree of anticipated starch hydrolysis is determined by the intended later use. The manufacture of coating compounds, surface sizing of paper and paints requires a colloidal starch solution with a final viscosity [2, 3]. Incomplete degradation of starch is brought about by α-amylase to attain a specific viscosity. During the production of glucose syrup, α-amylase is used in the first step of enzymatic degradation resulting in a mixture of glucose and fructose with high-fructose content. During the production of dextrans, the degradation of starch required is as high as 10 dextrose

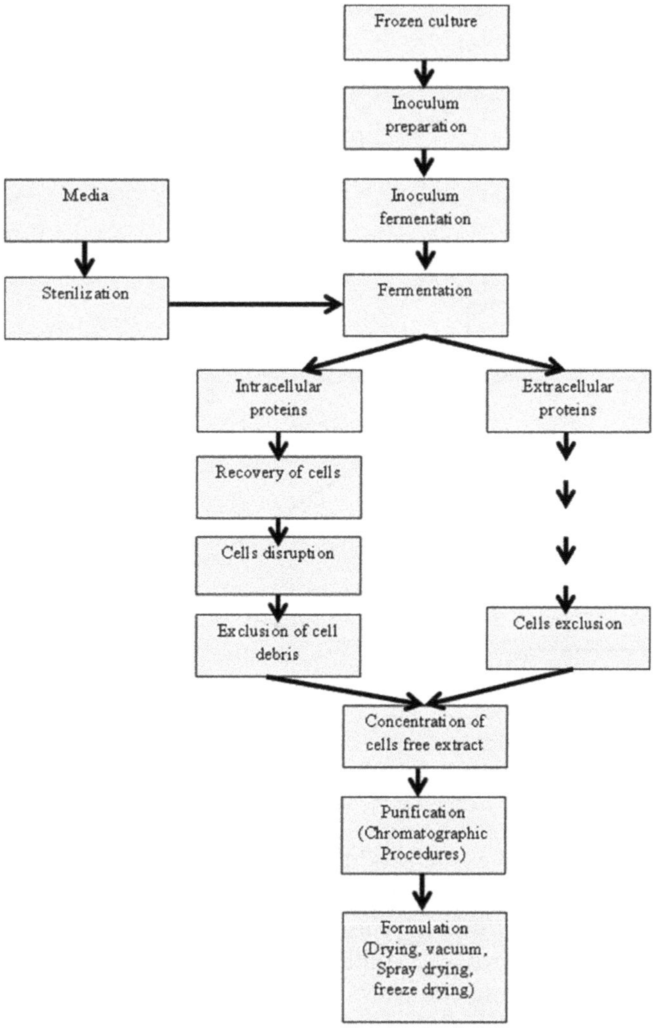

Figure 3.1. Basic outline of industrial production of enzymes.

equivalents. The increasing production of alcohol as a fuel from starch-containing raw material is opening greater prospects for the utilization of α-amylases and glucoamylases at large scales [2, 3].

3.3 Fungal α-amylases

Fungal amylases have been extensively used in large-scale industrial production owing to their many merits, for example, their cost effectiveness, consistency, smaller temporal and spatial requirements, and ease of process modification and optimization [4]. These enzymes account for around 30% of the world's enzyme production [3]. Amylases belong to a class of starch degrading enzymes that catalyze the breakdown of internal glycosidic bonds in polysaccharides with the retention of

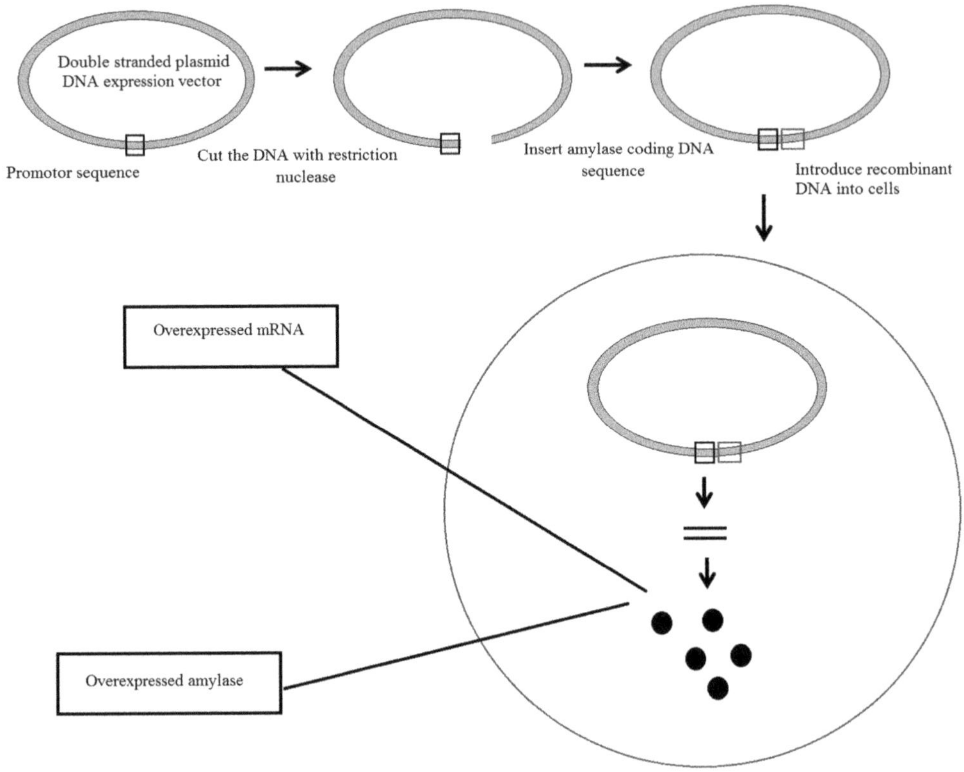

Figure 3.2. Amylase production using rDNA technology.

the anomeric configuration in the products. As mentioned above, most amylases are metalloenzymes, which require calcium ions (Ca^{2+}) for their activity, structural integrity and stability [5]. Endo-amylases, exo-amylases, debranching enzymes and glycosyltransferases are further examples of the starch degrading class of enzymes [6]. In the starch processing industry, enzymatic hydrolysis is favored over acid hydrolysis, owing to the specificity of the reaction, stability of the synthesized products, lower energy requirements and exclusion of neutralization steps [7]. Due to the high demand for these enzymes at large scales, there is great interest in developing enzymes with better properties, e.g. raw starch degrading amylases suitable for industrial applications and cost effective production techniques. Most amylases have been synthesized from soil fungi, e.g. *Aspergillus*, *Penicillium* and *Rhizopus* [7]. Only little evidence is available on amylases from endophytic fungi, which are primarily explored for beneficial secondary metabolites with different bioactivity [8]. Fungal α-amylases are mainly derived from the *Aspergillus* genus (*A. niger*, *A. oryzae*). Because of their low deactivation temperature, low optimum pH (pH 4–5) and high saccharifying action, they significantly differ from the bacterial amylases [3–8]. They are less fit for the liquefaction of starch than bacterial α-amylases. The synthesis of fungal α-amylase is accomplished in solid-substrate culture and occasionally in submerged culture for specific strains of *Aspergillus*.

The fermentation medium used remains the same as for bacterial α-amylases, but the concentration of glucose hinders the formation of amylase, thus the concentration of glucose in the fermentation medium must be measured and kept low [3–8]. Other fungal enzymes are produced in solid-substrate culture. The protease content is found to be higher when produced by solid-substrate culture in comparison to α-amylase produced by submerged culture [3–8]. The glucoamylase level is low in both methods. Fungal α-amylases are employed in the manufacturing of baked goods, where this enzyme is added to amylase-poor flour to reduce and standardize the proving time of dough. In addition, they are also used in the production of maltose-rich syrups and various other areas of the food industry [3–8].

3.4 Bacterial proteases

Microbial proteases are among the most important hydrolytic enzymes and have been studied widely. Proteases derived from micro-organisms have gained great attention in the last decade because of their biotechnological applications in various industrial processes, e.g. detergent, textile, leather, dairy and pharmaceutical preparations. Proteolytic enzymes derived from micro-organisms are preferred in industrial applications of enzymes owing to their technical and economic advantages [9]. The category of microbial proteases signifies one of the largest groups of industrial enzymes and accounts for approximately 60% of total enzyme sales in the world [9]. The majority of bacterial proteases are produced from *B. licheniformis*. The bacterium is cultured in a fermentation medium supplemented by protein and protein hydrolysate. The medium's pH is adjusted to neutral and the temperature is kept between 30–40 °C. Supplementation of carbohydrates *in situ* enhances the productivity of enzymes. Production begins when the culture passes into the stationary phase (after 10–20 h) and takes place at an almost constant rate for several hours, providing that the protein is still present in the medium. Starch hydrolysates are degraded and assimilated by *B. licheniformis,* however, at the end of the fermentation serine protease is virtually the only protein in the culture suspension. Proteases can also be synthesized by fermentation *of B. amyloquefaciens*, but the final yield of protease is higher in the case of *B. licheniformis* (at least 10% more protein added to the medium). In the case of *B. amyloliquefaciens,* the culture supernatant contains other enzymes, in particular, a large amount of α-amylase. Alkaline proteases derived from *Bacillus amyloliquefaciens* and *Bacillus licheniformis* represent the lead molecules for subtilisin. Serine proteases have been derived from species of *Bacillus*. These proteases have better pH stability (up to pH 12) compared to subtilisins (pH 8–11). The bulk of alkaline proteases are used as additives to detergents in the form of granulates, prills or marumerizer pellets. Proteases enhance the cleansing power of a detergent and, in addition, this allows the elimination of protein-containing soils, e.g. blood, egg yolk and chocolate, and enhances the detachment of soil from the fibers. The stability of subtilisins is independent of metal ions, hence complexing additives in detergents, e.g. EDTA or tripolyphosphates, do not significantly interfere. Currently, almost 80% of the detergents on the market contain enzymes in concentrations of about

0.015–0.025%. The high alkali resistance of alkaline proteases makes them suitable for the leather industry, for the dehairing (nuding) of skins and to reduce softening times. They exhibit much better stability in detergents when the complexing agent EDTA or tripolyphosphate is substituted by citrate or gluconate.

3.5 Fungal proteases

Fungal proteases have attracted the attention of researchers as fungi can grow on low-cost substrates and secrete large quantities of enzymes into the culture medium, which could ease downstream processing [10]. Fungal proteases can be employed for hydrolyzing protein and other components of soybeans and wheat in soy sauce production. Proteases can be synthesized in large amounts in a short time by established methods of fermentation [10]. Parameters such as the C/N ratio and the presence of some sugars, in addition to several other physical factors, are significant in the development of the fermentation process. Fungus-derived proteases can be produced cost effectively, and also have the advantages of fast production, ease of enzymes modification and the easy removal of mycelium by filtration. Protease production has mainly been achieved by using submerged fermentation, however, the conditions in solid-subtrate fermentation result in several potential merits for the production of fungal enzymes [10].

Fungal proteases are synthesized from the genus *Aspergillus*, in particular the two species *A. oryzae* and *A. niger* [11, 12]. When cultured on a solid substrate, large quantities of proteases are produced by these species. The yield can be further improved by the addition of inorganic nitrogen. Other than proteases, reasonable amounts of α-amylase, glucoamylase, cellulases and pectinases are also produced. As far as applications are concerned, *Aspergillus* proteases are employed for hydrolysis of soybean protein in the production of soya sauce, as a component of enzymatic preparations for the treatment of digestive disturbances and in the baking industry. The supplementation of protease into flour incompletely hydrolyzes the gluten, reduces the viscosity of dough and thus reduces the kneading time. Fungal proteases are deactivated at 50 °C, so a quick denaturation takes place in the beginning of the baking process and circumvents the complete hydrolysis of proteins. Acid proteases can be derived from *Mucor* strains, in particular from the species *M. pusillus* and *M. michei*. During the fermentation of *Mucor* strains, esterases and lipases are also formed, which are mainly used in the manufacture of cheese [13].

3.6 Glucose isomerase (D-xylose ketol-isomerase; EC. 5.3.1.5)

Glucose isomerase is the first intracellular enzyme used on a large scale. It catalyzes the reversible isomerization of D-glucose and D-xylose to D-fructose and D-xylulose, respectively [14]. This enzyme has the largest market in the food industry because of its application in the production of high-fructose corn syrup (HFCS) [14]. This mixture contains 42% fructose, 50% glucose and 8% other mono- and disaccharides used as liquid sugar in the food industry. The entire process is based on starch hydrolysis by α-amylase and glucoamylase, and then isomerization to a mixture of glucose and fructose. Fructose has a better sweetening power than sucrose and glucose. HFCS, an equilibrium mixture of glucose and fructose, is 1.3 times sweeter than sucrose and serves as a sweetener for use by diabetics [14]. Interconversion of xylose to xylulose by glucose isomerase acts as a nutritional requirement in saprophytic bacteria and has a potential application in the bioconversion of hemicellulose to ethanol. The enzyme is generally found in prokaryotes. The microbial strain development accomplished by utilizing xylan-containing raw materials for the development of or screening for constitutive mutants of glucose isomerase is expected to result in discontinuation of the use of xylose as an inducer for the production of the enzyme [14]. Removal of Co^{2+} from the fermentation medium is necessary to prevent health issues arising from human consumption of HFCS. Glucose isomerase is synthesized by submerged cultivation of the various organisms listed in table 3.3 in the pH range 6.5–8.5 at 30 °C. Enzyme isolation is typically omitted and whole or disintegrated cells are immobilized. Glucose isomerase immobilization requires adsorption or heat treatment together with cross linkage [14]. Glucose isomerase offers an efficient means for its easy recovery and reuse, and lowers the cost. X-ray crystallographic and genetic engineering reports support a hydride shift mechanism for the action of glucose isomerase. Glucose isomerase cloning in homologous as well as heterologous hosts has been performed, with the main aim of overproducing the enzyme and interpreting the genetic makeup of individual genes (xylA, xylB and xylR) in the xyl operon of different micro-organisms [14]. Using a low-cost inducer in the fermentation medium devoid of Co^{2+} and redesigning a tailor-made glucose isomerase with improved thermostability, greater affinity for glucose and lower optimum pH will contribute considerably to the growth of an economically viable industrial process for enzymatic isomerization of glucose to fructose [14]. Various glucose isomerase sources and immobilization techniques are listed in table 3.3.

Different varieties of reactor have been compared and it has been reported that continuous fixed-bed reactors are the most suitable [14]. The isomerization is in a pH range of 7.0–8.5 at 60–65 °C. The substrate solution contains 40–50% glucose and oligosaccharide and has a viscosity of 0.8–3 mPa s. A magnesium salt concentration 0.5–5 mM l^{-1} is supplemented to activate and stabilize the enzyme. Alkali-catalyzed transformation of fructose into mannose and psicose can be averted by decreasing

Table 3.3. Glucose isomerase sources and immobilization techniques.

Name of organism	Immobilization technique	Final form
Streptomyces olivaceus	Whole cell; cross-linked with glutaraldehyde	Granules
Streptomyces olivochromogenes	Aluminum oxide support	Powder
Arthrobacter	Immobilized in the cells with a flocculating agent	Cylindrical particles
Actinoplanes missouriensis	Mycelium encapsulated in gelatin cross-linked with glutaraldehyde	Microspheres

the pH of the product solution. The purification of fructose syrup is achived by treatment with activated charcoal and using ion exchangers. The fructose syrup is concentrated to 71% dry matter. The procedure for the production of fructose syrup is represented in figure 3.3.

3.7 Penicillinase

Penicillinases are produced on the inner surface of the bacterial cell membrane and perhaps pass through the membrane to the external surface while they are being synthesized [15]. Generally, a gram-positive bacterium produces an ample amount of penicillinase and the majority is released into the surrounding medium. This is called exopencillinase. On the other hand, gram-negative bacteria synthesize β-lactamase in smaller amounts. The production and processing of penicillinase by the gram-positive organism *Bacillus licheniformis* has been investigated. When the bacterium is exposed to inducers (e.g. penicillin), there is a sudden rise in penicinillase production [15]. This process is accompanied by the development of many vesicles with high enzymatic activity. There are considerable differences in the way that β-lactamases protect gram-positive and gram-negative bacteria from the β-lactam antibiotic. Generally, β-lactamase production in a gram-positive organism

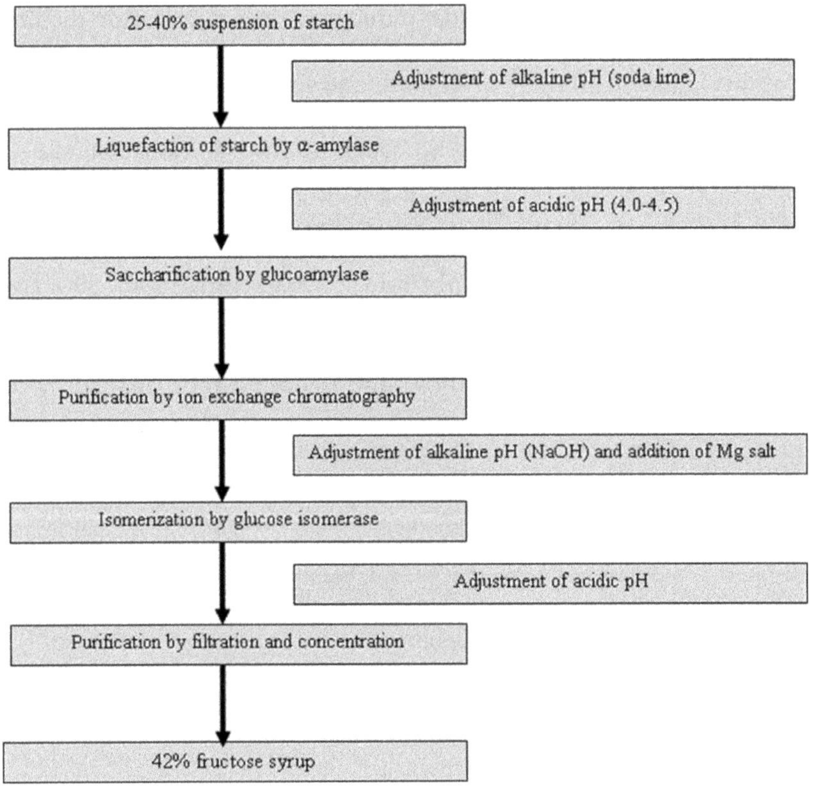

Figure 3.3. Flow diagram for the production of fructose syrup.

is inducible and a large quantity of penicillinase is released into the surrounding medium [15].

Because of its inactivation property against penicillin and certain cephalosporins, penicillinase (a bacterial enzyme) has gained wide attention. Several other hydrolytic enzymes which are also capable of inactivating penicillin have been explored, and a number of penicillinase producing pathogenic bacteria have been explored, e.g. staphylococcus species and some gram-negative bacteria of intestinal origin. This was first reported when some species showed resistance against penicillin. In particular, it was evidenced among patients infected by *Staphylococcus aureus*. Many common genera of fungi, such as *Penicillium*, *Aspergillus* and *Mucor*, synthesize the enzyme penicillin amidase [16–18]. Based on their activity penicillinase can be divided into two major classes:

- p-lactamase or penicillinase, and
- penicillin amidase or penicillin acylase.

Penicillinase reverses the therapeutic potential of penicillins. The activity of the enzyme penicillin amidase is based on a specific attack on the acyl group attached to the basic nucleus, i.e., 6-aminopenicillanic acid [16–18], therefore, it is also called penicillin acylase. Penicillin acylase specifically acts on the enzyme which is more specific to penicillin V and K. p-lactamase acts specifically on the basic nucleus itself, i.e., by cleaving the β-lactam bond and finally generating penicilloic acid. p-lactamase is more specific against penicillin G and X, and to a lesser extent against penicillin V [16–18].

As discussed earlier, certain bacteria have the capability to synthesize penicillinase in constitutents, whereas in others it may be inducible. The final effects can be established with the following case. If the antibiotic concentration is found to be non-lethal in a penicillin-treated patient, at that point, it can encourage the formation of penicillinase after which the bacteria becomes resistant to the antibiotic. A similar example is also reported when the penicillinase synthesizing bacteria are greater in number in comparison to the amount of antibiotic present in a certain time period. Under these conditions, although the bacteria lesser numbers are vulnerable to penicillin, they are resistant when present in greater numbers, as the concentration of penicillinase released by each bacterium adds up to an amount which can inactivate the antibiotic. In several clinically resistant strains of bacterium, the synthesis of β-lactamase is regulated by extra-chromosomal genetic elements. There are a number of notable differences between the enzymes synthesized by gram-positive and gram-negative bacteria. Considering molecular weight as a factor of comparison, gram-negative bacterium enzymes are generally small in molecular weight (22 000 Da), whereas gram-positive bacterium enzymes are large in molecular weight (28 000 Da) [19]. The concentration of β-lactamase synthesized by gram-positive bacteria is greater and, since the antibiotic is more specific against gram-positive bacteria than gram-negative bacteria, the enzyme β-lactamase is more sensitive against penicillin when synthesized by gram-positive bacteria. This is because the concentration of enzyme synthesized by gram-negative bacteria is far less and they do not secrete the enzyme into the external environment.

This can also be linked in an alternative way, i.e., as there is no β-lactamase secreted into the external environment, there is no dilution (loss) of β-lactamase in gram-negative bacteria, and therefore the need for production of β-lactamase by the cells is quenched.

3.8 Chloramphenicol acetyltransferase

Chloramphenicol has been widely used to treat typhoid fever since 1948 and is the drug of choice in the treatment of typhoid fever and certain infections caused by organisms such as *Haemophilus influenza* [20]. A major typhoid epidemic was first countered by chloramphenicol in Mexico [22]. During this period certain strains of the gram-negative bacterium *Salmonella typhi* came to light. It was later reported that the chloramphenicol resistance facilitated by R-factors in gram-negative bacteria and by transducible plasmids in *Staphylococcus aureus* is due to the occurrence of the enzyme chloramphenicol acetyltransferase (CAT), which acetylates the hydroxyl groups of the side chain of the chloramphenicol (see figure 3.4) [21].

The product derived from the two-stage inactivation process involving chloramphenicol acetyltransferase is 1,3-dicetoxychloramphenicol (which is inactive) [23]. Chloramphenicol acetyltransferase production in gram-positive bacteria is stimulated by chloramphenicol. Chloramphenicol acts as a protein synthesis inhibitor, thus producing an initial lag period in which the antibiotic prevents protein synthesis with simultaneous induction of chloramphenicol acetyltransferase. Therefore, an initial low amount of chloramphenicol acetyltransferase does not affect the antibiotic, however, at a later phase, it brings down the inhibitory level, i.e., when the amount of chloramphenicol acetyltransferase is high enough to break the drug moiety. As mentioned earlier for penicillinase, there exists a difference between enzymes released by gram-positive bacteria and gram-negative bacteria. However, in this case there is no such substantial difference. The molecular weight of chloramphenicol acetyltransferase produced by both gram-positive and -negative cells is almost the same (80 000 Da).

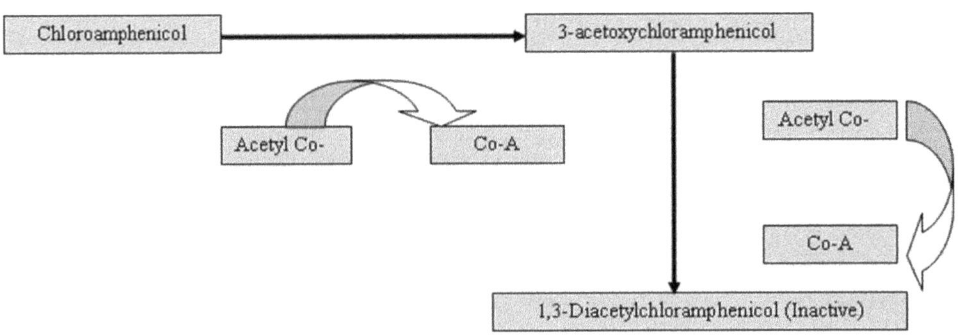

Figure 3.4. The two-stage chloramphenicol inactivation reaction involving chloramphenicol acetyltransferase.

3.9 Aminoglycoside antibiotic inactivating enzymes

Aminoglycoside antibiotics are also susceptible to inactivation. Different strains of *Streptomyces* have aminoglycoside-modifying enzymes. *Streptomyces kanamyceticus* synthesizes an enzyme that acetylates the 6′-amino group of kanamycin A and B, gentamicin C(1a) and neomycin. *Streptomyces spectabilis* synthesizes an enzyme that acetylates the 2′-amino group of the hexose ring of gentamicin C(1a). These enzymes catalyze reactions similar to those catalyzed by enzymes present in gram-negative bacteria having R(antibiotic resistance)-factors. The breakthrough of these enzymes proposes the likelihood of an evolutionary relationship between the aminoglycoside inactivating enzymes (synthesized by resistance determinants) in bacteria containing R-factors and similar enzymes found in the actinomycetes.

Despite the range of enzymes presenting full or partial resistance to antibiotics, there are only three types of inactivation reactions:

- N-acetylation of susceptible amino groups.
- Adenylation of hydroxyl groups. Phosphorylation of hydroxyl groups. Table 3.4 shows different enzymes that are accountable for the inactivation of aminoglycoside antibiotics.

Acetylation of the drug is performed by the co-factor, acetyl-CoA which serves as the source of the acetyl group, while the source of the phosphoryl and adenyl groups is adenosine triphosphate (ATP). Inactivation occurs in the periplasmic space (periplasm) or on the external surface of the cytoplasmic membrane. Intracellular accumulation of active aminoglycosides is determined by the induction of an energy-dependent process. The modified aminoglycoside is unable to stimulate intracellular accumulation after inactivation of the aminoglycosides by the enzymes. If somehow they succeed in entering a cell, they are incapable of restricting ribosomal function.

3.10 Fibrinolytic enzymes

Conversion of fibrinogen to fibrin inside blood vessels leads to thrombosis, resulting in myocardial infarction and other cardiovascular diseases. Generally, there are four therapy options: surgical operation or the intake of antiplatelets, anticoagulants or

Table 3.4. Different aminoglycoside antibiotic inactivating enzymes.

Enzymes	Substrates (aminoglycoside antibiotics)
Streptomycin/spectinomycin adenyltransferases	Streptomycin, spectinomycin
Streptomycin phosphotransferase	Streptomycin
Lividomycin phosphotransferase	Neomycin, paromomycin
Kanamycin/neomycin phosphotransferases I and II	Kanamycins, neomycins, gentamicin A
Kanamycin acetyltransferase	Kanamycin A and B, neomycin, gentamicin
Gentamicin adenyltransferace	Kanamycins, gentamicins
Gentamicin acetyltransferases I and II	Gentamicin

fibrinolytic enzymes [15]. Microbe-derived fibrinolytic enzymes have been gaining much attention compared to traditional thrombolytic agents because the latter are expensive and have side effects. Fibrinolytic enzymes have been successively derived from various micro-organisms, the most significant among which is the genus *Bacillus* [15]. Microbial fibrinolytic enzymes, particularly those from food-grade micro-organisms, have the capability to be established as functional food additives and drugs to avert or cure thrombosis and other associated diseases. There are numerous assay methods for these enzymes; this might be due to the insolubility of the substrate, fibrin. Current assay approaches can be categorized into three main groups:

- The first group involves the assay of fibrinolytic activity with natural proteins as substrates, such as fibrin plate methods.
- The second and third groups of assays are appropriate for kinetic investigations and are based on the determination of hydrolysis of synthetic peptide esters.

The fibrinolytic enzyme dissolves intravascular clots due to the action of plasmin (enzyme that digests fibrin) [15]. Plasminogen (the inactive precursor) is converted to plasmin by breakdown of a single peptide bond. For this reason, infusion of two types of enzymes can be considered, those with direct and general proteolytic activity, such as plasmin or trypsin, and those such as the plasminogen activators which increase the level of native plasmin [15]. In this respect, lyokinase plasminogen activators, such as streptokinase urokinase, and tissue plasminogen activators (t-PA) have gained more attention.

3.10.1 Streptokinase

Streptokinase is a proteolytic enzyme (47-Kda protein) derived from the culture of β-hemolytic streptococci of Lancefield group C. It is a potent activator of the fibrinolytic enzyme system in humans [26]. The protein molecule has mol. wt. 47 600 Da and an isoelectric point of pH 4.7. The thrombolytic capability of this enzyme catalyzes the transformation of human plasminogen to plasmin by hydrolytic cleavage of the Arg–Val bond. The enzymatic action of streptokinase is based upon a 1:1 noncovalent complex with a zymogen from which the active plasmin is released [26]. This protein complex can be dissociated into two parts, one with plasmin activity and the other with plasminogen activity. Figure 3.5 depicts a positive feedback mechanism, in which complexation first occurs between streptokinase and the plasmin contaminant of plasminogen. This complex is later free to convert a second plasminogen molecule [26].

As mentioned above, streptokinase can be utilized in the management of thromboembolic occlusive vascular disease. The major reasons behind acute arterial occlusion are embolism, thrombosis and trauma. The objective of thrombolytic treatment is to restore blood flow and to preserve life and limb. Antithrombotic therapy during chronic disease is considered to avoid the progression and development of thrombotic occlusion, and also to avoid thrombotic problems after vascular

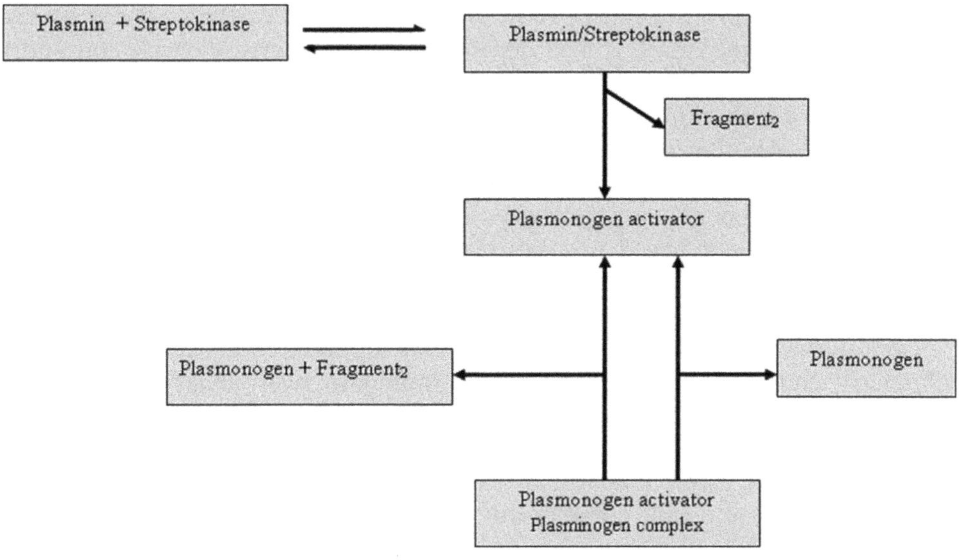

Figure 3.5. Proposed feedback mechanism attributed to streptokinase.

reconstructions. During acute arterial occlusion in the form of embolism or thrombosis, active anticoagulant treatment will avoid the spread of thrombi and also avoid re-occlusion after surgical or interventional procedures to re-establish flow; or, in the case of embolism, may prevent recurrence. In this thrombolytic treatment of occlusive vascular disease, both circulating and intrinsic thrombus plasminogen are activated. This is usually done by forming an activator complex (plasmin, as well as plasminogen, can form an activator complex with streptokinase) with the plasminogen molecule itself. Simultaneously, streptokinase can also encourage coagulation defects (depending on the dosage used) due to the induction of a hyperplasminaemic state [26]. So far the role of streptokinase therapy in the management of occlusive vascular disease is not clear. However, as per a recent report, streptokinase-based treatment can be utilized in acute massive or major pulmonary embolism, arterial occlusion and in deep-vein thrombosis (iliofemoral) and peripheral venous segments [26]. Based on a recent report it was suggested that zinc chelation (Ca-EDTA) can improve the efficacy of streptokinase in thrombolysis. Thus zinc reduced streptokinase-induced thrombolysis and zinc chelation can promote the effects of streptokinase-induced thrombolysis [30]. It was also observed that PEGylation of truncated streptokinase resulted in a formulation with increased *in vivo* half-life, reduced immune-reactivity and clot-specificity [31]. Additionally, streptokinase has also been employed in the management of chronic arterial occlusions and acute myocardial infarction. Streptokinase can also be utilized in the management of several other diseases associated with thromboembolic occurrences, however, more supporting research is essential before these conditions become definite indications for streptokinase treatment. Streptokinase-based thrombolytic treatment differs significantly with certain conditions. It has been observed that newly formed thrombi can be more easily broken down than old ones.

Similarly, venous thrombi can be more easily broken down than arterial thrombi. The action of streptokinase is also dependent on the degree of occlusion and the amount of blood supply in that particular region. Similarly to other thrombolytic agents, streptokinase also faces problems related to bleeding from sites of vascular puncture [26]. Another challenge is the development of antibodies after foreign protein (streptokinase) administration, which limits the course of treatment to 3–5 days. There should be two years between repeated administrations of streptokinase. Reports have also confirmed its cross reaction with anti-streptococcal antibodies. This may result in the development of resistance against therapy. However, this can be overcome by the administration of booster doses.

Streptokinase is superior in artificially dissolving induced coronary thrombi and pulmonary embolisms. The power of streptokinase over heparin in decreasing mortality in myocardial infection has also been well recognized. A streptokinase–plasminogen complex (antistreplase) in which lys-plasminogen is acylated at its catalytic site, serine is also used for coronary thrombolysis. Under *in vivo* conditions, the acyl group is allowing complex to bind to fibrin before activation, and this change confers some specificity towards clots during the fibrinolytic process. Owing to its bacterial origin, streptokinase is antigenic and may bind to antibodies and nonspecific inhibitors. The immune response presented by streptokinase affects the modification of urokinase.

3.10.2 Urokinase

Urokinase (molecular weight 54 500 Da) is a serine protease comprising 411 amino acid residues, synthesized and isolated from cultured human kidney cells [27]. This enzyme is single polypeptide and very stable against temperatures and extremes of pH [27]. Urokinase efficiently digests blood clots. Human urokinases often act as plasminogen activators by specifically binding to cleave peptide bonds adjacent to lysine and arginine side chains. As it is derived from humans, it is non-immunogenic and its action is very similar to that of trypsin, catalyzing the activation of plasminogen in a first order reaction with cleavage of the Arg–Val bond [27]. The current focus on the management of mortality due to thrombovascular disorders is providing openings for various cardiovascular agents, particularly plasminogen activators, to avoid strokes and heart attacks [27]. Therefore, urokinase, one of the most potent plasminogen activators, is gaining a great deal of attention. Progress in cell lines and bioprocessing technique has made it conceivable to synthesize urokinase from *in vitro* cell cultures. Efforts are now underway to increase urokinase synthesis from cell culture through medium optimization, bioreactor cultivation and advanced purification methods [27].

3.10.3 Tissue plasminogen activator (t-PA)

t-PA (527 amino acid residues) is a protein involved in the breakdown of blood clots. This enzyme belongs to the serine protease class and binds to fibrin via lysine binding sites at its amino terminus, and further activates bound plasminogens several hundred fold more quickly than it activates plasminogen in circulation.

In the absence of fibrin it acts as a poor plasminogen activator. Under biological conditions (5–10 ng ml^{-1}), the specificity of t-PA for fibrin restricts the systemic development of plasmin and the introduction of a systemic lytic state. It can successfully break down thrombi during treatment of acute myocardial infarction. t-PA is produced by recombinant DNA technology. Currently it is recommended for coronary thrombolysis. The major side effect is severe hemorrhage. The cost of t-PA is high, several times more than streptokinase per therapeutic dose.

The production of this drug in plants such as cucumber, one of the more common vegetables in the world, could reduce its production costs. In a study it was concluded that the presence of three expressions of regulatory factors (CaMV 35S, Kozak, NOS) and the KDEL signal in the construct caused the increase of t-PA gene expression in cucumber plants [28]. In another study a protease-deficient strain of *Aspergillus niger* was used as a host for the production of human tissue plasminogen activator (t-PA). Production was increased (up to 1.9 mg t-PA (g biomass)(−1)) by the addition of soy peptone to the defined medium [29]. However, the total t-PA (detected by enzyme-linked immunoassay) also eventually disappeared from culture supernatants, confirming significant extracellular proteolytic activity, even though the host strain was protease-deficient [29].

3.11 Biotechnological applications of enzymes

3.11.1 Algae and plant research

As mentioned above, there are various applications of enzymes in different disciplines, however, some of the applications are restricted to plant discipline. Recently, interest related with plant enzymes has augmented considerably. Plants are considered as the main source for secondary metabolites. Plant based enzymes such as peroxidases are extensively used in medicine as diagnostic tools and in the bioremediation and biobleaching industries, among others. Earlier these enzymes were derived from a natural source, a process that is sometimes difficult and influenced by environmental conditions and low yields. To prevent this obstacle, some inputs have been made to develop plant cell cultures *in vitro* to use the system as a continuous source of plant enzymes.

3.11.2 Immobilization

Different types of carriers and procedures have been implemented in the recent past to improve traditional enzyme immobilization targeted to increase enzyme loading, activity and stability to reduce the enzyme biocatalyst cost at large scale [32–42]. These include:

- recently nanoparticle-based immobilization of enzymes;
- microwave-assisted immobilization;
- mesoporous supports;
- cross-linked enzyme aggregates;
- click chemistry technology.

In nanotechnology method, mixture of the specific physical, chemical, optical and electrical properties of nanoparticles, especially catalytic properties of biomolecules, has resulted in the appearance of countless novel biotechnological applications [42–49]. Nanoparticles offer high surface-to-volume ratio resulting in an increase in the concentration of the immobilized entity that is considerably higher than that afforded by experimental protocols based on immobilization on planar 2-D surfaces. Enzymes immobilized on nanoparticles presented a broader working pH and temperature range and higher thermal stability than the native enzymes. In contrast with traditional procedures, nanoparticle based immobilization served three important features:

- nano-enzyme particles are easy to synthesize in high solid content without using surfactants and toxic reagents;
- homogeneous and well-defined core–shell nanoparticles with a thick enzyme shell can be obtained;
- particle size can be conveniently changed within utility limits. Moreover, with the growing attention paid to cascade enzymatic reaction and *in vitro* synthetic biology, it is possible that co-immobilization of multi-enzymes could be achieved on these nanoparticles.

References

[1] Anbu P, Gopinath Subash C B, Chaulagain B P, Tang T H and Citartan M 2014 Microbial enzymes and their applications in industries and medicine *Biomed. Res. Int.* **2017** 816419

[2] Souza P M and Magalhães P O 2010 Application of microbial α-amylase in industry—a review *Braz. J. Microbiol.* **41** 850–61

[3] van der Maarel M J, van der Veen B, Uitdehaag J C, Leemhuis H and Dijkhuizen L 2002 Properties and applications of starch-converting enzymes of the alpha-amylase family *J. Biotechnol.* **94** 137–55

[4] Burhan A, Nisa U, Gokhan C, Omer C, Ashabil A and Osman G 2003 Enzymatic properties of a novel thermostable, thermophilic, alkaline and chelator resistant amylase from an alkalophilic *Bacillus* sp. isolate ANT-6 *Process Biochem.* **38** 1397–403

[5] Bordbar A K, Omidiyan K C and Hosseinzadeh R 2005 Study on interaction of α-amylase from *Bacillus subtilis* with cetyltrimethylammoniumbromide *Colloids Surf.* B **40** 67–71

[6] Khajeh K, Shokri M M, Asghan S M, Moradian F and Ghasemi A 2006 Acidic proteolytic digestion of α-amylase from *Bacillus licheniformis* and *Bacillus amyloliquefaciens*: stability and flexibility analysis *Enzyme Microbiol. Technol.* **38** 422–8

[7] Pandey A, Nigam P, Soccol C R, Soccol V T, Singh D and Mohan R 2000 Advances in microbial amylases *Biotechnol. Appl. Biochem.* **31** 135–52

[8] Mariana Recco P, Molina G, Dionísio A P, Maróstica Junior M R and Pastore G M 2011 The use of endophytes to obtain bioactive compounds and their application in biotransformation process *Electronic J. Biotechnol. Res. Int.* **2011** 1–11

[9] Saran S, Isar J and Saxena R K 2007 A modified method for the detection of microbial proteases on agar plates using tannic acid *J. Biochem. Biophys. Methods* **70** 697–9

[10] Anitha T S and Palanivelu P 2013 Purification and characterization of an extracellular keratinolytic protease from a new isolate of *Aspergillus parasiticus Protein Expr. Purif.* **88** 214–20

[11] Ogawa A, Yasuhara A and Tanaka T *et al* 1995 Production of neutral protease by membrane-surface of *Aspergillus oryzae* IAM2704 *J. Ferment. Bioeng.* **80** 35–40
[12] Vishwanatha K S, Rao A G A and Singh S A 2009 Characterization of acid protease expressed from *Aspergillus oryzae* MTCC 5341 *Food Chem.* **114** 402–7
[13] Andrade V S, Sarubbo L A and Fukushima K *et al* 2002 Production of extracellular proteases by *Mucor circinelloides* using D-glucose as carbon source/substrate *Braz. J. Microbiol.* **33** 106–10
[14] Bhosale S H, Rao M B and Deshpande V V 1996 Molecular and industrial aspects of glucose isomerase *Microbiol. Rev.* **60** 280–300
[15] Fantin B, Pierre J, Castéla-Papin N, Saint-Julien L, Drugeon H, Farinotti R and Carbon C 1996 Importance of penicillinase production for activity of penicillin alone or in combination with sulbactam in experimental endocarditis due to methicillin-resistant *Staphylococcus aureus Antimicrob. Agents Chemother.* **40** 1219–24
[16] Hamilton-Miller J M T 1966 Penicillinacylase *Bacteriol. Rev.* **30** 761–71
[17] Cole M and Sutherland R 1966 The role of penicillin acylase in the resistance of gram-negative bacteria to penicillins *J. Gen. Microbiol.* **42** 345–56
[18] Holt R J and Stewart G T 1964 Production of amidase and, 8-lactamase by bacteria *J. Gen. Microbiol.* **36** 203–13
[19] Silhavy T J, Kahne D and Walker S 2010 The bacterial cell envelope *Cold Spring Harb. Perspect. Biol.* **2** a000414
[20] Maza L, Miller S I and Ferraro M J 1990 Use of commercially available rapid chloramphenicol acetyltransferase test to detect resistance in *Salmonella* species *J. Clin. Microbiol.* **28** 1867–9
[21] Shaw W V, Bentley D W and Sands L 1970 Mechanism of chloramphenicol resistance in *Staphylococcus epidermidis J Bacteriol.* **104** 1095–105
[22] Olarte J and Galindo E 1973 *Salmonella typhi* resistant to chloramphenicol, ampicillin, and other antimicrobial agents: strains isolated during an extensive typhoid fever epidemic in Mexico *Antimicrob. Agents Chemother.* **4** 597–601
[23] Derrick J P, Lian L Y, Roberts G C K and Shaw W V 1992 Analysis of the binding of 1, 3-diacetylchloramphenicol to chloramphenicol acetyltransferase by isotope-edited proton NMR and site-directed mutagenesis *Biochemistry* **31** 8191–5
[24] Benveniste R and Davies J 1973 Aminoglycoside antibiotic-inactivating enzymes in actinomycetes similar to those present in clinical isolates of antibiotic-resistant bacteria *Proc. Natl Acad. Sci. USA* **70** 2276-80
[25] Kotb E 2013 Activity assessment of microbial fibrinolytic enzymes *Appl. Microbiol. Biotechnol.* **97** 6647-65
[26] Brogden R N, Speight T M and Avery G S 1973 Streptokinase: a review of its clinical pharmacology, mechanism of action and therapeutic uses *Drugs* **5** 57–445
[27] Bansal V and Roychoudhury P K 2006 Production and purification of urokinase: a comprehensive review *Protein Expr. Purif.* **45** 1-14
[28] Asgari M, Javaran M J, Moieni A, Masoumiasl A and Abdolinasab M 2014 Production of human tissue plasminogen activator (tPA) in *Cucumis sativus Prep. Biochem. Biotechnol.* **44** 182-92
[29] Wiebe M G *et al* 2001 Production of tissue plasminogen activator (t-PA) in *Aspergillus niger Biotechnol. Bioeng.* **76** 164–74

[30] Wang Z, Yu X and Li Y V 2017 Zinc chelation promotes streptokinase-induced thrombolysis *in vitro Int. J. Physiol. Pathophysiol. Pharmacol.* **9** 137–46

[31] Sawhney P, Katare K and Sahni G 2016 PEGylation of truncated streptokinase leads to formulation of a useful drug with ameliorated attributes *PloS One* **11** e0155831

[32] Georgianna D R *et al* 2013 Production of recombinant enzymes in the marine alga *Dunaliella tertiolecta Algal Research* **2** 2–9

[33] Yan N, Fan C, Chen Y and Hu Z 2016 The potential for microalgae as bioreactors to produce pharmaceuticals *Int. J. Mol. Sci.* **17** 962

[34] Bhatia S, Rathee P, Sharma K, Chaugule B B, Kar N and Bera T 2013 Immuno-modulation effect of sulphated polysaccharide (porphyran) from *Porphyra vietnamensis Int. J. Biolog. Macromol.* **57** 50–6

[35] Bhatia S, Kumar V, Sharma K, Nagpal K and Bera T 2014 Significance of algal polymer in designing Amphotericin B nanoparticles *Sci. World J* **2014** 1

[36] Bhatia S, Sharma K, Nagpal K and Bera T 2015 Investigation of the factors influencing the molecular weight of porphyran and its associated antifungal activity *Bioact Carbohydr. Diet. Fiber* **5** 153–68

[37] Bhatia S, Sharma K, Nagpal K, Sharma S and Bera T 2015 Evaluation of pharmacognostical, phytochemical and anti-microbial properties of *Porphyra vietnamensis Int. J. Green Pharm.* **2** 1–5

[38] Bhatia S, Sharma K and Bera T 2015 Structural characterization and pharmaceutical properties of porphyran *Asian J. Pharm* **9** 1–3

[39] Bhatia S, Sharma K, Sharma A, Nagpal K and Bera T 2015 Anti-inflammatory, analgesic and antiulcer properties of *Porphyra vietnamensis Avicenna J. Phytomed.* **5** 1–7

[40] Bhatia S *et al* 2009 Novel algal polysaccharides from marine source: porphyran *Pharmacog. Rev.* **2** 271–6

[41] Bhatia S *et al* 2011 Mycosporine and mycosporine-like amino acids: A paramount tool against ultra violet irradiation *Pharmacogn Rev.* **5** 138–46

[42] Bhatia S *et al* 2010 Broad-spectrum sun-protective action of Porphyra-334 derived from *Porphyra vietnamensis. Pharmacog. Res.* **2** 45–9

[43] Bhatia S, Namdeo A G and Nanda S 2015 Factors affecting the gelling and emulsifying property of natural polymer *Sys. Pharm. Rev.* **9** 93–101

[44] Bhatia S, Sharma K, Sharma A, Namdeo A G and Chaugule B B 2011 Anti-oxidant potential of Indian Porphyra *Pharmacologyonline* **1** 248–57

[45] Bhatia S *et al* 2015 *Modern Applications of Plant Biotechnology in Pharmaceutical Sciences* (Amsterdam: Elsevier), pp 277–87

[46] Bhatia S 2016 *Nanotechnology in Drug Delivery: Fundamentals, Design, and Applications* (Boca Raton, FL: CRC Press), pp 121–7

[47] Bhatia S and Goli D 2016 *Leishmaniasis: Biology, Control and New Approaches for Its Treatment* (Boca Raton, FL: CRC Press), pp 177–9

[48] Bhatia S 2016 *Natural Polymer Drug Delivery Systems: Nanoparticles, Plants, and Algae* (Berlin: Springer), pp 117–27

[49] Bhatia S 2016 *Systems for Drug Delivery: Safety, Animal, and Microbial Polysaccharides* (Berlin: Springer), pp 122–7

[50] Bhatia S 2018 *Introduction to Pharmaceutical Biotechnology* Volume 1 (Bristol: IOP Publishing), pp 178–87

IOP Publishing

Introduction to Pharmaceutical Biotechnology, Volume 2
Enzymes, proteins and bioinformatics
Saurabh Bhatia

Chapter 4

Immobilization of enzymes

4.1 Introduction

In modern science, biologically active natural compounds, mainly enzymes, have gained significant attention in medicine and various biotechnological processes where they are being employed commercially. At present a number of ailments related to lysosomal enzyme activities have been recognized. These diseases could ideally be cured by supplementation or replacement of defective or deficient enzyme. Additionally, enzymes with very specific activity are usually involved in the catalysis of several reactions. Enzymes have various significant clinical applications as bioreactors, biosensors and in other enzyme-based reactions, however, their use is restricted for a number of reasons, such as high cost, low availability and quick inactivation under physiological conditions followed by fast clearance, therefore there is the requirement for a greater amount of enzyme during the course of treatment. Moreover, possible antigenicity, inactivation by different endogenous natural inhibitors and, lastly, the challenge to achieving higher levels for therapeutic activity, are limitations on the practical clinical use of enzymes. Regardless of these limitations, enzymes have a vast range of possible applications covering therapeutic, analytical and industrial aims. The application potential of an enzyme is established by the distinctive enzyme properties of precise specificity for the substrate and reaction catalyzed. In order to be practically suitable, an enzyme must have excellent catalytic behavior, with specificity, stability and reusability. According to research, the majority of enzymes involved in particular reactions become comparatively less active during the last phase because of the accumulation and detrimental effect of the reaction product.

Immobilized enzymes are currently the subject of significant interest because of their benefits over soluble enzymes. In addition to their use in industrial processes, immobilization methods are the basis for making a number of biotechnology products with applications in diagnostics, bioaffinity chromatography and biosensors [1]. Earlier, only single immobilized enzymes were employed, however, since the

doi:10.1088/978-0-7503-1302-5ch4

1970s more complex systems including two-enzyme reactions with co-factor regeneration and living cells were explored. The enzymes can be attached to the support by interactions ranging from reversible physical adsorption and ionic linkages, to stable covalent bonds. Although the choice of the most suitable immobilization method depends on the nature of the enzyme and the carrier, in the last few years immobilization technology has gradually become a matter of rational design [1]. As a result of enzyme immobilization, some important features, such as catalytic activity or thermal stability, are changed. These properties have been demonstrated and exploited [1]. The concept of stabilization has been an important driving force for immobilizing enzymes. Furthermore, true stabilization at the molecular level has been established, e.g. proteins immobilized through multipoint covalent binding [1].

Enzymes exist as bioproteins and thus are environmentally sensitive to different pH and temperature. Therefore, stability remains a challenge for the effective cure of problems associated with enzymes and enzyme technology. The major focus of immobilization is characteristically the economic application of enzyme systems. Additional merits of the immobilized enzyme system are ease of control and uniformity of conversion. The solution to qualifying the economic concerns is the evaluation of the cost of the constituents of the immobilized system against the value of its performance. During immobilization, the enzyme utilizes the carrier phase for stealth, and safe homing. The immobilization matrix or support facilitates exchange, but remains separate from the bulk phase in which substrate, effect or inhibitor molecules are present and isolated. This entails limiting soluble protein catalysts or the individual buoyancy of cells in reactor systems, which can require constant changes of fresh solution. Usually the immobilized system is placed in an appropriate column through which the charge is allowed to pass constantly. The definition may be expanded to cover intact cells and biocatalysts under the category of guest. The process of immobilization, i.e., the restrictive localization of enzymes, may in the immediate future have great potential in the development of biosensors, bioelectronic sensors in fermentation technology and enzyme therapeutics [2]. The carrier phase, also known as the enzyme phase, is classically water insoluble but can be combined with as hydrophilic porous polymeric matrix, e.g. cellulose, agarose and acrylamide. In other words, immobilization is a general term that designates the retention of a biologically active catalyst within a reactor or analytical system with the help of a suitable carrier support. A unique procedure is planned by using immobilization techniques through which fluid can pass without any difficulty, converting the substrate into the product under controlled enzymatic reaction, and all together allowing the easy removal of the catalyst from the product as it leaves the reactor. The immobilization, also referred to as forced homing, of enzymes is achieved by covalent coupling using adsorption or physical entrapment of enzymes within the enzyme phase (polymeric matrix). The activity of immobilized enzymes is usually dependent on the type of matrix and enzyme phase. One of the most common immobilization matrices which utilizes anchoring on the surface of an insoluble polymeric matrix is cellulose or polyacrylamide via covalent bond formation. The enzyme phase is available in the form of fine particulate, membranous or monolithic spheres. The enzyme in turn may be bound to another enzyme via cross-linking. This may render the overall composite as an insoluble, but active, polymeric enzyme system. Once a support or carrier is employed for the purpose of enzyme immobilization, the stability of that carrier during its use is equally important. A carrier (membrane or matrix) that is not stable at a suitable pH, ionic strength or under the solvent conditions of the process, may be disrupted or dissolved releasing the enzyme component. This obviously suggests that it is not essential for a carrier to be insoluble and durable at all pHs, ionic strengths or solvent conditions. The different procedures most frequently employed for enzyme immobilization using solid supports can be categorized into two main classes on the basis of the approaches used to restrain biocatalysts: those with catalysts within the support and those which hold the biocatalyst on the surface of the support. Each class is further divided on the basis of their usage of covalent and noncovalent coupling procedures for surface immobilization, with cross-linking entrapment and encapsulation techniques for immobilization in or within a support. The different types of supports used in immobilization are shown in figure 4.1 and table 4.1.

4.2 Types of immobilization

On-surface immobilization involves [3]:

- Covalent coupling with polymers employing their functional group for anchoring, moreover, the groups must be non-functional or non-essential for biological activity [4].
- Adsorption on an inert support or ion-exchange resins.
- Complexation and chelation.

In the procedure of within-support immobilization, the enzyme is immobilized by cross-linking by multifunctional reagents, followed by entrapment within a structure of known/defined geometry. Entrapment can be achieved by:

- encapsulation in microcapsules, hollow fibers, liposomes, etc.
- occlusion within a cross-linked gel.

Immobilization frequently causes obvious changes in the apparent parameters of the enzyme related to the reaction it catalyzes. These parameters include the reaction rate, effect of the inhibitor (which may be lost completely in the case of an immobilized enzyme), Michaelis–Menten kinetics, and the optimum temperature and pH corresponding to maximum activity. The degree of variation among these parameters may not only be related to the technique of immobilization selected, but also to some extent to the enzyme reaction. Iso *et al* have demonstrated the micro-encapsulation of lipase (*Pseudomonas fluorescens*), carried out using a (W/O)/W two-phase emulsion technique [5].

4.2.1 Surface immobilization by covalent coupling

Immobilization of enzymes by covalent bonding to an insoluble polymer has the supposed benefit of irreversible binding of the enzyme to the support matrix. Many

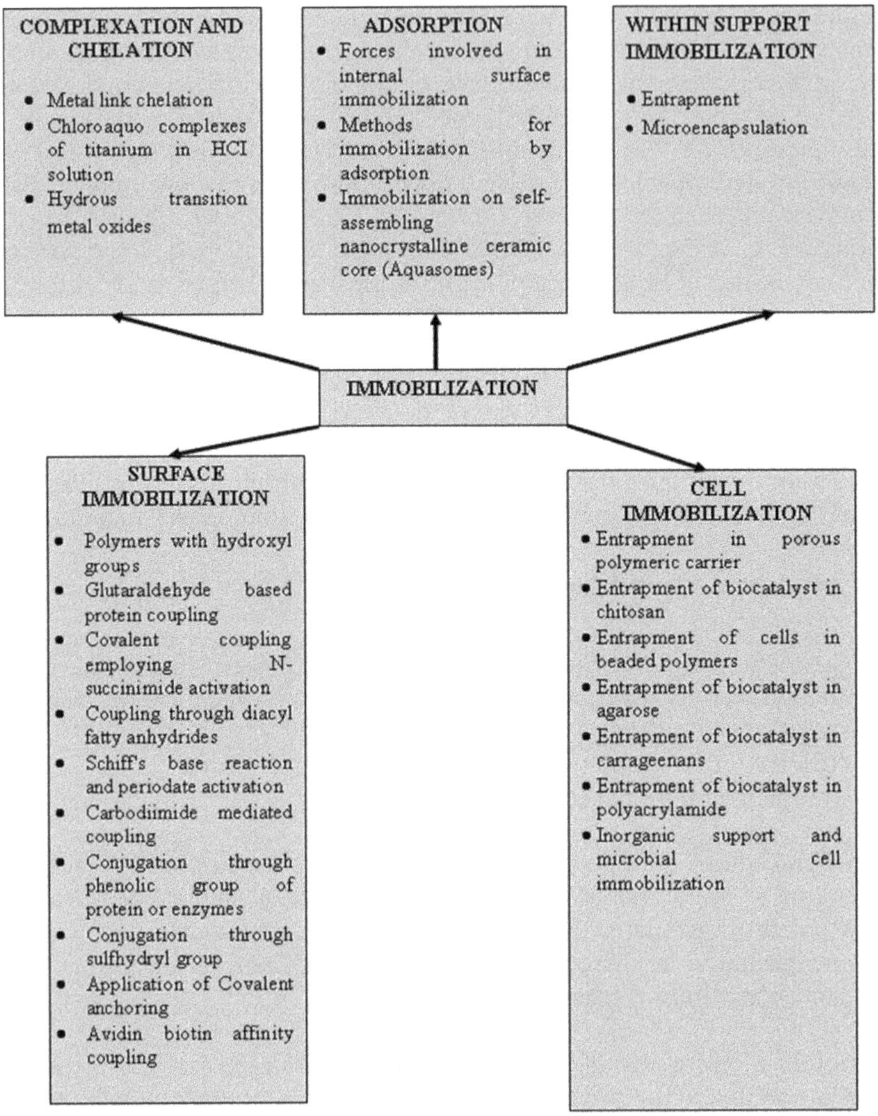

Figure 4.1. Schematic or immobilization techniques.

of the bonds usually employed are, however, not 100% stable in use. If insufficient care is taken to confirm that the enzyme can survive the reaction conditions, considerable loss of enzyme activity will occur. Even under promising conditions, 60% loss of activity is not uncommon [6]. The covalent bond between a biocatalyst and a support matrix results in a complex which is significantly stable throughout the functional use of the enzyme. However, the process subjects the delicate three-dimensional structure of the biopolymer to strong disruptive physical and chemical forces or stresses. Thus, where the superior form and stability of an immobilized enzyme are more important than the irreversible loss of activity that is certainly

Table 4.1. Different types of supports used in immobilization [16].

Classification of supports
Organic
Natural polymers
• Polysaccharides: cellulose, dextrans, agar, agarose, chitin, alginate
• Proteins: collagen, albumin
• Carbon synthetic polymers
• Polystyrene
• Other polymers: polyacrylate polymethacrylates, polyacrylamide, polyamides, vinyl and allyl-polymers
Inorganic
Natural minerals Bentonite Silica processed materials: glass (nonporous and controlled pore) Metals Controlled pore metal oxides

associated with immobilization processess, such enzymes are preferred over their soluble form. The functional groups found on enzyme proteins, by which a covalent bond with polymeric or other supports could be established, should be non-essential for enzymatic action. Moreover, the reactions which include comparatively mild conditions and principally utilize an aqueous medium are favored. This additionally requires broad information on the amino acids and their active groups involved in polymeric covalent bonding, the prospect of chemical amelioration or specific chemical modification on activity, the protection of active group(s) and the three-dimensional structure of the enzyme under consideration. The protein functional groups which could be employed in covalent coupling include:

- COOH: β and α of aspartic acid and glutamic acid, respectively.
- NH_2 lysine: N-terminal of polypeptide.
- OH: Phenol ring on tyrosine.
- SH: Group on cysteine.

The polymeric supports which have been considered as supports for covalent coupling based immobilization include [1]:

- Aldehyde and acetal groups of polymers.
- Amide group of polypeptides and polyamides (nylon).
- Amino and related groups of amino-ethyl coated polysaccharides and silica gel, poly(D-aminoglycine), etc.
- Carboxylic acid and related groups of poly(glutamic acid), maleic anhydride co-polymers, poly(acrylamide) and carboxymethyl cellulose.
- Hydroxyl groups of polysaccharides, PVA, polymethyl methacrylate (PMMA) and inorganic glasses.

The covalent coupling procedures usually employed for conjugation of enzymes with a solid/matrix support are cyanobromide, cyclic trans 2,3-carbonates and acid anhydrides. The approaches employed for covalent anchoring via support activation are: diazotization, Schiff's base formation, glutaraldehyde treatment, thiodisulfide and imido ester formation. With the help of these coupling agents and certain procedures, covalent bonds are developed between functional groups on enzymes and respective counter groups on polymeric support with a net result of an immobilized enzyme system formation.

4.2.2 Adsorption

Considering the functionality at the level of operation, adsorption appears to be a most economical and simple process, however, the forces involved are extremely complex. So far, practically no material has been reported which could be employed for adsorptive immobilization with its own surface contribution [1]. The nature of the carrier and the enzyme protein surface determines the bonds that exist between the enzyme protein and carrier, such as ionic, hydrogen, covalent, coordinated covalent or hydrophobic, or even a combination of any of these [1]. Immobilization can be achieved by coupling an enzyme to either the external or internal surface of a carrier. When an enzyme is immobilized externally, it is necessary that the particle size of the carrier must be small enough to offer a substantial surface for binding. These particles possibly have diameters ranging between 500 Å mm to 1 mm diameter [1]. The external surface binding employed during immobilization of the enzyme is beneficial as it does not include conditions such as pore diffusion. The drawbacks, however, include a comparatively small surface area for binding, exposure of enzymes against microbial attack, physical abrasion of the enzyme and inhibitory effects due to turbulence related to the bulk solution. Finally, the reduced particles cause a high pressure drop (gradient) in continuous packed-bed reactors [1]. While the internal surface participation of a porous carrier during enzyme immobilization has been a noticeable shortcoming, many adavatages can be gained from this approach. The main shortcoming of internal immobilization relates to pore diffusion. Therefore, the process calls for proper optimization of pore diameter, surface area, surface charges and other related parameters.

4.2.3 Complexation and chelation

Numerous techniques for the immobilization of enzymes have been developed. By choosing a suitable carrier material and coupling procedure, it is typically possible to obtain immobilized derivatives with good activity and stability [7]. However, the majority of these comprise irreversibly bonded enzymes, which means that the carrier cannot be regenerated and reused. This issue can be avoided by immobilization based on ion exchange, hydrophobic interaction or thiol disulfide interexchange [7].

4.2.3.1 Metal link chelation

This method provides an immediate coupling without chemical derivatization or activation of a support or matrix. The procedure is principally based on the chelation properties of transition metals, specifically titanium and zirconium, which are attractive due to their nontoxic nature [8]. A scheme based on titanium chloride–cellulose demonstrates the process of transition metal association–chelate formation. Titanium metal has the feature of octahedral co-ordination with ionic molecules which function as ligands. In the majority of cases these ligands are water molecules or chloride ions. In the event where the chloride ion behaves as a ligand, its surplus electron density is employed in the development of a partial-covalent bond with the metal ion (Ti), thus the whole positive charge of titanium is minimized by a unit.

4.2.3.2 Chloroaquo complexes of titanium in HCI solution

These ligands may be substituted by other water-based ligands or other electron donating groups. The strength of association depends on the chemical nature of the ligand [9]. Hydroxyl ions are active ligands for transition metals; consequently they may well have the potential to create complexes with polysaccharides, where new ligating hydroxyl ions may be substituted for pre-existing ones if any are present [9]. Certain polysaccharides, e.g. cellulose, contain a vicinal diol group which is not involved in glycosidic linkage and is therefore accessible for free chelation by transition metals. The chelate is an outcome of the substitution of two ligands from the titanium ion by polysaccharide hydroxyl groups. For steric reasons the number of ligands which can be substituted by cellulose hydroxyl groups is limited [9].

4.2.3.3 Hydrous transition metal oxides

During this procedure hydrous metal oxides are used as the support for enzyme or cell immobilization [9]. The oxides can be prepared by precipitation after hydrolysis of the corresponding chloride. The method of immobilization is mainly based on chelation. Therefore the precipitation is achieved to obtain hydrous oxide in the presence of the enzyme, which may result in effective and efficient immobilization. However, all safety measures needed to account for potential detrimental conditions should be taken in to consideration [9]. The metal chlorides which may be employed for transformation to hydrous oxides include cobalt(II), copper(II), iron(II), manganese(II), tin(II), zinc(II), chromium(II), vanadium(III), tin(IV) and zirconium(IV). The complexes retain a considerable portion of the activity profile of the chelated immobilized enzyme. Due to the low operational stability of enzymes immobilized on metal hydrous oxide, a cross-linking step based on glutaraldehyde treatment has been recommended. This method of modification could avert protein loss into the solution during operation.

4.2.4 Within-support immobilization

The entrapment or encapsulation of enzymes can be achieved via their inclusion in a matrix that is based on highly cross-linked polymers, encapsulation in microcapsules or in distinct non-aqueous phases [10]. The characteristic feature of these procedures, mainly matrix inclusion and encapsulation, is that the enzyme is not attached to the matrix, thus any issues, such as the steric blockades related to covalent or electrostatic binding, are not encountered. The procedure offers scope for developing biocatalysts in various physical shapes and forms. While beads are most common, fibers, sheets, emulsions and gels have been used in bioreactors [10]. In conclusion, immobilization by means of entrapment methods is an attractive option where the difference in size of the catalyst and substrate should be characteristically large. Practically, membrane reactors are well adapted for enzymes which act on a low molecular weight substrate, whereas three-dimensional gel is favored for immobilizing cellular/particular biocatalysts and therefore is the scheme of choice for immobilization of live cells [10]. Usually entrapment is achieved by dissolving an enzyme in a solution that is required either for dissolution or for preparation of the enzyme phase (carrier), and then treating this solution so that a distinctive and discrete biphasic system is created in the form of dispersion.

4.2.5 Cell immobilization

Enzyme immobilization has now become a mature technology and a number of immobilization procedures have been developed. The groundbreaking attention has moved onto the development of immobilization of whole plant/animal cells, which offers another route to highly specific enzyme processes aimed at multistage processing and production of fine chemicals against first generation biocatalysts. In the last few decades, experience and expertise in the immobilization of microbial cells has improved such that a range of gentle procedures are now accessible to produce fine chemicals, fabrics and membranes incorporating a high number of viable cells. Hydrogels are being studied for mammalian cell immobilization. Their material features can be designed or structured for biocompatibility, selective permeability, mechanical and chemical stability, and other necessities as specified by the application, including uniform cell distribution and a given membrane thickness or mechanical strength [11]. These aqueous gels are a good choice for analytical and tissue engineering applications and can be employed for immobilization in therapies for various diseases as well as to generate bioartificial organs [11].

4.2.5.1 Distinctive features of plant and animal cells

In cell immobilization, currently most research is focused on using the prokaryotic bacterial cells as a subject [12]. The differences between prokaryotic and eukaryotic cells are clearly distinguishable and involve size, form and function. Prokaryotes lack of organelles, have a different cell wall composition and are typically one tenth the size or one thousandth the volume of plant and animal cells. Eukaryotic cells have different subcellular, membrane bound organelles such as nuclei, mitochondria, lysosomes, vacuoles and various other plastids. As a result of these major differences the metabolic functions of eukaryotes proceed more slowly than those of bacteria. Eukaryotic cells can express a superior amount of genetic information in a number of ways, to produce distinct, differentiated (i.e., specific) cells which frequently function co-operatively to develop tissues, organs or even whole organisms. As a result, they develop complex mechanisms for both intracellular and intercellular control and informative transfer, involving hormones, cell mediators and specific cell surface receptors. They are hence capable of more sophisticated responses than the basic survival-related reactions of bacteria. The whole complexity of their metabolism, affecting the gene expression in individual cells, presents problems in the maintenance and use of eukaryotes, even when successful immobilization has been achieved. However, these different levels of control could clearly offer valuable opportunities for the manipulation of cellular activity. Unfortunately, at present these mechanisms are so poorly understood at the molecular level, that bacteria continue to be considered for any applications where either cell type could be employed.

4.2.5.2 Products from eukaryotes

There is developing interest in the large scale culture of plant and animal cells for the production of high value pharmaceuticals, flavors, hormones and immunological products, despite the difficulties associated with such cells. The complication of organization and regulation of biosynthetic pathways resulting in secondary products among eukaryotes is likely to be beyond the practical or economic limits, whereas the practical and economic limits of genetic manipulation of bacteria are yet to arise. In addition, certain operational and post-translational modifications, such as glycosylation, are unlikely to be accomplished by incorporating gene sequences into prokaryotes. It might also be claimed that the most satisfactory source of material for replacement therapy in the treatment of humans will be attained from genetically engineered human cell lines, and that the best way of synthesizing plant and animal cell products is by direct genetic manipulation of these cells and their use in an immobilized form.

4.2.5.3 Limitations of immobilization of eukaryotic cells

In contrast to enzymes, immobilized cells are less reactive to high substrate concentrations, surfactants, solvents and rapid changes in physical factors. Therefore, the process intensity and flexibility of response are likely to be lower. The substrate is the major nutrient of the cell and in a single-pass reactor, cells at the end of the process will either be exposed to nutrient starvation, or lower conversion must be accepted. The sensitivity of higher cells against environmental factors, e.g. osmotic shock, oxygen, shear, gas balance, pH, temperature and toxic chemicals, is a clear drawback to the procedures which might be employed for immobilization. Nevertheless, these issues are to some extent offset by the capacity of cells to regenerate after immobilization, if maintained under suitable conditions. The hormonal reactions of eukaryotes allow additional levels of control and flexibility compared to micro-organisms, which often only respond to nutrient levels or gross environmental changes. To achieve the maximum merits of immobilization, cells must be reused and thus must be retained in a healthy condition. The rich nutrient media in which both plant and animal cells must be maintained are highly susceptible to microbial contamination and so all operations must be carried out in an aseptic environment. This further restricts immobilization methods and adds to the cost and complexity of the techniques. Moreover, stored medium can deteriorate and may become increasingly unreliable when continuous operation of a process is considered. It can be seen that some of these limitations are in fact related to the cells rather than the immobilization process. If synthesis by eukaryotic cells is selected, as discussed above, the advantages ensuing from immobilization can be very high, as long as satisfactory process control is available. However, it is obvious that the complexity involved in using cultured plant and animal cells will in general limit their application in the production of high value compounds.

4.2.6 Commercial production of enzymes

Enzyme production still requires attention in the modern biotechnology industry. The arrival of genetic engineering has allowed the industrial scale production of enzymes and other proteins which are synthesized naturally in only minute quantities [13]. This progress is mainly significant with regard to the synthesis of enzymes and other proteins of therapeutic importance, which are now accessible in clinically useful quantities [13]. The level of downstream processing to which any enzyme is exposed depends on its future application. The production of enzymes in bulk usually requires little downstream processing, and therefore these are comparatively crude preparations. Enzymes intended for therapeutic applications are subject to a far higher grade of downstream processing, frequently introducing three to four chromatographic steps. Although enzymology is one of the oldest recognized branches of the biochemical sciences, it continues to be an area of constant, active research. The repeated detection of new enzymes, and better knowledge of previously discovered enzymes and their functional implications, has suggested several novel applications for these catalytic activities [13].

4.3 Genetic engineering for microbial enzyme production

Enzymes are the large biomolecules that are obligatory for the many chemical interactions that sustain life. Enzymes produced from genetically modified microorganisms play an important role in food technology [14]. Enzymes accelerate all the metabolic developments in the body and each carries out a specific task. They are highly efficient, and can trigger reaction rates 100 million to 10 billion times quicker than any normal chemical reaction. Owing to developments in recombinant technology and protein engineering, enzymes have developed as vital molecules that have been extensively employed for different industrial and therapeutical purposes. Microbial enzymes are currently gaining much attention with fast expansion of enzyme technology [14]. Microbial enzymes are poplular due to their economic feasibility, high yields, consistency, ease of product modification and optimization, regular supply due to absence of seasonal fluctuations, rapid growth of microbes on inexpensive media, stability and greater catalytic activity [14]. Microbial enzymes play a significant role in the diagnosis, treatment, biochemical investigation and monitoring of various dreaded diseases. Amylase and lipase are two key enzymes that have been closely studied and have great significance in different manufacturing and therapeutic industries. The expression of a gene always results in the synthesis of a specific protein in the form of an enzyme, hormone, etc, for a specific function. We can say that the enzymes are functional proteins that are synthesized after the expression of specific sequences present in the gene that encode for that particular protein (enzyme) [14]. Currently genetic engineers are paying more attention to the production of these therapeutic proteins using rDNA technology. Due to their many advantages microbes are considered as an excellent source for the production of these therapeutic proteins. Recent developments in rDNA technology have allowed the efficient microbial production of commercial enzymes [14]. After the identification of an enzyme, especially those are with potential use in industry, multiple copies can be produced by cloning a suitable gene and introducing it into a suitable host.

4.3.1 Cloning methods

There are several cloning methods for the commercial production of enzymes, as mentioned in figure 4.2. cDNA templates are considered the best for the production of enzymes. cDNA, which can be synthesized from mRNA to establish a cDNA library, is required for the development of a suitable probe for a particular enzyme. The specific cDNA clones required can be identified on hybridization with oligonucleotide probes. This desired gene should be inserted into a suitable host such as *Aspergillus oryzae* for the expression of a desirable enzyme. This approach can be utilized to produce high quality industrial enzymes

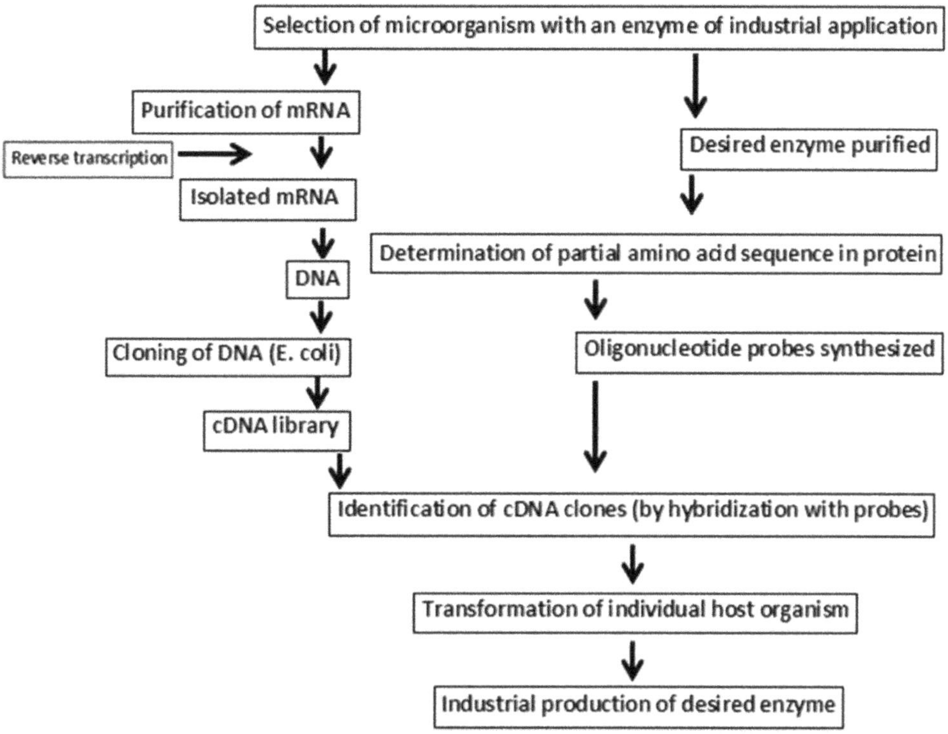

Figure 4.2. Cloning for the industrial production of enzymes.

with better yield. There have been several reports based on enzyme synthesis using cloning methods:

- The fungal enzyme lipolase, present in *Humicola languinosa*, was recently shown to eliminate fat stains in fabrics [15]. However, commercial production of this fungal enzyme from this organism is not possible as it synthesizes in very low amounts. Thus the gene that encodes for this lipolase was initially identified and then separated, cloned and finally incorporated into *Aspergillus oryzae* to produce multiple copies [15]. By following this approach, commercial production of this enzyme can be positively attained. One of the most appreciated features of lipolase is its stability and resistance against degradation by the proteases that are usually employed in the detergent industry. These features make lipolase a good candidate for applications in the fabric and detergent industry.
- Rennet (major component is chymosin) is a milk clotting enzyme that can be isolated from different sources, in particular the stomach of a young calf. Generally, rennet is considered as combination of various enzymes, in particular chymosin, which is added to raw milk in the cheese industry to produce cheese. Usually this process involves curd cutting and fractional exclusion of the whey. Afterwards the prepared cubes are allowed to ripen in the whey until they change into slim blocks. Chymosin is extensively utilized commercially, in particular in cheese manufacturing [16]. Its commercial preparation also includes other proteases; bovine pepsin is usually present in varying concentrations. To meet the market requirements commercial production should be high. Unfortunately, using earlier or existing approaches it is not possible to produce the high amounts of rennet for the market demand. Therefore, using a cloning method, in particular by exploring the genes responsible for the production of chymosin, can be considered as a potential approach for its industrial production.

4.4 Protein studies for modification of commercial enzymes

By protein engineering and site-directed mutagenesis, it is now possible to modify the structure of a protein/enzyme. In enzyme research, most of the modifications in the enzymes are carried out to enhance enzyme stability, which can further improve its catalytic function. Modification also allows the development of degradation and oxidation resistant enzymes. Moreover, structural modification of enzymes can change their substrate preference and improve tolerance against alkali and organic solvents [14]. There are number of approaches available to achieve the aforementioned objective, one of the most common is site-directed mutagenesis. Enzymes can be redesigned by using site-directed mutagenesis. This requires the systematic alteration of protein structure by rDNA technology and chemical synthesis of DNA fragments to allow the superficial modification of proteins by site specific mutagenesis of their genes. Thus mutated genes can be utilized to produce enzymes with improve yield and novel properties. Kinetic investigation of mutant enzymes with high-resolution structural data from protein x-ray crystallography can be utilized to draw the relationships between structure and function [18]. In particular, the strength and nature of enzyme–substrate interactions and their roles in catalysis and specificity may be examined [18].

During mutagenesis a number of amino acids present in the target enzyme at a particular site are allowed to modify to produce an enzyme with desirable features, e.g. structural modification of phospholipase A2 by using fundamental knowledge of protein engineering to resist elevated concentrations of acid. This modification allows its efficient utilization as a food emulsifier. Commercial preparation of phospholipase A2 is mainly produced from the pancreas for the further production of lysolecithin. This allows its extensive utilization as an excellent emulsifier for the food, cosmetics and pharmaceutical industries. Most eukaryotic proteins are N-terminally altered by one or more processing enzymes. Enzymes acting on the very first amino acid of a polypeptide include different peptidases, transferases and ligases [17].

The tyrosyl–tRNA synthetase from *Bacillus stearothermophilus* is being scientifically examined by site-directed mutagenesis. A fine-structure inspection is revealing the subtle roles of hydrogen bonding in catalysis and specificity [18]. Alteration of the residues that hydrogen-bond with ATP and tyrosine display how the energetics must be examined in terms of an exchange reaction with solvent water. Based on this knowledge, and structural data, an enzyme of vastly improved enzyme–substrate affinity has been engineered [18].

4.5 Enzyme and cell immobilization

Usually, enzymes in free solutions (i.e., in soluble or free form) react with substrates to yield final products. This type of enzyme utilization is inefficient, mainly for large -scale production, as enzymes are not stable and they cannot be recovered for reuse.

Enzymes or cells immobilization emerged as a technique of confining/anchoring the enzymes or cells in or on an (inert) support for their stability and functional reuse. By means of this technique, the enzymes produced are more effective and economical for large-scale use. A number of researchers referred to immobilization as 'the goose that lays the golden egg' in enzyme technology. Immobilized enzymes retain their structural conformation necessary for catalysis. There are various benefits to immobilized enzymes:

- They are stable and more efficient in function.
- They can be reused repeatedly.
- Their products are enzyme-free.
- They are ideal for multi-enzyme reaction systems.
- The control of enzyme function is easy.
- They are appropriate for large scale and therapeutic use.
- They reduce effluent disposal problems.

There are, however, certain drawbacks related to immobilization:

- There is a possibility of loss of biological activity of an enzyme
- Immobilization is a costly process that frequently requires sophisticated equipment.

Immobilized enzymes are usually chosen over immobilized cells owing to their ability to produce the products in a pure form. However, there are numerous benefits to using immobilized multi-enzyme systems, e.g. organelles and whole cells, over immobilized enzymes.

4.6 Immobilization methods

The frequently used procedures for immobilization of enzymes are adsorption, entrapment, covalent binding and cross-linking.

4.6.1 Adsorption methods

Adsorption includes the physical binding of enzymes (or cells) on the surface of an inert support. The support materials may be inorganic (e.g. alumina, silica gel, calcium phosphate glass) or organic (starch, carboxymethyl cellulose, DEAE-cellulose, DEAE-sephadex). Adsorption of enzyme molecules on the inert support involves weak forces, e.g. van der Waals forces and hydrogen bonds (figure 4.3). Consequently, the adsorbed enzymes can be removed by minor changes in pH, ionic strength or temperature. This is a shortcoming for the large scale utilization of enzymes.

4.6.2 Nonspecific adsorption

The immobilization procedure of nonspecific adsorption is generally based on physical adsorption or ionic binding [19, 20]. During physical adsorption the

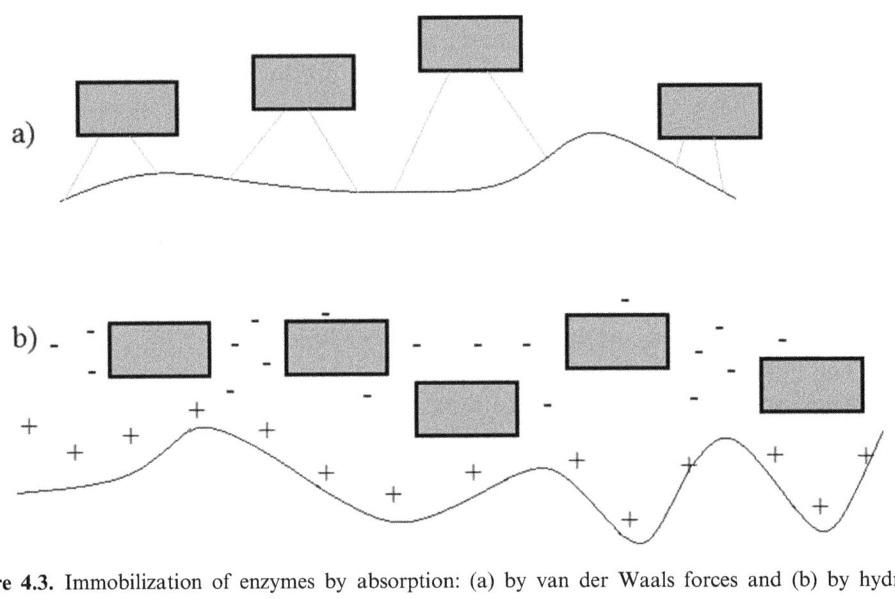

Figure 4.3. Immobilization of enzymes by absorption: (a) by van der Waals forces and (b) by hydrogen bonding.

enzymes are attached to the matrix via hydrogen bonding, van der Waals forces, or hydrophobic interactions; while in ionic bonding the enzymes are bound via salt linkages. The types of forces present in noncovalent immobilization can be reversed by varying the conditions that affect the strength of the interaction (e.g. pH, ionic strength, temperature, or polarity of the solvent). Immobilization by adsorption is a gentle, easy to perform process, and generally conserves the catalytic activity of the enzyme. Such procedures are consequently economically attractive, but may suffer from limitations, e.g. enzymes escape from matrix when the interactions are comparatively weak.

4.6.3 Ionic binding

One method for the reversible immobilization of enzymes is to use protein–ligand interactions based on the principles employed in chromatography. For instance, one of the first claims of chromatographic principles in the reversible immobilization of enzymes was the use of ion-exchangers [21, 22]. The procedure is simple and reversible, but it is generally challenging to discover conditions under which the enzyme remains both firmly bound and fully active. Currently, the use of immobilized polymeric-ionic ligands has permitted the modification of protein–matrix interactions and has thus enhanced the properties of the derivative. Some patents have been published on the use of polyethyleneimine to bind a rich variety of enzymes and whole cells [22]. However, issues may rise from the use of an extremely charged support when the substrates or products themselves are charged; the kinetics can be distorted resulting in partition or diffusion phenomena. Consequently, enzyme features, e.g. pH optimum or pH stability, may change [23, 24]. Although this could be a problem it could also be valuable to change the optimal conditions of a certain enzyme against more alkaline or acidic conditions, depending on the application [25].

4.6.4 Hydrophobic adsorption

Another approach is the use of hydrophobic interactions. During this procedure, it is not the development of chemical bonds but rather an entropically driven interaction that occurs. Hydrophobic adsorption has been employed as a chromatographic principle for more than three decades. It relies on well-known experimental variables such as pH, salt concentration and temperature [26]. The power of interaction relies on both the hydrophobicity of the adsorbent and the protein. The hydrophobicity of the adsorbent can be controlled by the degree of substitution of the support and by the size of the hydrophobic ligand molecule. The effective reversible immobilization of β-amylase and amyloglucosidase to hexyl-agarose carriers has been demonstrated [27, 28]. Several other examples of strong reversible binding to hydrophobic adsorbents have also been demonstrated [29, 30].

4.6.5 Affinity binding

The basis of affinity between complementary biomolecules has been applied to enzyme immobilization. The outstanding selectivity of the interaction is a key benefit of the procedure. However, the method often needs the covalent binding of a costly affinity ligand (e.g. antibody, or lectin) to the matrix [31, 32].

4.6.6 Entrapment method

By means of physical entrapment enzymes can be easily immobilized inside a polymer or a gel matrix. The size of the matrix pores is maintained in such a manner that the enzyme is retained, whereas the substrate and product molecules can easily pass through. During this procedure, also known as lattice entrapment, the enzyme (or cell) is not exposed against strong binding forces or structural distortions. During an immobilization process, the level of deactivation might be possible. This happens due to the fluctuations in pH or temperature or the addition of solvents. The matrices employed for entrapment of enzymes include polyacrylamide gel, collagen, gelatin, starch, cellulose, silicone and rubber. Enzymes can be entrapped in numerous ways. There are different approaches to entrapping enzymes such as gel [33] or fiber entrapping [34, 35] and micro-encapsulation [36, 37].

4.6.6.1 Enzyme inclusion in gels

This is achieved by enzyme entrapment inside gels (figure 4.4). The poly(vinyl alcohol) bearing styrylpyridinium groups (PVA-SbQ), a soluble pre-polymer bearing photo-cross-linkable group, has been used extensively to entrap enzymes, and numerous bioassays based on this immobilization matrix have been demonstrated [38, 39]. The sol–gel process was established on the capability to form solid metal or semi-metal oxides through the aqueous process of hydrolytically labile precursors. Enzymes can also be entrapped in an agarose gel [39]. In contrast to synthetic polymers, e.g. polyacrylamide, this matrix is biocompatible, nontoxic, provides a

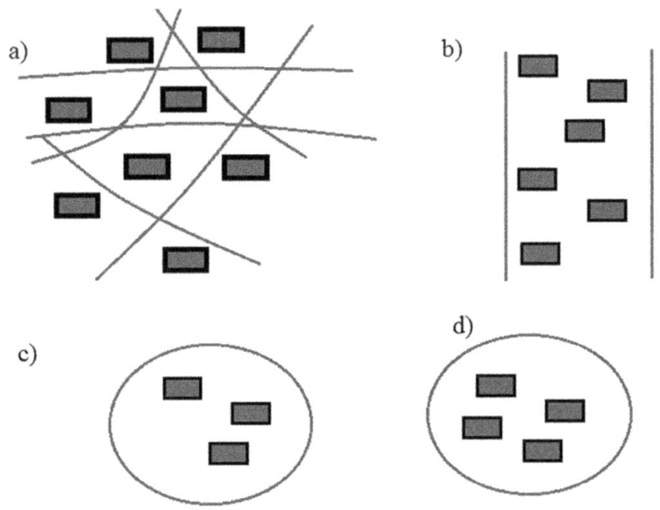

Figure 4.4. Immobilization of enzymes by entrapment. (a) Inclusion in gels. (b) Inclusion in fibers. (c) Inclusion in microcapsules.

natural microenvironment to the enzyme and also offers adequate accessibility to electrons to shuttle between the enzyme and the electrode. The entrapment approaches are easy to perform, and allow the deposition of enzymes, mediators, and additives in the same sensing layer. Furthermore, the activity of the enzyme is conserved during the immobilization process, as the biological element is not altered [39]. Biosensors based on physically entrapped enzymes are frequently described by increased operational and storage stability.

4.6.6.2 Enzyme inclusion in fibers
Entrapment involves inclusion of an enzyme in a polymer network (gel lattice) such as an organic polymer, a silica sol–gel or a membrane device such as a hollow fiber or a microcapsule. Entrapment requires the synthesis of the polymeric network in the presence of the enzyme. The last step involves cross-linking of enzyme aggregates or crystals, using a bifunctional reagent, to prepare carrier-free macroparticles [40]. The enzymes are trapped in the fibers of the matrix.

4.6.6.3 Enzyme inclusion in microcapsules
During this procedure enzymes are trapped inside a microcapsule matrix (figure 4.4). The hydrophobic forms of the matrix polymerize to yield a microcapsule containing enzyme molecules [41]. The major limitation for entrapment of enzymes is the leakage from the matrix. Most of the researchers suggested utilizing this method of entrapment for immobilization of whole cells. Entrapped cells can be used in large scale production of amino acids such as L-isoleucine, L-aspartic acid, L-malic acid and hydroquinone [41].

4.6.6.4 Micro-encapsulation
Micro-encapsulation is a kind of entrapment in which spherical particles in a liquid or suspension are enclosed in a semipermeable membrane [43]. The membrane may be polymeric, lipoidal, lipoprotein-based or non-ionic in nature. The three steps of micro-encapsulation are:

- Building superior membrane reactors.
- Formation of the emulsion.
- Stabilization of the emulsion to yield microcapsules [42].

Currently micro-encapsulation is employed for immobilization of enzymes and mammalian cells. For example, cultured pancreatic cells can be immobilized by micro-encapsulation; immobilized hybridoma cells have also been evidenced by this technique.

4.6.7 Covalent binding

Enzyme immobilization can be accomplished by formation of covalent bonds between the chemical groups of enzymes and the chemical groups of the support (figure 4.4). This procedure is extensively used. However, covalent binding is frequently linked with loss of some enzyme activity. Pretreatment (to form pre-activated support) of the inert support is typically required before it binds to an enzyme. The following are the common procedures of covalent binding. Immobilizations of proteins by procedures based on the formation of covalent bonds are among the most extensively used. A benefit of these procedures is that, because of the stable nature of the bonds developed between enzyme and the matrix, the enzyme is not released into the solution upon use. However, in order to attain high levels of bound activity, the amino acid residues vital for catalytic activity must not be involved in the covalent linkage to the support; this may prove a challenging condition to accomplish in some cases. A simple method that occasionally increases the activity yield is to perform the coupling reaction in the presence of substrate analogs [44]. Covalent methods for immobilization are used when there is a strict obligation for the absence of the enzyme in the product. An extensive range of reactions have been established depending on the functional groups accessible on the matrix [45]. Coupling procedures can generally be categorized in two main classes:

- Activation of the matrix by addition of a reactive function to a polymer.
- Alteration of the polymer backbone to synthesize an activated group.

The activation procedures are usually intended to make electrophilic groups on the support which, in the coupling step, react with the strong nucleophiles on the proteins. The basic principles governing the course of covalent coupling to the matrices are similar to those used for the chemical alteration of proteins. The most often used reactions include the following side chains of the amino acids: lysine (ε-amino group), cysteine (thiol group), and aspartic and glutamic acids (carboxylic group).

- *Cyanogen bromide activation:* The inert support materials such as cellulose, sepharose and sephadex enclosing glycol groups are activated by CNBr, which at that time binds to enzymes and immobilizes them (figure 4.5).
- *Diazotation:* A number of the support materials (aminobenzyl cellulose, amino derivatives of polystyrene, aminosilanized porous glass) are exposed to diazotation on treatment with $NaNO_2$ and HCI. They, in return, bind covalently to tyrosyl or histidyl groups of enzymes (figure 4.5).
- *Peptide bond formation*: Enzyme immobilization can also be achieved by the development of peptide bonds between the amino (or carboxyl) groups of the support and the carboxyl (or amino) groups of enzymes (figure 4.5). The support material is initially chemically treated to form active functional groups.

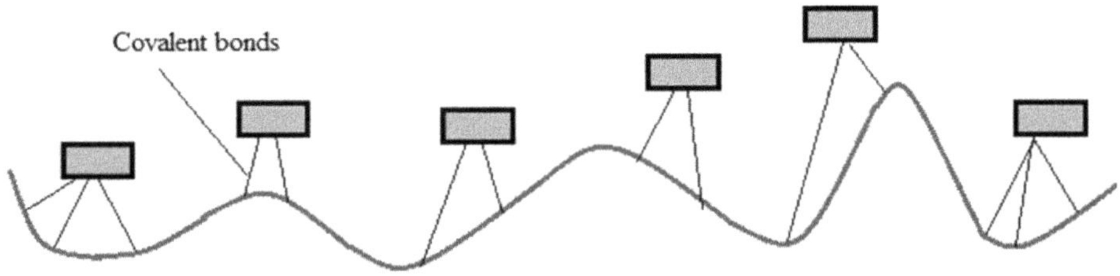

Figure 4.5. A general representation of the immobilization of enzymes by covalent binding.

- *Activation by bi- or polyfunctional reagents*: A number of reagents, e.g. glutaraldehyde, can be used to build bonds between the amino groups of the enzymes and the amino groups of the support (such as aminoethylcellulose, albumin, amino alkylated porous glass). This is represented in figure 4.6.

4.6.8 Cross-linking

A characteristic feature of the immobilization of enzymes by cross-linking is the lack of a solid support. The enzyme molecules are immobilized by building cross-links between them, and by the participation of polyfunctional reagents. In fact, these

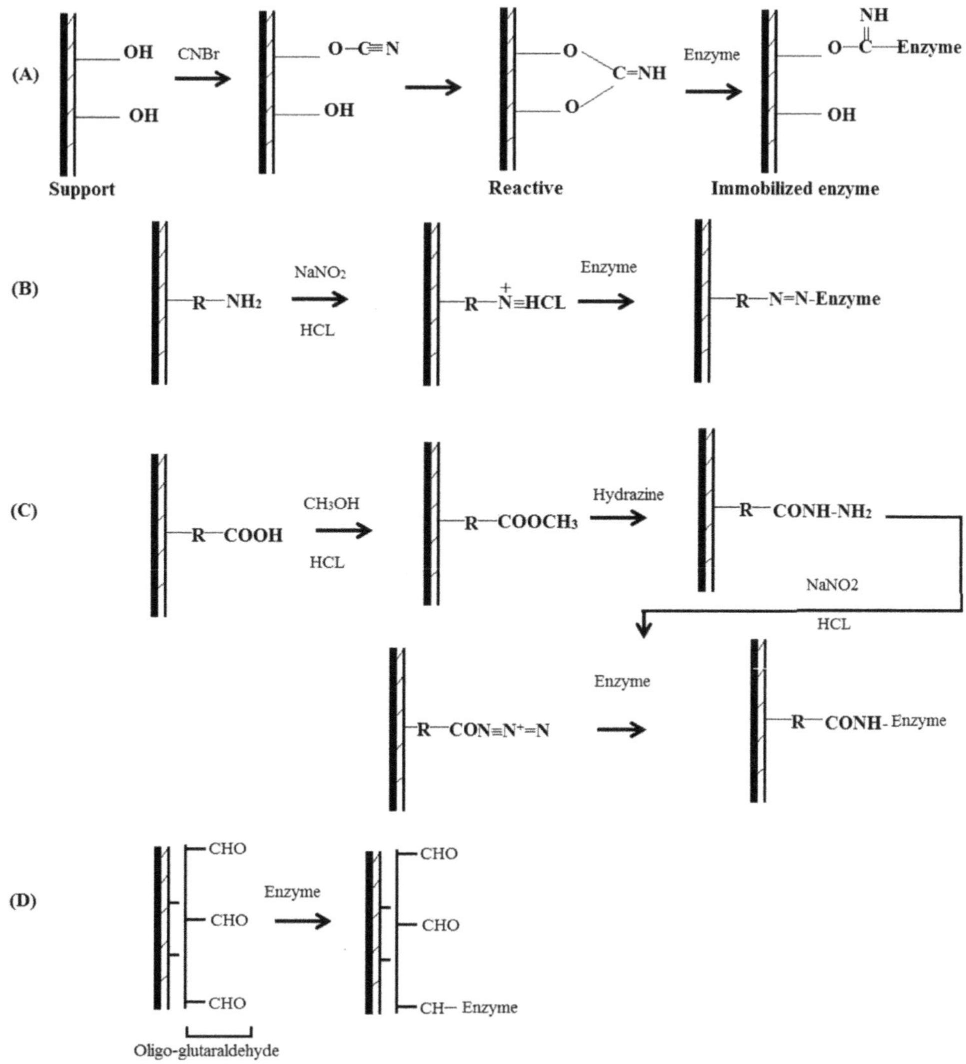

Figure 4.6. Immobilization of enzymes by covalent binding: (a) cyanogen bromide activation, (b) diazotation, (c) peptide bond formation and (d) activation by bifunctional agent.

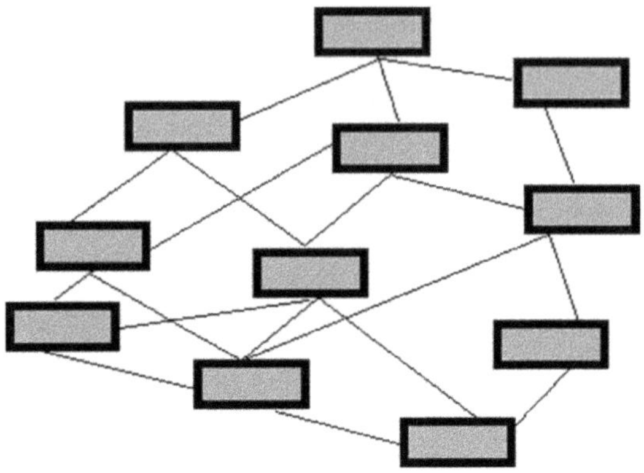

Figure 4.7. Immobilization of enzyme molecules by cross-linking.

reagents react with the enzyme and make bridges which further form the support to hold enzyme molecules (figure 4.7). There are a number of reagents used for cross-linking. These include glutaraldehyde, diazobenzidine, hexamethylene diisocyanate and toluene di-isoth iocyanate. Glutaraldehyde is the most frequently used cross-linking reagent [38]. It successfully reacts with lysyl residues of the enzymes and finally yields a Schiff's base. The cross-links developed between the enzyme and glutaraldehyde are irreversible and can tolerate extreme pH and temperature. Glutaraldehyde cross-linking has been effectively utilized to immobilize a number of industrial enzymes such as glucose isomerase and penicillin amidase [38]. The method of cross-linking is very simple and cost-effective, but the drawback is that it includes the possibility of enzyme denaturation by the polyfunctional reagent.

4.7 Choice of immobilization technique

Selection of a particular procedure for immobilization of enzymes is based on a trial and error approach to select the perfect one. The enzyme catalytic activity, stability, regenerability and cost are the most important factors that decide a technique.

4.7.1 Immobilization of L-amino acid acylase

Japanese researchers [46] discovered the first enzyme immobilized on a matrix, known as L-amino acid acylase. Later, a number of immobilization procedures were attempted by these researchers, but only three showed promising results. These were:

- Covalent binding to iodoacetyl cellulose.
- Ionic binding to DEAE-sephadex.
- Entrapment with *n* polyacrylamide.

4.7.2 Stabilization of soluble enzymes

Enzyme stabilization is essential for any biomedical or industrial application of enzymes. In several applications, the objective is to deliver a prolonged active lifetime under normal environmental conditions with traditional substrates at low concentrations in buffered solutions [47]. However, as enzymes are being employed for more and more applications, there is a need to use them under extreme environmental conditions (i.e., high temperatures), in high substrate concentrations and in nontraditional solvent systems [47]. Certain enzymes cannot be immobilized as they should be used in soluble form, for example, enzymes used in liquid detergents, some diagnostic reagents and food additives. With the aid of a number of additives or by chemical modifications these enzymes can be stabilized. The stabilized enzymes have longer half-lives, although they cannot be recycled. Several significant procedures for enzyme stabilization are briefly described in the following [47]:

- *Solvent stabilization*: Some solvents at low concentrations stabilize the enzymes, whereas at high concentrations the enzymes become denatured, e.g. acetone (5%) ethanol (5%) can stabilize benzyl alcohol dehydrogenase.
- *Substrate stabilization*: By adding substrates the active site of an enzyme can be stabilized, e.g. starch stabilizes α-amylase; glucose stabilizes glucose isomerase.
- *Stabilization by polymers*: This type of stabilization can be achieved by the addition of polymers, such as gelatin, albumin and polyethylene, particularly at increased temperatures.
- *Stabilization by salts*: Metalloenzymes stability of can be attained by supplementing salts, e.g. Ca, Fe, Mn, Cu and Zn. Proteases are commonly stabilized by the addition of calcium.
- *Stabilization by chemical modifications*: During this procedure, enzymes can be stabilized by appropriate chemical modifications without loss of biological activity. There are different types of chemical modifications:
    - Addition of polyamino side chains, such as polytyrosine.
    - Acylation of enzymes by adding groups, e.g. acetyl, propionyl and succinyl.
- *Stabilization by rebuilding*: Ideally, enzyme stability is dependent on the hydrophobic interactions in the core of the enzyme. It was thus suggested that enzymes can be stabilized by increasing hydrophobic interactions. For this reason, the enzyme is initially unfolded and later rebuilt in one of the following ways (see figure 4.8):
    - The enzyme can be chemically treated (e.g. using urea and a disulfide) and then refolded.
    - The refolding can be done in the presence of low molecular weight ligands.
    - For certain enzymes, refolding at higher temperatures (around 50 °C) stabilizes them.
- *Stabilization by site-directed mutagenesis*: Site-directed mutagenesis has been fruitfully used to synthesize more stable and functionally more efficient enzymes such as subtilisin E.

4.8 Immobilization of cells

A single-step reaction procedure is suitable for the immobilization of an individual cell, but is not appropriate for multienzymes or for reactions demanding co-factors. During this procedure whole cells or cellular organelles can be immobilized, yielding

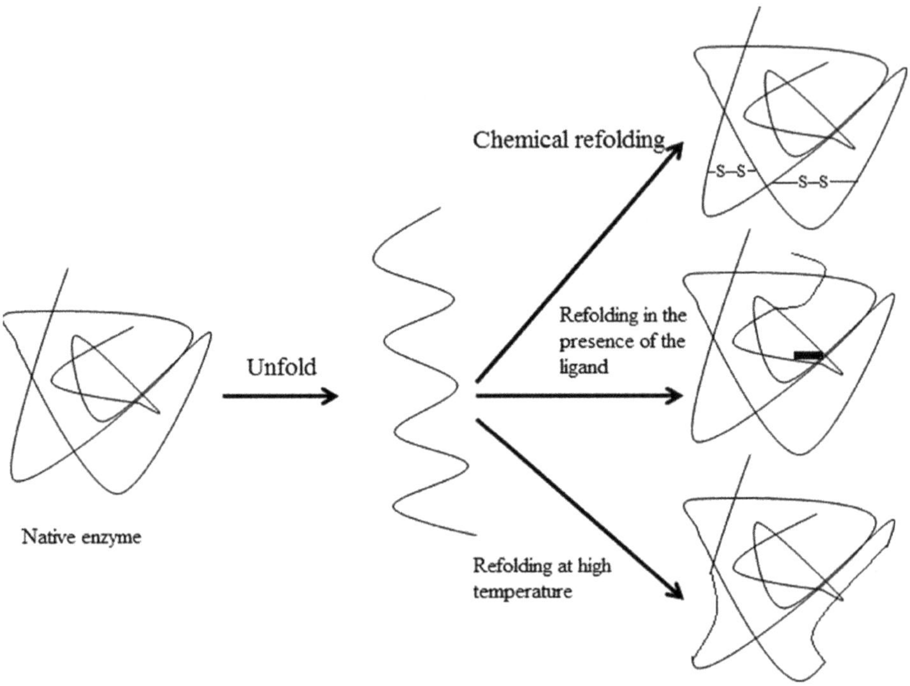

Figure 4.8. Stabilization of an enzyme by refolding.

a system which acts as a source for various enzymes, and hence is referred to as a multienzyme system. Furthermore, immobilized cells rather than enzymes are often favored even for single reactions, owing to the cost factor in isolating enzymes. For those enzymes which are dependent on the spacial arrangement of the membrane for their activity, cell immobilization is ideal. There are several significant applications of immobilized enzymes, for example, they are traditionally used for the treatment of sewage. The procedures used for the immobilization of cells are almost the same as those employed for the immobilization of enzymes, with appropriate modifications. Techniques such as entrapment and surface attachment are frequently used. Gels, and to some extent membranes, are also used.

4.8.1 Immobilization of viable cells

By means of gentle immobilization it is possible to preserve the viability of the cells. These immobilized cells are mainly useful for fermentation. Occasionally, mammalian cell cultures are made to function as immobilized viable cells.

4.8.2 Immobilized non-viable cells

In most cases, immobilized non-viable cells are considered over enzymes or even viable cells. This is primarily because of the expensive isolation and purification procedures. An excellent example is the immobilization of cells producing glucose isomerase for the large-scale production of high-fructose syrup.

4.8.3 Drawbacks of immobilizing eukaryotic cells

Prokaryotic cells (mainly bacterial) are generally used for immobilization. According to various reports it is also possible to immobilize eukaryotic plant and animal cells. Owing to the presence of cellular organelles, the metabolism of eukaryotic cells is slow. Consequently, for the large-scale production of biochemicals, prokaryotic cells are considered. However, for the synthesis of complex proteins (e.g. immuno-globulins) and for proteins that experienced post-translational modifications, eukaryotic cells may be used.

4.8.4 The effect of immobilization on enzyme properties

Cell immobilization is often linked to modifications in enzyme properties, particularly the kinetic properties of enzymes. Some of these are as follows:

- A considerable reduction in enzyme specificity. This may be because of conformational changes that take place when the enzyme becomes immobilized.
- The kinetic constants (K_m and V_{max}) of an immobilized enzyme differ from those of the innate enzyme. This is because of the conformational modifications of the enzyme which may further affect the affinity between the enzyme and substrate.

4.8.5 Immobilized enzyme reactors

Immobilized enzyme cells are exploited for large-scale production in the form of enzyme reactors [43]. They are generally of two types: batch reactors and continuous reactors. The frequently used enzyme reactors are depicted in figure 4.9.

Batch reactors. In these types of reactors, the immobilized enzymes and substrates are employed, and the reaction is allowed to start under constant stirring. As the reaction is finished, the product is isolated from the enzyme (typically by denaturation). Soluble enzymes are frequently used in batch reactors. It is somewhat challenging to isolate the soluble enzymes from the products, therefore, there is the drawback of their reuse. Special procedures have been developed for the recovery of soluble enzymes, which may occasionally result in loss of enzyme activity [48].

Stirred tank reactors. The most modest form of batch reactor is the stirred tank reactor (figure 4.9). It is composed of a reactor fixed with a stirrer that permits good mixing, and a suitable temperature and pH control. However, in this system there may be loss of some enzyme activity. The basket reactor is a modification of the stirred tank reactor. In this method, the enzyme is retained over the impeller blades. Both the stirred tank reactor and basket reactor have a well-mixed flow pattern [48].

Plug flow type reactors. These reactors are alternatives to flow pattern type reactors. The flow rate of fluids can be regulated by a plug system. Plug flow type reactors are available in the form of packed-bed or fluidized-bed reactors (figure 4.9). These reactors are mainly useful when flow type reactors furnish insufficient product formation. Moreover, plug flow reactors are also beneficial for obtaining kinetic data on the reaction systems [48].

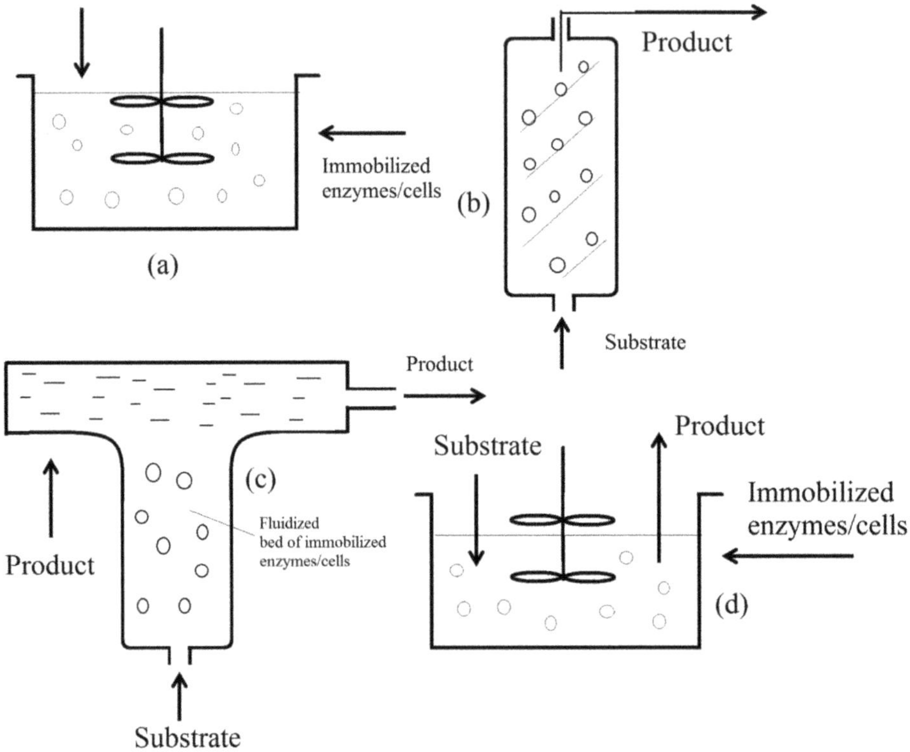

Figure 4.9. Immobilized enzyme/cell reactors: (a) batch stirred tank reactor, (b) packed-bed reactor, (c) fluidized-bed reactor and (d) continuous stirred tank reactor.

Continuous reactors. During this procedure the substrate is supplemented constantly, whereas the product is eliminated. Immobilized enzymes can also be employed for continuous operation. Continuous reactors have certain merits over batch reactors. These entail regulation over the product formation, suitable operation of the system and easy automation of the whole procedure. There are two main types of continuous reactors: the continuous stirred tank reactor (CSTR) and plug reactor (PR) [49]. A schematic depiction of CSTR is shown in figures 4.9 and 4.10. CSTR is suitable for good quality product formation.

Membrane reactors. A number of membranes with an array of chemical compositions can be used. Polysulfone, polyamide and cellulose acetate are the most often used materials for membranes. The biocatalysts (enzymes or cells) are usually retained on the membranes of the reactor. The substrate is added into reactor while the product passes out. Good mixing in the reactor can be attained using a stirrer (figure 4.10). In a continuous membrane reactor, the biocatalysts are held on membrane layers over which substrate molecules are passed (figure 4.10). In a recycle model membrane reactor, the contents, i.e., the solution containing enzymes, co-factors, and substrates, along with freshly released product, are recycled by using a pump (figure 4.10). The product passes out of the system and can be recovered [43–45].

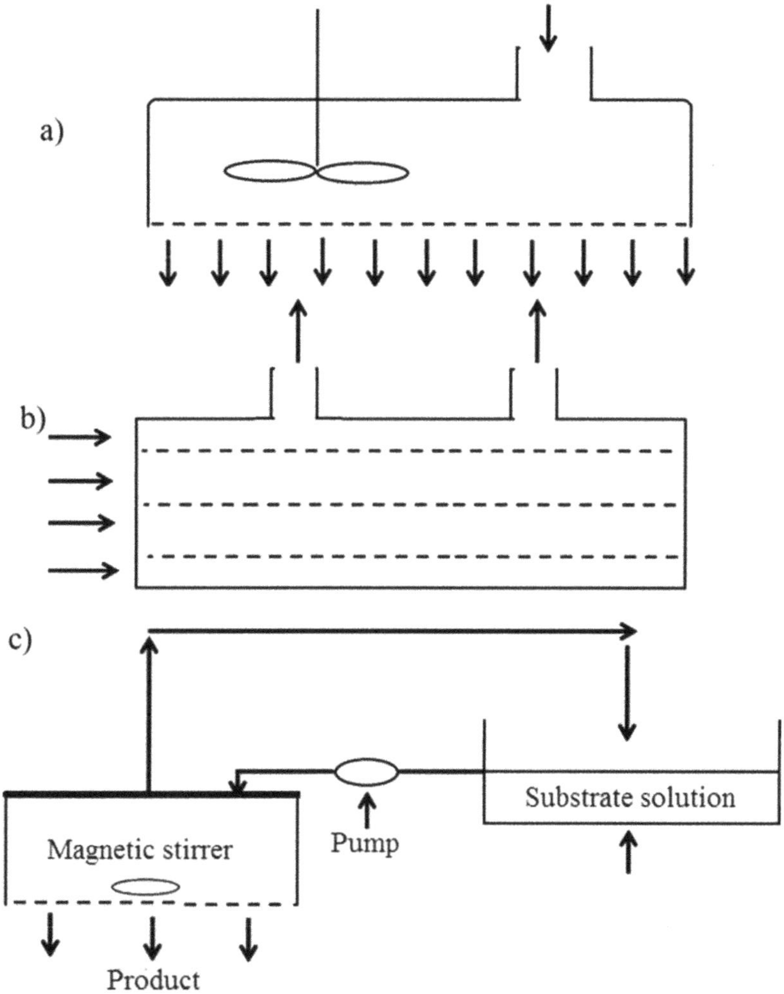

Figure 4.10. Membrane reactors: (a) batch membrane reactor, (b) continuous membrane reactor and (c) recycle membrane reactor (coloured lines indicate membranes).

4.8.6 Applications of immobilized enzymes and cells

Immobilized enzymes and cells are extensively utilized for industrial, analytical and therapeutic applications, including their utilization in food production and research in biochemistry, microbiology and other allied sciences [50]. A summary of industrial applications of immobilized cells is provided in table 4.2.

4.9 Manufacture of commercial products

A list of significant immobilized enzymes and their industrial applications is mentioned in table 4.2. Further details on the production of L-amino acids and high-fructose syrup are mentioned below.

Table 4.2. A selected list of important immobilized enzymes and their industrial applications.

Immobilized enzyme	Applications
β-galactosidase	Splitting of lacrosse to glucose and galactose.
Ribonuclease	Synthesis of nucleotides from RNA.
Penicillin acylase	Commercial production of semi-synthetic penicillins.
Nitrilase	Production of acrylamide from acrylonitrile.
Invertase	Splitting of sucrose to glucose and fructose.
Histidine ammonia lyase	Production of urocanic acid from histidine.
Glucose isomerase	Production of high-fructose syrup from glucose (or starch).
Fumarase	Synthesis of malic acid from fumaric acid.
Aspartase	Production of aspartic acid from fumaric acid.
Amylase	Production of glucose from starch.
Aminocyclase	Production of L-amino acids from D, L-acyl amino acids.

4.9.1 Production of L-amino acids

L-amino acids (and not D-amino acids) are very important for use in food and feed supplements and therapeutic applications [49, 51]. The chemical methods used for their production result in a racemic mixture of D- and L-amino acids. They can be acylated to yield D, L-acyl amino acids. The immobilized enzyme aminocyclase (frequently immobilized on DEAE-sephadex) can selectively hydrolyze D, L-acyl amino acids to produce L-amino acids.

$$\text{D, L-acyl amino acids} \longrightarrow \text{L-amino acids} + \text{D, L-acyl amino acids}$$

The free L-amino acids can be separated from the unhydrolysed D-acyl amino acids. The latter can be racemized to D, L-acyl amino acids and recycled via an enzyme reactor containing immobilized amnocyclase. Large amounts of L-methionine, L-phenylalanine, L-tryptophan and L-valine are produced worldwide using this method.

4.9.2 Production of high-fructose syrup

Among the monosaccharaides, fructose is the sweetest, and has doubled the sweetening strength of sucrose. Glucose is about 75% as sweet as sucrose. Thus, glucose (the most abundant monosaccharide) cannot be a good replacement for sucrose for sweetening. Therefore, there is a great demand for fructose which is very sweet, but has a similar calorific value to that of glucose or sucrose. High-fructose syrup (HFS) contains about equivalent amounts of glucose and fructose. From the nutritional point of view high-fructose syrup is similar to sucrose [52]. High-fructose syrup is a good replacement for sugar in the manufacturing of soft drinks, processed foods and baking. High-fructose syrup can be manufactured from glucose by using an immobilized glucose isomerase. The starch-containing raw materials (wheat, potato, corn) are exposed to hydrolysis to produce glucose. Glucose isomerase then isomerises glucose to fructose (figure 4.11). The ultimate product formed is high-fructose syrup comprising 50% fructose.

4.9.2.1 Glucose isomerase

Glucose isomerase is an intracellular enzyme synthesized by several micro-organisms. Species of *Arthrobacter, Bacillus* and *Streptomyces* are the preferred sources. As it is an intracellular enzyme, the separation of glucose isomerase without the loss of biological activity involves specialized and expensive methods. Often, whole or partly broken cells are immobilized and used [18].

4.9.3 Immobilized enzyme and cell analytical applications

4.9.3.1 In biochemical analysis

Immobilized enzymes (or cells) can be employed for the advancement of precise and specific analytical methods for the determination of several compounds. The main

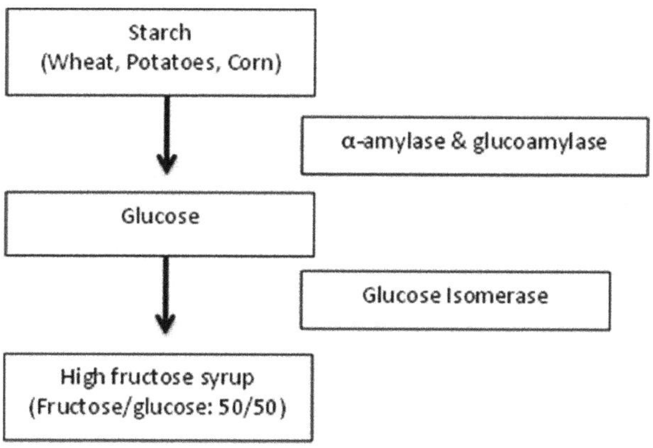

Figure 4.11. The production of high-fructose syrup from starch with glucose isomerase as the immobilized enzymes.

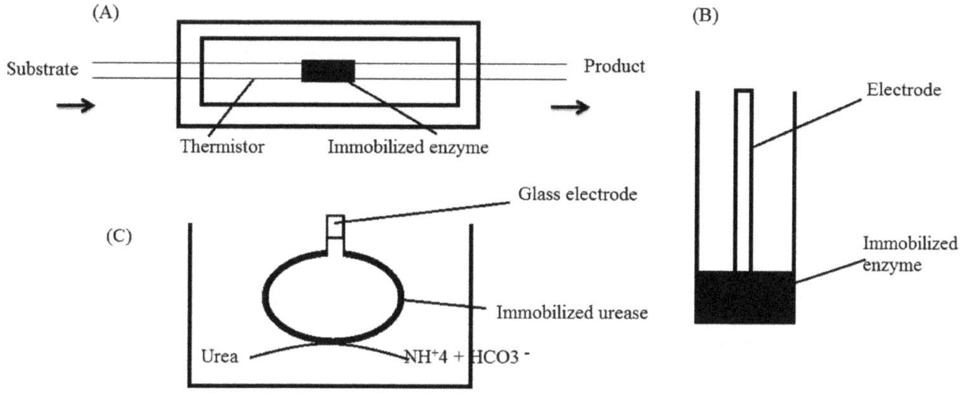

Figure 4.12. Immobilized enzymes or cells in analytical biochemistry. (a) Enzyme thermistor. (b) Enzyme electrode. (c) Urease electrode.

principle of analytical assays involves the action of the enzyme on the substrate. A decrease in the substrate concentration, an increase in the product or an alteration in the co-factor concentration can be used for the assay. Two types of detectors are commonly employed, thermistors and electrode devices.

Thermistors are heat measuring devices which can analyze the heat produced in an enzyme-calibrated reaction. Electrode devices are employed for determining potential differences in the reaction systems. In figure 4.12 an enzyme thermistor, enzyme electrode and a urease electrode are shown.

4.9.3.2 Affinity chromatography and purification

Proteins purification by traditional methods is often tedious and partial, and the yields are often low. Earlier isolation of enzyme is limited to certain concepts such as:

- its specific biological property;
- its reversible association capability with specific substrates;
- its reversible association capability with specific inhibitors.

During affinity chromatography, the desired enzyme considered for the purification is allowed to pass through a column which usually contains a cross-linked polymer or gel. This specific competitive enzyme is covalently attached with polymer or gel present in the column. Enzymes without considerable affinity for the bound inhibitor present in the polymeric or gel matrix will pass directly or spontaneously through the column, whereas those that have substantial affinity, recognize the inhibitor and form in a complex way and are eventually retained by the matrix. Elution of complex or bound enzymes can be achieved by altering certain factors such as salt concentration or pH, or by addition of a competitive inhibitor in solution. Generally, the matrix is usually inert in nature, i.e., it shows minimal interaction with proteins, both before and after coupling to the specific binding group such as substrate or inhibitors. For affinity chromatography, the adsorbent must have a number of promising characteristics. A loose and porous matrix or gel allows easy entry and exit of macromolecules. Additionally, it also retains suitable flow properties during use. The chemical nature of the matrix must allow the suitable and extensive attachment of the particular ligand under mild conditions. This will only happen with chemical bonds that are stable to the conditions of adsorption and elution.

Affinity chromatography can be employed in several applications such as purification of nucleic acid and protein from cell free extracts and blood. Based on the property of affinity it is possible to purify several compounds e.g. antigens, antibodies and co-factors.

4.10 Application of various immobilization techniques for algal bioprocesses

Algal bioprocesses are beneficial in handling wastewater treatment processes with valuable biomass production [53–71]. Algal biomass can be utilized for different reasons such as:

- slow-release fertilizers or soil conditioners;
- pharmaceuticals;
- food additives;
- cosmetics;
- biohydrogen or biogas;
- biofuel generation as biodiesel;
- other valuable chemicals.

Microalgal cells have numerous merits in not needing several raw materials to produce their biomass, if procedure is at lower costs, with the dissolved oxygen produced by the algae being valued to raise the oxygen levels of water effluents, and can be used for the reduction of carbon dioxide emissions by CO_2 for their biomass or energy generation. Collection and procurement, especially drying part of the algal biomass is one of the major challenges of algal bioprocesses. Numerous harvesting procedures have been developed [53, 54]. Algal cells immobilization has been introduced to prevent these challenges faced during harvesting and dewatering stages [55, 56]. Immobilization of various cells in polymeric matrix has numerous merits over their free-cell counterparts, as immobilized cells occupy less space, are easier to handle, and can be used recurrently for product production [57]. Cells immobilization has been projected to enhance the biosorption capacity and bioactivity of the biomass [58, 59]. It encourages bioprocesses with higher cell densities and offers ease during harvesting of biomass from its liquid environment [60]. Immobilization techniques have numerous merits such as:

- remaining unaffected against unfavorable conditions such as salinity, metal toxicity, and pH;
- protecting the old cultures against the detrimental effects of photoinhibition;
- Offering higher biomass concentrations;
- recovering the cells in a less-destructive way;
- increasing the cost-effectiveness of the process by reusing the regenerated biomass.

Considering scale-up of algal biomass at the industrial level, one of the most important challenges is related with algal biomass recovery. Immobilization systems have now become attractive replacements for scale-up processing [61]. Various types of immobilization systems are in use today, such as:

- entrapment within polymers immobilization system;
- covalent coupling immobilization system;
- confinement in liquid–liquid emulsions immobilization system;
- capturing with semipermeable membranes immobilization system;
- adsorption immobilization system.

Entrapment of the cells within polymeric matrices and self-adhesive attachment of cells onto the surfaces of solid-supports is considered as one of the most common immobilization processes [62]. Different polymers (natural and synthetic) can be utilized as the immobilization matrix [63]. Methods for effective entrapment are to set the algal cells free within their partition, while pores inside the gel matrix allow the diffusion of substrates and the metabolic products toward and from the cells [64]. However, entrapment immobilization systems still have some disadvantages in decreasing the mass transfer kinetics of the uptake of metal ions [65]. Nevertheless, these can be circumvented by carefully selecting the immobilization procedure or matrix [53–71]. One of the most important applications for immobilized algal cells involves:

- removal of nutrients from aqueous solutions;
- biodiesel production;
- biosorption of heavy metals from wastewaters;
- photoproduction of hydrogen and photopigments;
- providing an alternative technique to the common cryopreservation processes;
- toxicity testing.

References

[1] Brena B, González-Pombo P and Batista-Viera F 2013 Immobilization of enzymes: a literature survey *Methods Mol. Biol.* **1051** 15–31

[2] Mehrotra P 2016 Biosensors and their applications—a review *J. Oral Biol. Craniofac. Res.* **6** 153–9

[3] Mohamad N R, Nur Haziqah Che M, Buang N A, Huyop F and Wahab R A 2015 An overview of technologies for immobilization of enzymes and surface analysis techniques for immobilized enzymes *Biotechnol. Biotechnol. Equip.* **29** 205–20

[4] Kamra T *et al* 2016 Covalent immobilization of molecularly imprinted polymer nanoparticles on a gold surface using carbodiimide coupling for chemical sensing *J. Colloid Interface Sci.* **461**(1) 1–8

[5] Iso M, Shirahase T, Hanamura S, Urushiyama S and Omi S 1989 Immobilization of enzyme by microencapsulation and application of the encapsulated enzyme in the catalysis *J. Microencapsul.* **6** 165–76

[6] Trevan M D 1988 Enzyme immobilization by covalent bonding *New Protein Tech.* **3** 495–510

[7] Coulet P R, Carlsson J and Porath J 1981 Immobilization of enzymes on metal-chelate regenerable carriers *Biotechnol. Eng.* **23** 663–8

[8] Kennedy J F and Cabral J M S 1986 Use of titanium etc. Species for the immobilization of bioactive compounds—enzymes *Transition Met. Chem.* **11** 41–66

[9] Kennedy J F, Cabral J M, Kosseva M R and Paterson M 1997 Transition metal methods for immobilization of enzymes and cells *Immobilization of Enzymes and Cells* (Methods in Biotechnology vol 1) (Berlin: Springer), 345–9

[10] Datta S, Christena L R, Rani Y and Rajaram S 2013 Enzyme immobilization: an overview on techniques and support materials *Biotechnology* **3** 1–9

[11] Jen A C, Wake M C and Mikos A G 1996 Review: hydrogels for cell immobilization *Biotechnol. Bioeng.* **50** 357–64

[12] Hall R D, Holden M A and Yeoman M M 1988 Immobilization of higher plant cells *Medicinal and Aromatic Plants* (Biotechnology in Agriculture and Forestry vol 4) (Berlin: Springer), pp 136–56

[13] Headon D R and Walsh G 1994 The industrial production of enzymes *Biotechnol. Adv.* **12** 635–46

[14] Gurung N, Ray S, Bose S and Rai V 2013 A broader view: microbial enzymes and their relevance in industries, medicine, and beyond *Biomed. Res. Int.* **2013** 329121

[15] Hasan F, Shah A, Javed S and Hameed A 2010 Enzymes used in detergents: lipases *African J. Biotechnol.* **9** 4836–44

[16] Fernandes P 2010 Enzymes in food processing: a condensed overview on strategies for better biocatalysts *Enzyme Res.* **2010** 862537

[17] Varland S, Osberg C and Arnesen T 2015 1,2 N-terminal modifications of cellular proteins: the enzymes involved, their substrate specificities and biological effects *Proteomics* **15** 2385–401

[18] Fersht A R and Winter G P 1985 Redesigning enzymes by site-directed mutagenesis *Ciba Found. Symp.* **111** 204–18

[19] Messing R A 1976 Adsorption and inorganic bridge formations *Methods in Enzymology* ed K Mosbach vol 44 (New York: Academic), pp 148–69

[20] Woodward J 1985 Immobilized enzymes: adsorption and covalent coupling *Immobilized Cells and Enzymes: A Practical Approach* ed J Woodward (Oxford: IRL), pp 3–17

[21] Tosa T, Mori T, Fuse N and Chibata I 1967 Studies on continuous enzyme reactions I. Screening of carriers for preparation of water insoluble aminoacylase *Enzymologia* **31** 214–24

[22] Sharp A K, Kay G and Lilly M D 1969 The kinetics of β-galactosidase attached to porous cellulose sheets *Biotechnol. Bioeng.* **11** 363–80

[23] Goldstein L 1972 Microenvironmental effects on enzyme catalysis. A kinetic study of polyanionic and polycationic derivatives of chymotrypsin *Biochemistry* **11** 4072–84

[24] Goldman R, Kedem O, Silman I, Caplan S and Katchalski-Katzir E 1968 Papain–collodion membranes. I. Preparation and properties *Biochemistry* **7** 486–500

[25] Guisan J M, Alvaro G, Rosell C M and Fernandez-Lafuente R 1994 Industrial design of enzymic processes catalysed by very active immobilized derivatives: utilization of diffusional limitations (gradients of pH) as a profitable tool in enzyme engineering *Biotechnol. Appl. Biochem.* **20** 357–69

[26] Porath J 1987 Salting-out adsorption techniques for protein purification *Biopolymers* **26** S193–204

[27] Caldwell K, Axén R, Bergwall M and Porath J 1976 Immobilization of enzymes based on hydrophobic interaction. I. Preparation and properties of a betaamylase adsorbate *Biotechnol. Bioeng.* **18** 1573–88

[28] Caldwell K, Axén R, Bergwall M and Porath J 1976 Immobilization of enzymes based on hydrophobic interaction. II. Preparation and properties of an amyloglucosidase adsorbate *Biotechnol. Bioeng.* **18** 1589–604

[29] Cashion P, Lentini V, Harrison D and Javed A 1982 Enzyme immobilization on trityl-agarose: Reusability of both matrix and enzyme *Bioechnol. Bioeng.* **24** 1221–4

[30] Yon R 1974 Enzyme purification by hydrophobic chromatography: an alternative approach illustrated in the purification of aspartate transcarbamoylase from wheat germ *Biochem. J.* **137** 127–30

[31] Dixon J, Andrews P and Butler L 1979 Hydrophobic esters of cellulose: properties and applications in biochemical technology *Biotechnol. Bioeng.* **21** 2113–23

[32] Solomon B, Hollaander Z, Koppel R and Katchalski-Kazir E 1987 Use of monoclonal antibodies for the preparation of highly active immobilized enzymes *Methods in Enzymology*vol 135 ed K Mosbach (London: Academic), pp 160–70

[33] Bernfeld P and Wan J 1963 Antigens and enzymes made insoluble by entrapping them into lattices of synthetic polymers *Science* **142** 678–9
[34] Dinelli D, Marconi W and Morisi F 1976 Fibre-entrapped enzymes *Methods in Enzymology* ed K Mosbach vol 44 (New York: Academic), pp 227–43
[35] Wadiack D T and Carbonell R G 1975 Kinetic behavior of microencapsulated *β*-galactosidase *Biotechnol. Bioeng.* **17** 1157–81
[36] Mattiasson B and Kaul R 1991 Determination of coupling yields and handling of labile proteins in immobilization technology *Protein Immobilization. Fundamentals and Applications* ed R F Taylor (New York: Marcel Dekker), pp 161–79
[37] Scouten W H 1987 A survey of enzyme coupling techniques *Methods in Enzymology* vol 135 ed K Mosbach (London: Academic), pp 30–65
[38] López-Gallego F, Guisán J M and Betancor L 2013 Glutaraldehyde-mediated protein immobilization *Methods Mol. Biol.* **1051** 33–41
[39] Sassolas A, Hayat A and Marty J L 2013 Enzyme immobilization by entrapment within a gel network *Methods Mol. Biol.* **1051** 229–39
[40] Cao L, van Langen L and Sheldon R A 2003 Immobilised enzymes: carrier-bound of carrier-free? *Curr. Opin. Biotechnol.* **14** 387–94
[41] Trevan M D 1988 Enzyme immobilization by entrapment *Methods Mol. Biol.* **3** 491–4
[42] Klibanov A M 1979 Enzyme stabilization by immobilization *Anal. Biochem.* **93** 1–25
[43] Sirkar K K, Shanbhag P V and Kovvali A S 1999 Membrane in a reactor: a functional perspective *Ind. Eng. Chem. Res.* **38** 3715–37
[44] Wöltinger J, Karau A, Leuchtenberger W and Drauz K 2005 Membrane reactors at Degussa *Adv. Biochem. Eng. Biotechnol.* **92** 289–316
[45] Chang H N and Furusaki S 1991 Membrane bioreactors: present and prospects *Adv. Biochem. Eng. Biotechnol.* **44** 27–64
[46] Tosa T, Mori T, Fuse N and Chibata I 1969 Studies on continuous enzyme reactions *Agric. Biol. Chem.* **33** 1047–59
[47] Moehlenbrock M J and Minteer S D 2011 Introduction to the field of enzyme immobilization and stabilization *Methods Mol. Biol.* **679** 1–7
[48] Roig M G, Bello J F, Velasco F G, Celis C D D and Cachaza J M 2010 Biotechnology and applied biology section applications of immobilized enzymes *Biochem. Educ.* **15**
[49] Lortie R 1994 Evaluation of the performance of immobilized enzyme reactors with Michaelis–Menten kinetics *J. Chem. Technol. Biotechnol.* **60** 189–93
[50] Freed L E, Vunjak-Novakovic G V, Drinker P A and Langer R 1993 Bioreactor based on suspended particles of immobilized enzyme *Ann. Biomed. Eng.* **21** 57–65
[51] Chibata I, Tosa T, Sato T and Mori T 1976 Production of L-amino acids by aminoacylase adsorbed on DEAE-Sephadex *Methods Enzymol.* **44** 746–59
[52] Jin L Q, Xu Q, Liu Z Q, Jia D X, Liao C J, Chen D S and Zheng Y G 2017 Immobilization of recombinant glucose isomerase for efficient production of high fructose corn syrup *Appl. Biochem. Biotechnol.* **183** 293–306
[53] Moreno-Garrido I 2013 Microalgal immobilization methods *Meth Mol Biol.* **1051** 327–47
[54] Yan N, Fan C, Chen Y and Hu Z 2016 The potential for microalgae as bioreactors to produce pharmaceuticals *Int. J. Mol. Sci.* **17** 962
[55] Bhatia S, Rathee P, Sharma K, Chaugule B B, Kar N and Bera T 2013 Immuno-modulation effect of sulphated polysaccharide (porphyran) from *Porphyra vietnamensis Int. J. Biolog. Macromol.* **57** 50–6

[56] Bhatia S, Kumar V, Sharma K, Nagpal K and Bera T 2014 Significance of algal polymer in designing Amphotericin B nanoparticles *Sci. World J.* **2014** 1
[57] Bhatia S, Sharma K, Nagpal K and Bera T 2015 *Investigation of the factors influencing the molecular weight of porphyran and its associated antifungal activity* Bioact Carbohydr. Diet. Fiber **5** 153–68
[58] Bhatia S, Sharma K, Nagpal K, Sharma S and Bera T 2015 Evaluation of pharmacognostical, phytochemical and anti-microbial properties of *Porphyra vietnamensis Int. J. Green Pharm* **2** 1–5
[59] Bhatia S, Sharma K and Bera T 2015 Structural characterization and pharmaceutical properties of porphyran *Asian J. Pharm* **9** 1–3
[60] Bhatia S, Sharma K, Sharma A, Nagpal K and Bera T 2015 Anti-inflammatory, analgesic and antiulcer properties of *Porphyra vietnamensis Avicenna J. Phytomed.* **5** 1–7
[61] Bhatia S *et al* 2009 Novel algal polysaccharides from marine source: porphyran *Pharmacog. Rev.* **2** 271–6
[62] Bhatia S *et al* 2011 Mycosporine and mycosporine-like amino acids: A paramount tool against ultra violet irradiation *Pharmacogn Rev.* **5** 138–46
[63] Bhatia S *et al* 2010 Broad-spectrum sun-protective action of Porphyra-334 derived from *Porphyra vietnamensis. Pharmacog. Res.* **2** 45–9
[64] Bhatia S, Namdeo A G and Nanda S 2015 Factors affecting the gelling and emulsifying property of natural polymer *Sys. Pharm. Rev.* **9** 93–101
[65] Bhatia S, Sharma K, Sharma A, Namdeo A G and Chaugule B B 2011 Anti-oxidant potential of Indian Porphyra *Pharmacologyonline* **1** 248–57
[66] Bhatia S *et al* 2015 *Modern Applications of Plant Biotechnology in Pharmaceutical Sciences* (Amsterdam: Elsevier), pp 277–87
[67] Bhatia S 2016 *Nanotechnology in Drug Delivery: Fundamentals, Design, and Applications* (Boca Raton, FL: CRC Press), pp 121–7
[68] Bhatia S and Goli D 2016 *Leishmaniasis: Biology, Control and New Approaches for Its Treatment* (Boca Raton, FL: CRC Press), pp 177–9
[69] Bhatia S 2016 *Natural Polymer Drug Delivery Systems: Nanoparticles, Plants, and Algae* (Berlin: Springer), pp 117–27
[70] Bhatia S 2016 *Systems for Drug Delivery: Safety, Animal, and Microbial Polysaccharides* (Berlin: Springer), pp 122–7
[71] Bhatia S 2018 *Introduction to Pharmaceutical Biotechnology*Volume 1 (Bristol: IOP Publishing), pp 178–87

IOP Publishing

Introduction to Pharmaceutical Biotechnology, Volume 2
Enzymes, proteins and bioinformatics
Saurabh Bhatia

Chapter 5

Biosensors

5.1 Introduction

The field of biosensors has developed hugely from the initial demonstration of the concept of biosensors by Leland C Clark Jr in 1962. Currently, the biosensor market is dominated by glucose biosensors, mass-produced enzyme electrodes for the quick self-diagnosis of blood-glucose levels by diabetes patients [1–3]. A biosensor is an analytical device containing an immobilized biological material (enzyme, antibody, nucleic acid, hormone, organelle or whole cell) which can specifically interact with an analyte and produce physical, chemical or electrical signals that can be measured. An analyte is a compound (e.g. glucose, urea, drug, pesticide) whose concentration needs to be measured [1–3]. Biosensors mainly consist of the quantitative analysis of different substances by altering their biological actions into quantifiable signals. The majority of biosensors have immobilized enzymes. The efficiency of biosensors is typically dependent on two important factors, the specificity and sensitivity of the biological reaction, in addition to the stability of the enzyme. The category of biosensors that utilizes enzymes as the recognition element is known as enzyme-based chemical biosensors. Their mechanism is based on biological recognition. Typically it is based on a mixture of a biocatalyst and a supporting material to transduce the biological signal, thus enzymes are allowed to participate in the biochemical reaction. During their participation and after completion of biochemical reaction, the enzyme should be stable under the normal functional conditions of the biosensor. Information about the target analyte always determines the structure of biosensors along with the type of matrix in which the analyte has to be introduced [1–3].

The two different components of a biosensor are (see figure 5.1):

- the biological component (enzyme, cell, etc) and
- the physical component (transducer, amplifier, etc).

The biological component identifies and interacts with the analyte to produce a physical change (a signal) that can be easily examined by the transducer. In practice,

doi:10.1088/978-0-7503-1302-5ch5

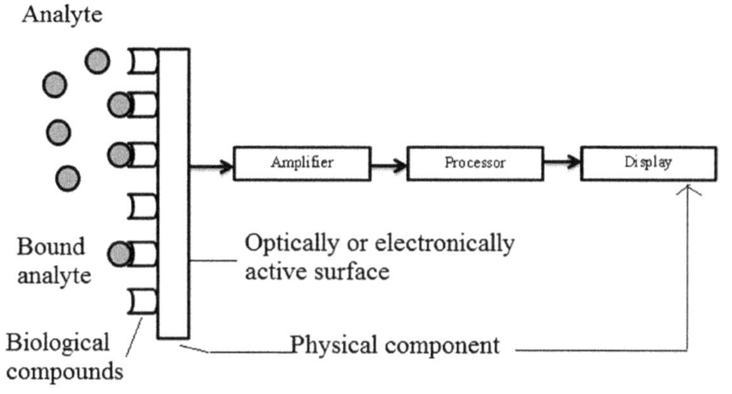

Figure 5.1. A schematic representation of a biosensor.

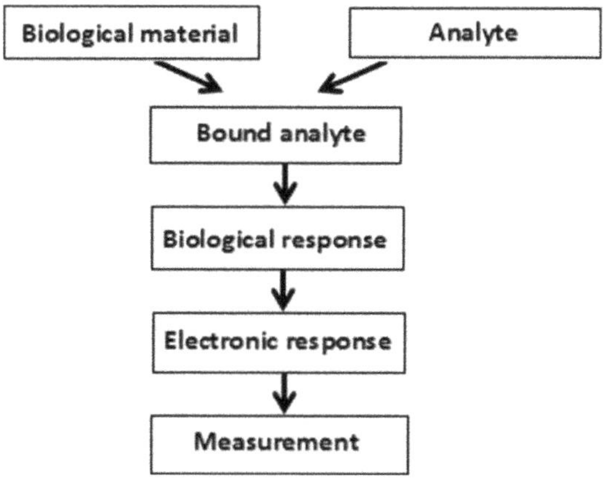

Figure 5.2. Principles of a biosensor.

the biological material is properly immobilized on the transducer and the so prepared biosensors can be used several times (maybe about 10 000 times) for an extensive period (e.g. several months).

5.2 Principles of a biosensor

Biosensors are valuable analytical devices that can be assimilated with on-line process monitoring schemes [4]. The basic principle of a biosensor is dependent on the affinity between the biological material and analyte (figure 5.2) [4]. The selected biological material (usually a specific enzyme) is immobilized by conventional procedures (physical or membrane entrapment, non-covalent or covalent bonding). This immobilized biological material is in close interaction with the transducer. The analyte binds to the biological material to form a bound analyte which in turn produces an electronic response that can be examined.

In a number of cases, the analyte is transformed into a product which may be related to the release of heat, gas (oxygen), electrons or hydrogen ions. The transducer can transform the product-associated changes into electrical signals which can be amplified and measured. The manufacture of biosensors, their resources, transducing devices and immobilization techniques entails multidisciplinary research in chemistry, biology and engineering. The resources used in biosensors are divided into three groups based on their mechanisms:

- the biocatalytic group comprising enzymes;
- the bioaffinity group including antibodies and nucleic acids; and
- the microbe-based group containing micro-organisms.

A biosensor comprises three main parts:

- the biological recognition elements that differentiate the target molecules in the presence of various chemicals;
- a transducer that converts the biorecognition event into a measurable signal; and
- a signal processing system that converts the signal into a readable form [5–7].

The molecular recognition elements comprise receptors, enzymes, antibodies, nucleic acids, micro-organisms and lectins [8, 9]. The five principal transducer classes are electrochemical, optical, thermometric, piezoelectric and magnetic [10]. Most currently available glucose biosensors are of the electrochemical type, due to their better sensitivity, reproducibility and easy maintenance, as well as their low cost. Electrochemical sensors may be further divided into potentiometric, amperometric or conductometric types [10–12]. Enzymatic amperometric glucose biosensors are the most common devices commercially available, and have been extensively investigated over the last few years. Amperometric sensors monitor currents generated when electrons are exchanged either directly or indirectly between a biological system and an electrode [13, 14].

5.3 Different types of biosensors

Biosensors were developed in the 1960s by the pioneers Clark and Lyons. Some types of biosensors currently in use are enzyme-based, tissue-based, immunosensors, DNA biosensors, thermal biosensors and piezoelectric biosensors. Enzyme biosensors have been developed based on immobilization procedures, i.e., adsorption of enzymes by van der Waals forces, ionic bonding or covalent bonding. The enzymes frequently employed for this function are oxidoreductases, polyphenol oxidases, peroxidases and aminooxidases [15–18]. The principal microbe-based or cell-based sensor was presented by Diviès. The tissues for tissue-based sensors are obtained from plant and animal sources. The analyte of interest can be an inhibitor or a substrate of these procedures. Rechnitz [19] established the first tissue-based sensor for the detection of the amino acid arginine. Organelle-based sensors were introduced, incorporating membranes, chloroplasts, mitochondria and microsomes. For this type of biosensor, stability is high, however, the recognition time is longer and the specificity is reduced. Immunosensors were based on the well-known fact that antibodies have high affinity towards their corresponding antigens, i.e., the antibodies exactly bind to pathogens or toxins, or interact with components of the host's immune system. DNA biosensors were developed based on the feature that a single-strand nucleic acid molecule is able to identify and bind to its opposite strand in a sample. This interaction is due to the development of stable hydrogen bonds between the two nucleic acid strands [20]. Magnetic biosensors and miniaturized biosensors distinguish magnetic micro- and nanoparticles in microfluidic channels by means of the magnetoresistance effect and have great efficiency in terms of sensitivity and size [21]. Thermal biosensors or calorimetric biosensors are devised by assimilating biosensor materials into a physical transducer. Piezoelectric biosensors are of two types: the quartz crystal microbalance and the surface acoustic wave devices. They are based on the extent of changes in the resonance frequency of a piezoelectric crystal because of mass changes in the crystal structure. Optical biosensors comprise a light source, as well as many optical components to produce a light beam with specific properties and to focus this light to a modulating agent, a modified sensing head along with a photodetector [22]. Green fluorescent protein and the subsequent autofluorescent protein (AFP) alternatives and the development of genetic fusion have supported the growth of genetically encoded biosensors [23–30]. This kind of biosensor is user-friendly, and easy to design, manipulate and transfer into cells. A single-chain Förster resonance energy transfer (FRET) biosensor is another example. They comprise a pair of AFPs, which are efficient in transferring fluorescence resonance energy between themselves when brought close together. Various procedures may be used to control variations in FRET signals based on the intensity, ratio or lifetime of the AFPs. Peptide and protein biosensors are simply fabricated via synthetic chemistry followed by enzymatic labeling with synthetic fluorophores. Owing to the individuality of the genetically encoded AFPs, they are easily employed to regulate target activity and constitute attractive alternatives to more traditional methods. They have an additional benefit of being able to improve the signal-to-noise ratio and the sensitivity of response via introduction of chemical quenchers and photoactivatable groups. Based on the sensor devices and the type of materials used, there are different types of biosensors. A number of them are discussed below.

5.3.1 Electrochemical biosensors

Electrochemical biosensors are simple devices based on electric current, ionic or conductance changes caused by bioelectrodes. Enzyme-based electrochemical biosensors are employed extensively in our day-to-day lives, such as in healthcare, food safety and environmental monitoring. Healthcare is a key area for biosensor applications, such as checking blood-glucose levels in diabetics using glucose biosensors. Moreover, the reliable determination of urea has potential applications for patients with renal disease, for use either at home or in the hospital. Large-scale applications for biosensors comprise observing fermentation broths or food processing events via monitoring amounts of glucose and other fermentative end products. The sensitive recognition of phenolic compounds is a significant topic for environmental research, as phenolic compounds are frequently present in the wastewaters of various industries, and can cause harm to our living environment, as many of them are very toxic [31].

5.3.1.1 Amperometric biosensors

Enzyme-based, amperometric electrochemical biosensing facilitates a highly selective and sensitive response in a complex environment. In comparison to other methods, such as microdialysis and nuclear magnetic resonance (NMR) spectroscopy, microsensors can rapidly and precisely measure extracellular low analyte concentrations within the tissue in near real time [32]. The principle of amperometric biosensors is based on the movement of electrons (i.e., determination of electric current) as a result of enzyme-catalyzed redox reactions (figure 5.3). Usually, a continuous voltage passes between the electrodes which can be easily measured. During an enzymatic reaction, the substrate or product can transfer an electron with the electrode surface to be oxidized or reduced. This results in an improved current flow that can be measured. The intensity of the current is proportional to the substrate concentration. The Clark oxygen electrode, which controls reduction of O_2, is one of the simplest forms of amperometric biosensor and glucose determination by glucose oxidase is a common example of this type of biosensor. There is a direct transfer of the electrons released to the electrode, which may pose some practical problems. After the first-generation amperometric biosensors, second-generation amperometric biosensors have been established wherein a mediator (e.g. ferrocenes) takes up the electrons and then transfers them to the electrode. These biosensors, however, are yet to become prevalent. One drawback of this sensor principle is the need for oxygen, which has inadequate solubility in aqueous fluids, only about 200 μM under physiological conditions. It is accessible only in lower concentrations within tissues due to the restricted supply by blood vessels and permanent cellular consumption [32, 33]. Thus when one is measuring higher analyte concentrations, the objective is to attain a diffusion-limited regime by restricting analyte diffusion using an additional membrane. Preferably, this membrane limits the transport of oxygen to a much lesser degree than analyte diffusion, and thus permits an unaltered enzymatic reaction even at

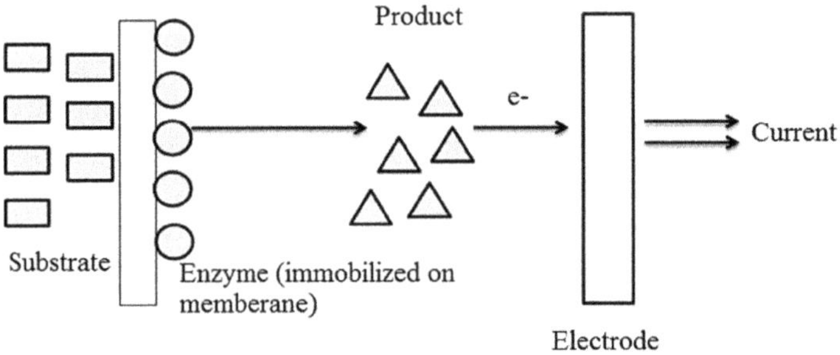

Figure 5.3. A schematic representation of an amperometric biosensor.

high analyte concentrations. With the aim to eradicate oxygen dependence, second-generation reagentless biosensors use a surplus electron acceptor (mediator) as a replacement for oxygen [32, 33].

5.3.1.1.1 Blood-glucose biosensor

The blood-glucose biosensor is a good example of amperometric biosensors, and is extensively used all over the world by diabetic patients. The blood-glucose biosensor is a watch-like device and has a single use disposable electrode. This electrode consist of a Ag/AgCI reference electrode and a carbon working electrode with glucose oxidase and ferrocene derivative (as a mediator). For uniform dispersal of a blood drop, the electrodes are covered with a hydrophilic mesh gauze. The disposable test strips, packed in aluminum foil, have a shelf-life of about six months.

Different types of amperometric biosensor can be used for measuring the freshness of fish. Compared to other nucleotides, the accumulation of inosine and hypoxanthine designate the freshness of fish, i.e., how long it has been dead and stored. For this purpose, a biosensor using immobilized nucleoside phosphorylase and xanthine oxidase over an electrode has been developed.

Usually, glucose measurements are based on interactions with one of three enzymes: hexokinase, glucose oxidase (GOx) or glucose-1-dehydrogenase (GDH) [34, 35]. The basic idea of the glucose biosensor is based on the fact that the immobilized GOx catalyzes the oxidation of β-D-glucose by molecular oxygen, generating gluconic acid and hydrogen peroxide [36]. To work as a catalyst, GOx requires a redox co-factor–flavin adenine dinucleotide (FAD). FAD works as the initial electron acceptor and is reduced to $FADH_2$:

$$\text{Glucose} + \text{GOx} - \text{FAD} + \rightarrow \text{Glucolactone} + \text{GOx} - \text{FADH}_2.$$

The co-factor is redeveloped by reacting with oxygen, resulting in the production of hydrogen peroxides:

$$\text{GOx} - \text{FADH}_2 + \text{O}_2 \rightarrow \text{GOx} - \text{FAD} + \text{H}_2\text{O}_2.$$

Hydrogen peroxide is oxidized at a catalytic, characteristically platinum (Pt) anode. The electrode easily identifies the number of electron transfers, and this electron flow is directly related to the number of glucose molecules present in the blood [37]:

$$\text{H}_2\text{O}_2 \rightarrow 2\text{H} + +\text{O}_2 + 2\text{e}.$$

Three overall strategies are employed for the electrochemical sensing of glucose: by determining oxygen consumption; by determining the concentration of hydrogen peroxide formed by the enzyme reaction; or by employing a diffusible or immobilized mediator to transfer the electrons from the GOx to the electrode. The number and types of GDH-based amperometric biosensors have been increasing. The GDH family includes GDH-pyrroquinolinequinone (PQQ) [38–40] and GDH-nicotinamide-adenine dinucleotide (NAD) [40–43]. The enzymatic reaction of GDH is independent of the dissolved oxygen. The quinoprotein GDH recognition element uses PQQ as a co-factor:

$$\text{Glucose} + \text{PQQ(ox)} \rightarrow \text{gluconolactone} + \text{PQQ(red)}.$$

This mechanism necessitates neither oxygen nor NAD^+. GDH–PQQ is a largely efficient enzyme system, with a rapid electron transfer rate, however, it is comparatively expensive.

GDH with NAD as a co-factor produces NADH rather than H_2O_2. NAD is a major electron acceptor in the oxidation of glucose, during which the nicotinamide ring of NAD^+ accepts a hydrogen ion and two electrons, equivalent to a hydride ion. The reduced form of this carrier generated in this reaction is called NADH, which can be electrochemically oxidized:

$$\text{Glucose} + NAD^+ \rightarrow \text{gluconolactone} + \text{NADHNADH} \rightarrow NAD^+ + H^+ + 2e.$$

5.3.1.2 Potentiometric biosensors

A potentiometric device determines the accumulation of a charge potential at the working electrode compared to the reference electrode in an electrochemical cell when zero or no considerable current flows between them [44–46]. In particular, potentiometry offers information about the ion activity in an electrochemical reaction [44–46]. The principle of this bioreactor is based on fluctuations in ionic concentrations which are determined using ion-selective electrodes (figure 5.4). The pH electrode is the most frequently used ion-selective electrode, as many enzymatic reactions include the release or absorption of hydrogen ions. Other significant electrodes are ammonia-selective and CO_2-selective electrodes. The potential difference found between the potentiometric electrode and the reference electrode can easily be determined. It is relative to the concentration of the substrate. One of the major limitations of potentiometric biosensors is the sensitivity of enzymes to ionic concentrations such as H^+ and NH^{4+}.

Potentiometry is also employed as a technique to electrically measure the point in a (bio)chemical reaction at which equal concentrations of opposing solutions reach a state of equilibrium (e.g. 0.1 mol HCl and 0.1 mol NaOH). This is called determining a titration end-point; the procedure is known as potentiometric titration. By carrying out a titration at constant or zero current, the end-point is

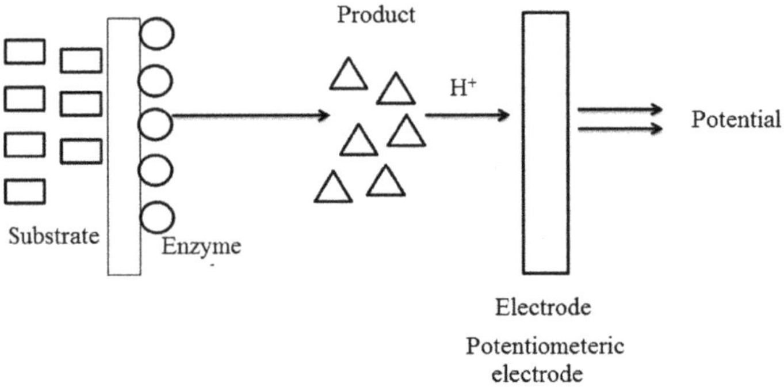

Figure 5.4. A schematic representation of a potentiometric biosensor.

recognized from the variations in electrode potential, which are produced by variations in solution concentration of the potential-determining ion. Numerous potentiometric devices are also based on many forms of field-effect transistor devices to determine pH variations, selective ion concentrations and the kinetics of biocatalytic reactions encompassing enzymes [47]. An additional example and novel optical/electrochemical hybrid method is called the light addressable potentiometric sensor [47–53]. This is a silicon-based detector that takes advantage of the photovoltaic effect to selectively detect the point of measurement. By scanning with a focused light source, it can measure the spatially resolved surface potential distribution along the interface of the sample and substrate surfaces [54].

5.3.1.3 Ion-selective field-effect transistors

For several years, much attention has been given to silicon-based biosensors in the area of bio-analytical applications owing to their promising features, which include sensitivity, speed, miniaturization and low cost [55–57]. This attention is obvious in the many investigations that have observed biological events, such as nucleic acid hybridization, protein–protein interaction, antigen–antibody binding and enzyme–substrate reactions, by means of these silicon-based biosensors. Among these, the ion-sensitive field-effect transistor (ISFET) is one of the most common electrical biosensors and was presented as the first miniaturized silicon-based chemical sensor [55–57]. The ISFET, usually known as a pH sensor, has been employed to determine ion concentration (H^+ or OH^-) in a solution through the effect of an interface potential on the gate insulator. The ISFET is a type of potentiometric device that functions similarly to a metal oxide semiconductor field-effect transistor. Consequently, so as to evaluate the performance of an ISFET, it makes sense to first recognize the overall principles behind the setup of the potentiometric sensor. ISFETs are the economic devices that can be employed for miniaturization (to manufacture ever smaller mechanical, optical and electronic products and devices) of potentiometric biosensors. An excellent example is the ISFET biosensor employed to monitor intramyocardial pH during bypass surgery [55–57].

5.3.1.4 Conductimetric biosensors

Conductometric biosensors were first introduced in 1961 to measure urea. The procedure is based on electrical conductivity variation. Formaldehyde, pesticides, insecticides and nitrate biosensors using conductometry were also established [58]. The urea biosensor was upgraded using a platinum electrode as a matrix for urease immobilization [59]. A conductimetric biosensor determines small variations in the conductivity of a solution by employing a conductimetric transducer, i.e., a conductivity meter. Conductivity determination is dependent on the biocatalytic reaction of the sample on an electrode. The reaction will generate ions which will result in the variation in conductivity [60]. The conductimetric transducer comprises two electrodes, a reference and a working electrode. Both electrodes are coated with a membrane referred to as the nata de coco membrane. The enzyme is immobilized on the working electrode, however, not on the reference electrode. Throughout the reaction, CO_2 is formed on the working electrode, which is soluble in water to form

HCO_3^- and H_3O^+. Furthermore, no reaction takes place on the reference electrode, so the mobility of ions on the two electrodes is different and the conductivity is changed. There are a number of reactions in the biological systems that bring about changes in the ionic species. These ionic species change the electrical conductivity which can be determined. An outstanding example of conductimetric biosensor is the urea biosensor using immobilized urease. Urease catalyzes the following reaction:

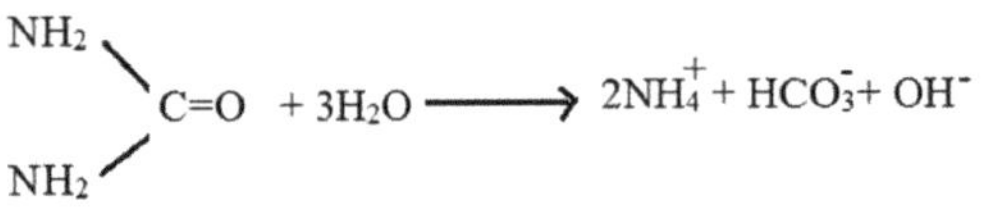

The above reaction is related to sudden modification in ionic concentration which can be used for monitoring urea concentration. Actually, urea biosensors are very fruitfully employed in dialysis and renal surgery.

5.3.2 Thermometric biosensors

Thermometric or calorimeteric biosensors utilize an essential property of biological reactions, i.e., absorption or emission of heat [60]. This is represented as a variation in the temperature within the reaction medium. In previous reports on calorimetry, the variation in heat was directly observed to measure the level of reaction (for catalysis) or structural dynamics of biomolecules in the dissolved state [61]. Its utilization in biosensors results in the development of thermometric devices [62]. These mainly determine the variation in temperature of the circulating fluid following the reaction of a appropriate substrate with the immobilized enzyme molecules. Thermometry basically means the determination of temperature. The most basic type of such a device is a thermometer, regularly used for determination of body or ambient temperature. However, simple mercury based-thermometers are restricted by their temperature sensitivity in addition to the toxicity of metallic mercury. Based on similar methods, in thermometric devices the heat is determined using sensitive thermistors [62]; such devices are generally called enzyme thermistors [63]. Calorimetric devices for regular use were restricted by the cost of the process and the comparatively long experimental processes. The creation of an enzyme thermistor based on flow injection examination in combination with an immobilized biocatalyst and heat-sensing element circumvented several of these limitations [64, 65]. A number of instruments were introduced in the past two decades that combined the principles of calorimetry, enzyme catalysis, immobilization on suitable matrices and flow injection analysis.

A schematic representation of a thermal biosensor is shown in figure 5.5. It includes a heat insulated box fitted with a heat exchanger (an aluminum cylinder). The reaction occurs in a small enzyme packed bed reactor. As the substrate comes into the bed, it is transformed into a product and heat is generated. The temperature difference between the substrate and product is determined by thermistors. Even a

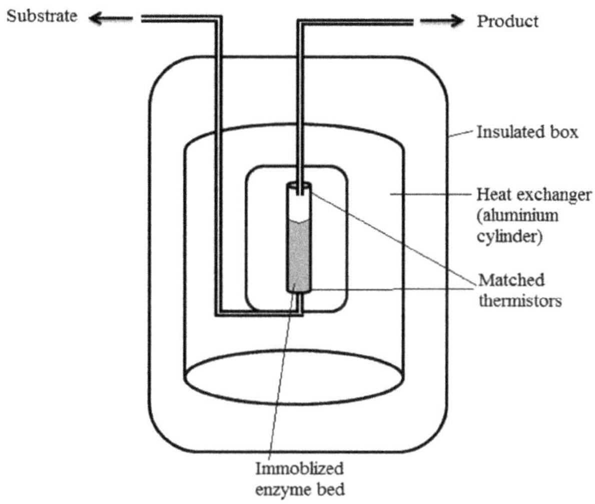

Figure 5.5. A schematic representation of a thermometric biosensor.

small change in temperature can be measured by thermal biosensors. Thermometric biosensors are usually employed for the estimation of serum cholesterol. When cholesterol is oxidized by the enzyme cholesterol oxidase, heat is produced which can be determined. Similarly, measurements of glucose (enzyme glucose oxidase), urea (enzyme-urease), uric acid (enzyme-uricase) and penicillin G (enzyme β lactamase) can be achieved by these biosensors. Generally, however, their efficacy is limited. Thermometric biosensors can be employed as part of enzyme linked immunoassays (ELISA) and this technique is called thermometric ELISA.

5.3.3 Optical biosensors

The principle of this biosensor is based on optical measurements (absorbance, fluorescence, chemiluminescence, etc). Its utilization is based on the quality of the fiber optics and optoelectronic transducers used in fabricating these biosensors. The word optrode is used as a condensation of the words optical and electrode. Optical biosensors mainly involve enzymes and antibodies as the transducing elements. Optical biosensors facilitate safe non-electrical sensing of materials. An additional merit of optical biosensors is that these biosensors usually do not require reference sensors, as the comparative signal can be produced by means of the same source of light as the sampling sensor. Certain important optical biosensors are discussed below.

5.3.3.1 Fiber optic lactate biosensor

Figure 5.6 show a fiber optic lactate biosensor. Its functioning is based on the determination of changes in molecular O_2 concentrations by detecting the quenching effect of O_2 on a fluorescent dye. The subsequent reaction is catalyzed by the enzyme lactate mono-oxygenase:

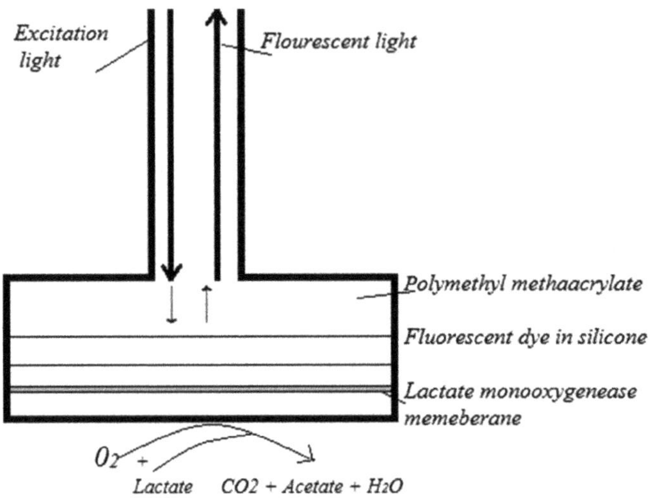

Figure 5.6. Schematic representation of a fiber optic lactate biosensor.

$$O_2 + \text{Lactate} \xrightarrow[\textit{Lactate monooxygenase}]{} CO_2 + \text{ACETATE} + H_2O$$

The extent of fluorescence produced by the dyed film is dependent on O_2, because O_2 has a reducing effect on the fluorescence. As the concentration of lactate in the reaction mixture rises, O_2 is utilized, and as a result there is a proportionate decline in the quenching or reducing effect. Finally, there is an increase in fluorescent production which can be determined.

5.3.3.2 Optical biosensors for blood glucose

Optical biosensors exploit light and the selective nature of biological components to measure specific analytes [66–69]. Vast studies have been reported on optical biosensors [69], and great developments have been made [70] from the time when the first optical biosensor was introduced by Lübbers and Opitz [71]. The utilization of optical sensors can help circumvent many of the issues produced by electrochemical sensors [67, 72]. Unlike several other analytical approaches, such as electrochemical procedures, the light used is typically not damaging to the body or the system it is used in, and in theory the excitation and computing can be performed noninvasively from out the body [66]. Using fluorescence can be very sensitive [73] and the signal can travel great distances, making it possible to determine glucose concentrations in hard to reach places [72]. The device used is fairly cheap, convenient to use and is not disturbed by electrical or magnetic fields [72]. An additional advantage is that there are various procedures which can be utilized within the field of fluorescence; using a steady-state estimate by determination of the intensity differences, time-resolved fluorescence, Förster resonance energy transfer (FRET) and other procedures that can offer information about the micro-environment and structure of the molecules

[66]. The first glucose biosensor was introduced in 1962 by Clark and Lyons [73] by means of glucose oxidase (GOx) entrapped over an oxygen electrode with a dialysis membrane. In 1975 the first available commercial glucose sensor was introduced by Yellow Spring Instruments Inc. [74]. A demonstration of *in vivo* glucose monitoring was presented in 1982 by Shichiri *et al* [75] and a wearable and noninvasive glucose sensor was created by Cygnus Inc. in 2000 [76]. In December 2011 Microsoft Research revealed that they were employed in making noninvasive glucose sensing contact lenses [77, 78]. After a couple of years Google started working on a similar project [79]. The fact that two multi-billion dollar companies like Microsoft and Google are investing time and money into noninvasive glucose sensors is evidence of how significant this field of research is, and that noninvasive sensors are the future.

Assessment of blood glucose is very significant for the monitoring of diabetes. This technique involves paper strips saturated with reagents. The strips contain glucose oxidase, horseradish peroxidase and a chromogen (e.g. toluidine). The following reactions occur:

$$\text{Glucose} \xrightarrow{\textit{Glucose oxidase}} \text{Gluconic acid} + H_2O_2$$

$$\text{Chromogen} + 2\,H_2O_2 \xrightarrow{\textit{Peroxidase}} \text{Colour dye} + 2H_2O$$

The intensity of the color (of the dye) can be determined by using a portable reflectance meter. Glucose strip production has now been commercialized worldwide. Colorimetric test strips of cellulose coated with suitable enzymes and reagents are also in use for the estimation of several blood and urine parameters.

5.3.3.3 Luminescent biosensors to detect urinary infections

Biosensors are being used widely in the medical field to detect infectious diseases. Various favorable biosensor technologies for urinary tract infection diagnosis along with pathogen identification and anti-microbial susceptibility are under development. The cause of urinary tract infections, i.e., the micro-organisms in the urine, can be identified by employing luminescent biosensors. For this, an immobilized (or even free) enzyme called luciferase is used. The amount of light production can be determined by electronic devices. The principles of luminescent biosensors for detecting urinary infections are shown in figure 5.7. On lysis, the micro-organisms release ATP.

$$\text{Leuciferin} + \text{ATP} + O_2 \xrightarrow{\textit{Luciferase}} \text{Oxyluciferin} + CO_2 + \text{AMP} + \text{Pyrophosphate} + \text{Light}$$

Figure 5.7. Principle of luminescent biosensors for detecting urinary infections.

5.3.3.4 Other optical biosensors

Optical fiber sensing devices are in use for determining pH, pCO_2 and pO_2 in critical care and surgical monitoring.

5.3.4 Piezoelectric biosensors

The piezoelectric effect is not an entirely new idea as it has been recognized since the nineteenth century with comprehensive scientific applications since the beginning of the twentieth century. The innovation of the piezoelectric effect is associated with the famous physicists Jacques Curie and Pierre Curie who documented the first anisotropic crystals, i.e., crystals without a center of symmetry that can produce an electric dipole when mechanically squeezed. The designated effect can work in the opposite way when an anisotropic crystal become distorted due to a voltage imposed on it [80]. This phenomenon is depicted figure 5.8.

The principle of piezoelectric biosensors is based on acoustics (sound vibrations), therefore, they are also known as acoustic biosensors. Piezoelectric crystals form the basis of these biosensors. These particles have characteristic frequencies and cystals with positive and negative charges vibrate with typical frequencies. Developed resonance frequencies are modified by the adsorption of certain molecules on the crystal surface. This resonance can be easily detected by electronic devices. Some

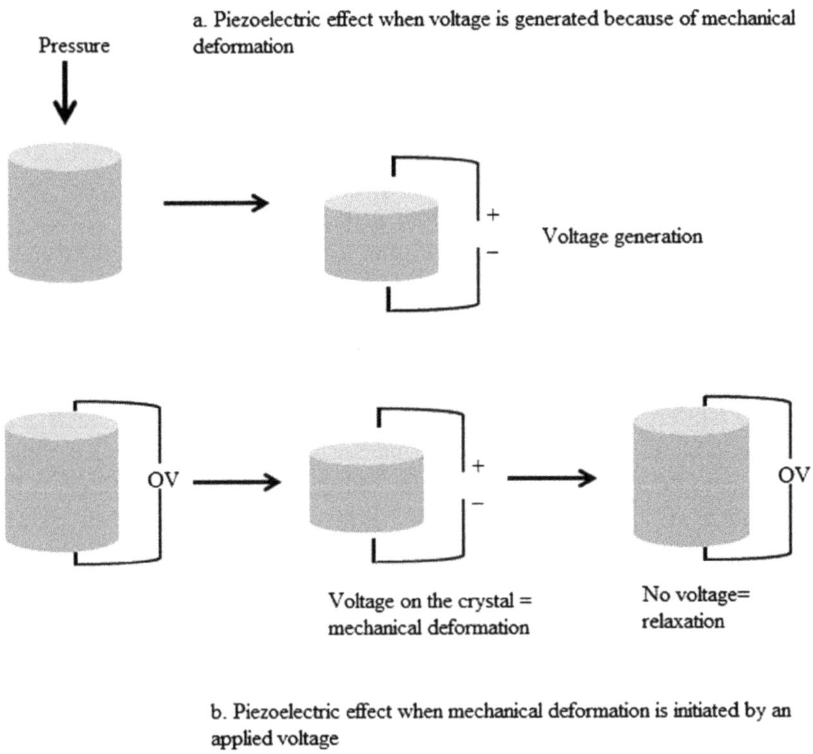

Figure 5.8. The principle of piezoelectric biosensors.

other biological compounds in different forms, such as enzymes with gaseous substrates or inhibitors, can also be attached to these crystals. By the introduction of acetylcholine esterase a piezoelectric biosensor for organophosphorus insecticide has been developed. Similarly, by introduction of formaldehyde dehydrogenase, a biosensor for formaldehyde has been created. Another example of a biosensor for cocaine (in the gas phase) has been developed by attaching cocaine antibodies to the surface of a piezoelectric crystal.

5.3.4.1 Drawbacks of piezoelectric biosensors

These biosensors cannot be utilized for the determination of substances in a solution. This is because the crystals may stop oscillating fully in viscous liquids.

5.3.5 Whole-cell biosensors

Whole-cell biosensors are mainly beneficial for multi-step or co-factor requiring reactions. These biosensors can be utilized for live or dead microbial cells. Certain examples of organisms along with the analytes and the types of biosensors used are listed in table 5.1.

Advantages of microbial cell biosensors. Microbial cells are economical as they have extended half-lives. Furthermore, in contrast to isolated enzymes, they are less sensitive to changes in pH and temperature.

Disadvantages of microbial cell biosensors. In general, whole cells entail extended periods of catalysis. Moreover, the specificity and sensitivity of whole-cell biosensors may be lower in comparison to that of enzymes.

5.3.6 Immunobiosensors

Immunosensors are compact analytical devices in which the development of antigen–antibody complexes is identified and converted, by means of a transducer, to an electrical signal, which can be processed, recorded and displayed [81]. Various transducing mechanisms are used in immunological biosensors, based on signal generation (such as an electrochemical or optical signal) or changes in properties

Table 5.1. A number of biosensor organisms along with the analytes and the types of biosensors.

Organism	Analyte	Type of biosensor
Sarcina flava	Glutamine	Potentiometric (NH_3)
Proteus morganii	Cysteine	Potentiometric (H_2S)
Nitrosomanas spp.	Ammonia	Amperometric (O_2)
Many organisms	Biological oxygen demand (BOD)	Amperometric (O_2)
Lactobacillus fermenti	Thiamine	Amperometric (mediated)
Lactobacillus arabinosus	Nicotinic acid	Potentiometric (H^+)
Escherichia coil	Glutamate	Potentiometric (CO_2)
Desulfovibrio desulfuricans	Sulfate	Potentiometric (SO_3)
Cyanobacteria	Herbicides	Amperometric (mediated)

(such as mass changes) following the formation of antigen–antibody complexes [81]. These are the biosensors which are based on the principle of immunological specificity, coupled with measurement (mostly) by amperometric or potentiometric biosensors. Immunobiosensors or immunochemical sensors have numerous possible configurations. Selected configurations are depicted in figure 5.9, and briefly explained in the following:

- It may contain an immobilized antibody to which the antigen can bind directly.
- It may also contain an immobilized antigen that binds to antibody which in turn can bind to a second, free antigen.

Enzyme immunosensors can be used for therapeutic applications. Some examples are as follows:

- Immobilized streptokinase and urokinase (on sephadex) can be employed for the treatment of thrombosis.
- Some success has been achieved in the management of congenital defects by employing immobilized enzymes, e.g. phenylalanine hydroxylase and lysosomal α-1, 4 glucosidase to correct type II glycogen storage disease.
- Immobilized enzymes have been demonstrated in the treatment of congenital defects, e.g. phenylalanine in the treatment of phenylketonuria.

5.4 Applications of biosensors

Biosensors have become very popular in recent years. Due to their small size, easy handling, high specificity and sensitivity, and low cost, they are extensively used in

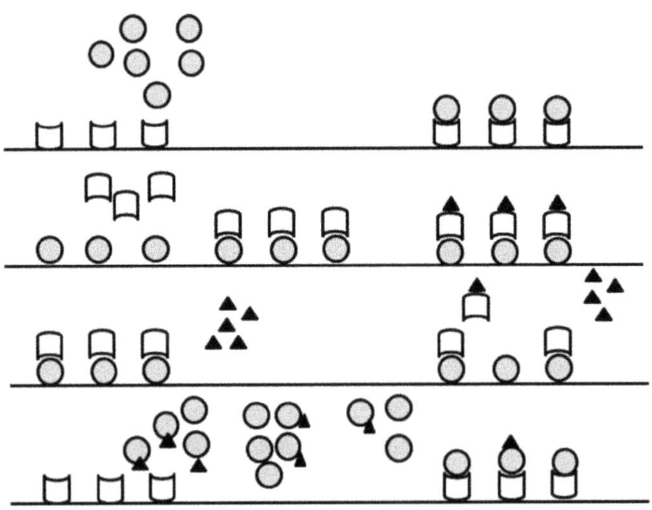

Figure 5.9. Diagrammatic representation of selected immunobiosensors. (a) Direct binding of an antigen to an immobilized antibody. (b) Antigen–antibody sandwiches (an immobilized antigen binds to an antibody and then to a second antigen). (c) An antibody binds to an immobilized antigen which is partially released by a competitive free antibody binding to a free antigen and an enzyme labeled antigen (in competition).

various sectors. Certain important applications of biosensors are broadly described below.

5.4.1 Applications in medicine and health

Biosensors are effectively used for the quantitative assessment of several biologically vital substances in body fluids, e.g. glucose, cholesterol and urea. The glucose biosensor is of great benefit for diabetic patients for regular monitoring of blood glucose. Blood gas monitoring for pH, pCO_2 and pO_2 is performed during critical care and surgical monitoring of patients. The mutagenicity of numerous chemicals can be assessed by employing biosensors. Numerous toxic substances synthesized in the body can also be identified.

5.4.2 Applications in industry

Biosensors can also be employed for monitoring fermentation products and the assessment of various ions. Thus, biosensors assist in improving the fermentation conditions for superior yield. Currently, biosensors are used to detect the odor and freshness of foods. For example, the freshness of stored fish can be examined by ATPase. ATP is not found in spoiled fish and this can be examined by using ATPase. Some pharmaceutical companies have developed immobilized cholesterol oxidase system for the estimation of cholesterol concentration in foods (e.g. butter).

5.4.3 Applications in pollution control

Biosensors are very helpful in measuring environmental pollutants. Pesticides and biological oxygen demand can be measured by biosensors.

5.4.4 Applications in the military

Biosensors have been developed to detect toxic gases and other chemical agents used during war.

5.4.5 Immobilized enzymes and cell therapeutic applications

The industrial and analytical applications of immobilized enzymes are described in detail in an above section. There are several limitations on the direct utilization of enzymes for therapeutic purposes:

- Poor availability of the enzyme at the active site.
- Sensitivity to natural inhibitors.
- Interference by endogenous proteases.
- Immunogenicity of certain enzymes.

Further important therapeutic applications of immobilized enzymes and cells are briefly described in the following.

5.4.5.1 Improved drug delivery by using immobilized enzymes

Urea-urease modulated system. Drug delivery can be improved by using immobilized urease enzymes along with the substrate urea (figure 5.10). This procedure is dependent on the action of urease. Urease splits urea which may result in an increase in pH owing to the formation of ammonium hydroxide. Thus a therapeutic drug present in a pH sensitive biodegradable polymer can be successfully released to perform its function (figure 5.11).

Glucose oxidase-glucose based system. A biodegradable polymeric system comprising insulin has been developed for efficient insulin delivery to the human body. By the help of immobilized glucose oxidase enzyme insulin delivery can be controlled (figure 5.11). After the action of glucose oxidase, gluconic acid is produced which can ultimately decrease the pH. The low pH, in turn, causes the release of insulin from the bio-erodible polymeric system.

5.4.5.2 Immobilization of artificial cells

An artificial cell mainly comprises a spherical semipermeable membrane with similar dimensions to a living cell. The biological materials, e.g. enzymes, enclosed within the artificial cells can be immobilized. The thus-immobilized compact artificial cells can function as artificial organs. Using this procedure, the most important artificial organs developed include artificial kidney, artificial liver, blood detoxifiers and immunosorbents. Their operation is, however, very restricted. Through multiple reactions, a multienzyme system can be immobilized in the form of artificial cells for the transformation of a substrate to a product.

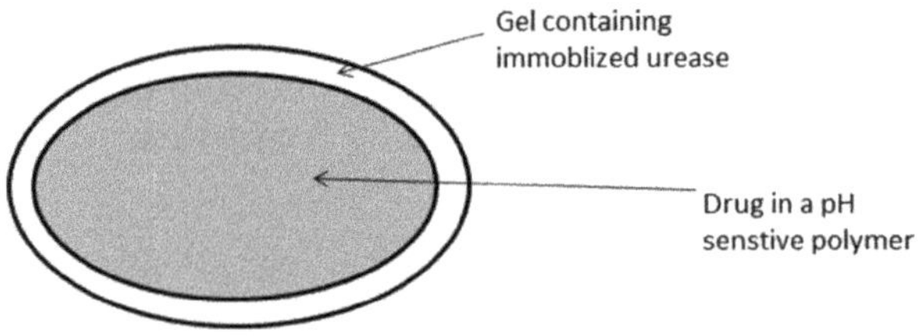

Figure 5.10. A schematic representation of immobilized urease for drug delivery.

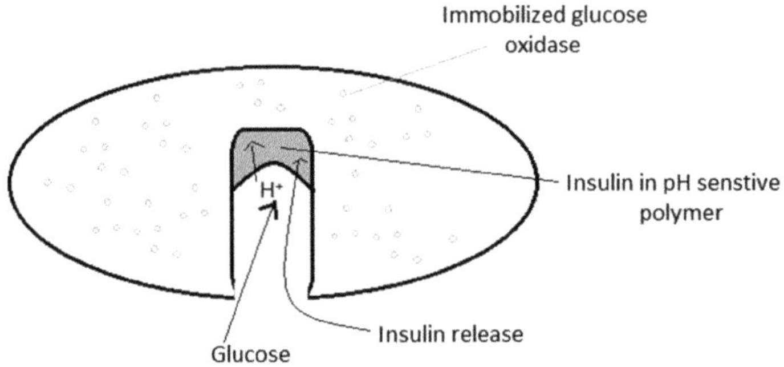

Figure 5.11. A schematic representation of immobilized glucose oxidase for insulin delivery.

References

[1] Suekane M 1982 Immobilization of glucose isomerase *Z. Allg. Mikrobiol.* **22** 565–76

[2] Newman J D and Setford S J 2006 Enzymatic biosensors *Mol. Biotechnol.* **32** 249–68

[3] Rocchitta G *et al* 2016 Enzyme biosensors for biomedical applications: strategies for safeguarding analytical performances in biological fluids *Sensors* **16** E780

[4] Mulchandani A and Bassi A S 1995 Principles and applications of biosensors for bioprocess monitoring and control *Crit. Rev. Biotechnol.* **15** 105–24

[5] Clark L C and Lyons C 1962 Electrode systems for continuous monitoring in cardiovascular surgery *Ann. NY Acad. Sci.* **102** 29–45

[6] Updike S J and Hicks G P 1967 The enzyme electrode *Nature* **214** 986–8

[7] Hiratsuka A, Fujisawa K and Muguruma H 2008 Amperometric biosensor based on glucose dehydrogenase and plasma-polymerized thin films *Anal. Sci.* **24** 483–6

[8] Chambers J P, Arulanandam B P, Matta L L, Weis A and Valdes J J 2008 Biosensor recognition elements *Curr. Issues Mol. Biol.* **10** 1–12

[9] Iqbal S S, Mayo M W, Bruno J G, Bronk B V, Batt C A and Chambers J P 2000 A review of molecular recognition technologies for detection of biological threat agents *Biosens. Bioelectron.* **15** 549–78

[10] Newman J D and Turner A P 1992 *Biosensors: Principles and Practice*vol 27 (London: Portland), pp 147–59

[11] Habermuller K, Mosbach M and Schuhmann W 2000 Electron-transfer mechanisms in amperometric biosensors *Fresenius J. Anal. Chem.* **366** 560–8

[12] Pearson J E, Gill A and Vadgama P 2000 Analytical aspects of biosensors *Ann. Clin. Biochem.* **37** 119–45

[13] Thevenot D R, Toth K, Durst R A and Wilson G S 2001 Electrochemical biosensors: recommended definitions and classification *Biosens. Bioelectron.* **16** 121–31

[14] Turner A P, Chen B and Piletsky S A 1999 *In vitro* diagnostics in diabetes: meeting the challenge *Clin. Chem.* **45** 1596–601

[15] Wang J 2008 Electrochemical glucose biosensors *Chem Rev.* **108** 814–25

[16] Akyilmaz E, Yorganci E and Asav E 2010 Do copper ions activate tyrosinase enzyme? A biosensor model for the solution *Bioelectrochemistry* **78** 155–60

[17] Venugopal V 2002 Biosensors in fish production and quality control *Biosens. Bioelectron.* **17** 147–57

[18] Diviès C 1975 Remarques sur l'oxydation de l'éthanol par une electrode microbienne d'acetobacter zylinum *Ann. Microbiol.* **126A** 175–86

[19] Rechnitz G A 1978 Biochemical electrodes uses tissues slice *Chem. Eng. News* **56** 16–21

[20] Wang J 1998 DNA biosensors based on peptide nucleic acid (PNA) recognition layers. A review *Biosens. Bioelectron.* **13** 757–762

[21] Scognamiglio V, Arduini F, Palleschi G and Rea G 2014 Biosensing technology for sustainable food safety *Trends Anal. Chem.* **62** 1–10

[22] Leatherbarrow R J and Edwards P R 1999 Analysis of molecular recognition using optical biosensors *Curr. Opin. Chem. Biol.* **3** 544–7

[23] Zhang J, Campbell R E, Ting A Y and Tsien R Y 2002 Creating new fluorescent probes for cell biology *Nat. Rev. Mol. Cell Biol.* **3** 906–18

[24] Lippincott-Schwartz J and Patterson G H 2003 Development and use of fluorescent protein markers in living cells *Science* **300** 87–91

[25] Shaner N C, Steinbach P A and Tsien R Y 2005 A guide to choosing fluorescent proteins *Nat. Methods* **2** 905–9

[26] Tsien R Y 2005 Breeding and building molecules to spy on cells and tumors *FEBS Lett.* **579** 927–32

[27] Giepmans B N, Adams S R, Ellisman M H and Tsien R Y 2006 The fluorescent toolbox for assessing protein location and function *Science* **312** 217–24

[28] Ibraheem A and Campbell R E 2010 Designs and applications of fluorescent protein-based biosensors *Curr. Opin. Chem. Biol.* **14** 30–6

[29] Wu B, Piatkevich K D, Lionnet T, Singer R H and Verkhusha V V 2011 Modern fluorescent proteins and imaging technologies to study gene expression, nuclear localization, and dynamics *Curr. Opin. Cell Biol.* **23** 310–7

[30] Aye-Han N N, Qiang N and Zhang J 2009 Fluorescent biosensors for real-time tracking of post-translational modification dynamics *Curr. Opin. Chem. Biol.* **13** 392–7

[31] Zhao Z and Jiang H 2010 Enzyme-based electrochemical biosensors *Biosensors* ed P A Serra (Rijeka: InTech)

[32] Wang J 2008 Electrochemical glucose biosensors *Chem. Rev.* **108** 814–25

[33] Weltin A, Kieninger J and Urban G A 2016 Microfabricated, amperometric, enzyme-based biosensors for *in vivo* applications *Anal. Bioanal. Chem.* **408** 4503–21

[34] Price C P 2003 Point-of-care testing in diabetes mellitus *Clin. Chem. Lab. Med* **41** 1213–9

[35] D'Costa E J, Higgins I J and Turner A P 1986 Quinoprotein glucose dehydrogenase and its application in an amperometric glucose sensor *Biosensors* **2** 71–87

[36] Weibel M K and Bright H J 1971 The glucose oxidase mechanism. Interpretation of the pH dependence *J. Biol. Chem.* **246** 2734–44

[37] Guilbault G G and Lubrano G J 1973 An enzyme electrode for the amperometric determination of glucose *Anal. Chim. Acta* **64** 439–55

[38] Jin W, Wollenberger U and Scheller F W 1998 PQQ as redox shuttle for quinoprotein glucose dehydrogenase *Biol. Chem.* **379** 1207–1211

[39] Zayats M, Katz E, Baron R and Willner I 2005 Reconstitution of apo-glucose dehydrogenase on pyrroloquinoline quinone-functionalized Au nanoparticles yields an electrically contacted biocatalyst *J. Am. Chem. Soc.* **127** 12400–6

[40] Raitman O A, Patolsky F, Katz E and Willner I 2002 Electrical contacting of glucose dehydrogenase by the reconstitution of a pyrroloquinoline quinone-functionalized polyaniline film associated with an Au-electrode: an *in situ* electrochemical SPR study *Chem. Commun.* 1936–7

[41] Bartlett P N and Whitaker R G 1987 Strategies for the development of amperometric enzyme electrodes *Biosensors* **3** 359–79

[42] Bartlett P N, Simon E and Toh C S 2002 Modified electrodes for NADH oxidation and dehydrogenase-based biosensors *Bioelectrochemistry* **56** 117–122

[43] Gorton L and Dominguez E 2002 Electrocatalytic oxidation of NAD(P) H at mediator-modified electrodes *J. Biotechnol.* **82** 371–92

[44] Eggins B 2002 *Analytical Techniques in the Sciences: Chemical Sensors and Biosensors* (Chichester: Wiley)

[45] Chaubey A and Malhotra B D 2002 Mediated biosensors *Biosens. Bioelectron.* **17** 441–56

[46] D'Orazio P 2003 Biosensors in clinical chemistry *Clin. Chim. Acta* **334** 41–69

[47] Bakker E and Pretsch E 2005 Potentiometric sensors for trace-level analysis *Trac-Trends Anal. Chem.* **24**(3) 199–207

[48] Caras S and Janata J 1980 Field-effect transistor sensitive to penicillin *Anal. Chem.* **52** 1935–7
[49] Hafeman D G, Parce J W and McConnell H M 1988 Light-addressable potentiometric sensor for biochemical systems *Science* **240** 1182–5
[50] Mourzina Y, Yoshinobu T, Schubert J, Luth H, Iwasaki H and Schoning M J 2001 Ion-selective light-addressable potentiometric sensor (laps) with chalcogenide thin film prepared by pulsed laser deposition *Sens. Actuators* B **80**(2) 136–40
[51] Poghossian A, Yoshinobu T, Simonis A, Ecken H, Luth H and Schoning M J 2001 Penicillin detection by means of field-effect based sensors: EnFET, capacitive EIS sensor or LAPS? *Sens. Actuators* B **78** 237–42
[52] Xu G, Ye X, Qin L, Xu Y, Li Y, Li R and Wang P 2005 Cell-based biosensors based on light-addressable potentiometric sensors for single cell monitoring *Biosens. Bioelectron.* **20** 1757–63
[53] Kloock J P, Moreno L, Bratov A, Huachupoma S, Xu J, Wagner T, Yoshinobu T, Ermolenko Y, Vlasov Y G and Schoning M J 2006 PLD-prepared cadmium sensors based on chalcogenide glasses—ISFET, LAPS and μISE semiconductor structures *Sens. Actuators* B **118** 149–55
[54] Stein B, George M, Gaub H E and Parak W J 2004 Extracellular measurements of averaged ionic currents with the light-addressable potentiometric sensor (LAPS) *Sens. Actuators* B **98** 299–304
[55] Lee C S, Kim S K and Kim M 2009 Ion-sensitive field-effect transistor for biological sensing *Sensors* **9** 7111–31
[56] Schöning M and Poghossian A 2002 Recent advances in biologically sensitive field-effect transistors (BioFETs) *Analyst* **127** 1137–51
[57] Jaffrezic-Renault N and Dzyadevych S V 2008 *Sensors* **8** 2569–88
[58] Ghourchian H, Moulaie Rad A, Elyasvandi H and Iran I 2004 *Chem. Chem. Eng.* **23** 55–63
[59] Mulchandani A and Rogers K R 1998 *Enzyme and Microbial Biosensors: Techniques and Protocols* (Totowa, NJ: Humana), pp 157
[60] Spink C and Wadso I 1976 Calorimetry as an analytical tool in biochemistry and biology *Methods Biochem. Anal.* **23** 1–159
[61] Grime J K 1985 *Analytical Solution Calorimetry* (New York: Wiley)
[62] Mosbach K and Danielsson B 1974 An enzyme thermistor *Biochim. Biophys. Acta* **364** 140–5
[63] Danielsson B, Mattiasson B and Mosbach K 1981a Enzyme thermistor devices and their analytical applications *Appl. Biochem. Bioeng.* **3** 97–143
[64] Danielsson B and Mosbach K 1988 Enzyme thermistors *Methods Enzymol.* **137** 181–97
[65] Danielsson B and Mosbach K 1986 Theory and application of calorimetric sensors *Biosensors: Fundamentals and Applications* ed A P F Turner, I Karube and G S Wilson (Oxford: Oxford University Press), pp 575–95
[66] Steiner M S, Duerkop A and Wolfbeis O S 2011 Optical methods for sensing glucose *Chem. Soc. Rev.* **40**(9) 4805-39
[67] Pickup J C, Hussain F, Evans N D, Rolinski O J and Birch D J 2005 Fluorescence based glucose sensors *Biosens. Bioelectron.* **20** 2555-65
[68] Ligler F S and Taitt C A R 2002 *Optical Biosensors: Present and Future* (Amsterdam: Elsevier)
[69] Lakowicz J R 2006 *Principles of Fluorescence Spectroscopy* (Berlin: Springer)
[70] Tan W H 1998 Optical measurements on the nanometer scale *Trac-Trends Anal. Chem.* **17** 501-13

[71] Lübbers D W and Opitz N 1974 The pCO_2/pO_2-optode: a new probe for measurement of pCO_2 or pO in fluids and gases *Z. Naturforsch.* C **30** 532–3
[72] Chudobová I, Vrbová E, Kodíček M, Janovcová J and Káš J 1996 Fibre optic biosensor for the determination of d-glucose based on absorption changes of immobilized glucose oxidase *Anal. Chim. Acta* **319** 103–10
[73] Weiss S 1999 Fluorescence spectroscopy of single biomolecules *Science* **283** 1676-83
[74] Clark L C and Lyons C 1962 Electrode systems for continuous monitoring in cardiovascular surgery *Ann. NY Acad. Sci.* **102** 29-45
[75] YSI Inc 1975 New products *Anal. Chem.* **47** 787–91
[76] Shichiri M, Kawamori R, Yamasaki Y, Hakui N and Abe H 1982 Wearable artificial endocrine pancreas with needle-type glucose sensor *Lancet* **2** 1129–31
[77] Tierney M J, Tamada J A, Potts R O, Eastman R C, Pitzer K, Ackerman N R and Fermi S J 2000 The GlucoWatch biographer: a frequent automatic and noninvasive glucose monitor *Ann. Med.* **32**(9) 632-41
[78] Yao H, Liao Y, Lingley A R, Afanasiev A, Lahdesmak I, Otis B P and Parviz B A 2012 A contact lens with integrated telecommunication circuit and sensors for wireless and continuous tear glucose monitoring *J. Micromech. Microeng.* **22** 1–10
[79] Web page http://googleblog.blogspot.no/2014/01/introducing-our-smartcontact-lens.html Accessed: 25.02.2014
[80] Pohanka M 2017 The piezoelectric biosensors: principles and applications, a review *Int. J. Electrochem. Sci.* **12** 496-506
[81] Moina C and Ybarra G 2012 Fundamentals and applications of immunosensors *Advances in Immunoassay Technology* ed H L C Norman (Rijeka: InTech)

IOP Publishing

Introduction to Pharmaceutical Biotechnology, Volume 2
Enzymes, proteins and bioinformatics
Saurabh Bhatia

Chapter 6

Biotransformation and enzymes

6.1 Introduction

Micro-organisms have the tendency to enzymatically alter a wide range of organic compounds. Biotransformation (bioconversions or microbial transformations) widely refers to the processes in which micro-organisms convert organic substances into structurally related products. In particular, biotransformation deals with microbial (enzymatic) transformation of a substrate into a product with a limited number of (one or a few) enzymatic reactions. This is in comparison to fermentation which includes a number of reactions (often complex in nature). Although a number of biotransformations are recognized, only particular reactions are valuable for the production of commercially important products. The implication of bioconversion reactions becomes evident when the synthesis of a certain compound is either difficult or costly using chemical approaches. In addition, biotransformations are usually considered for chemical reactions because of substrate specificity, stereo-specificity and mixed reaction conditions (pH, temperature, pressure). The environmental pollution due to biotransformation is virtually insignificant or negligible. Moreover, it is easy to apply rDNA technology to make suitable changes in biotransformations. Another practical benefit of biotransformations is that it is very easy to scale-up the procedures due to the restricted number of reactions.

6.2 Types of biotransformation reactions

Different types of chemical reactions occur during biotransformations, such as oxidation, reduction, hydrolysis, condensation, isomerization, formation of new C–C bonds, production of chiral compounds and reversal of hydrolytic reactions. Among these, oxidation, isomerization and hydrolysis reactions are more commonly observed in biotransformations. Other biotransformations involve more than one type of reaction. Some examples of significant biotransformation reactions along with the micro-organisms involved are listed in table 6.1. The conversion time required for biotransformation is related to the type of reaction, the substrate

doi:10.1088/bk978-0-7503-1302-5ch6

Table 6.1. A selected list of important biotransformation reactions.

Type of reaction	Example	Commonly used micro-organism(s)
Reduction	Benzaldehyde → benzyl alcohol nitropentachlorobenzol → pentachloroaniline	*Saccharomyces cerevisiae,* *Streptomyces aureofaciens*
Oxidation	Tryptophan → 5 hydroxytryptophan naphthalene → salicylic acid	*Bacillus subtitis,* *Corynebacterium* spp.
Hydrolysis	Anhydrotetracycline → tetracycline menthyl laureate → menthol	*Streptomyces aureofaciens,* *Mycobacterium phlei*
Condensation	Streptomycin → streptomycin-phosphate	*Streptomyces griseus*

concentration and the micro-organism used. Usually, oxidation, hydrolysis and dehydration reactions are accomplished in a few hours.

6.3 Sources of biocatalysts and techniques for biotransformation

There are number of techniques available to enhance biotransformation and bioconversion processes, which may include immobilization techniques, genetic engineering and the use of biocatalysts that can tolerate organic solvents [6]. The utilization of enzymes to conduct biotransformation reactions is often challenging, as it faces difficulties such as:

- Reactant or product toxicity.
- Inhibition.
- High dilution.
- Conformational changes.
- Stability issues.
- pH and temperature dependence.

Nevertheless, biocatalysts are the only source for conducting biotransformation reactions [1, 2]. Another challenge is its dependence on the substrate; if the selected substrate is toxic then it can kill the micro-organism and may further delay the biotransformation reaction. The micro-organism also utilizes the substrate as a nutrient and energy source, which can affect product recovery. Limitations related to process time and the choice of biocatalysts makes biotransformation more complicated in the manufacturing of a small molecule pharmaceutical [3]. Another challenge is our inadequate understanding of complex biological systems and the need to increase the yield of the desirable product. However, the slow action (in comparison to chemical transformation) and specific nature of enzymes makes them more suitable for biotransformation. An extensive range of biological catalysts is available for biotransformation reactions [4]. These include growing cells, resting cells, killed cells, immobilized cells, cell-free extracts, enzymes and immobilized enzymes. The most significant sources of biocatalysts and the protocol used for biotransformation are briefly described below.

6.3.1 Growing cells

Suitable cells are cultured in a suitable medium. As the desired growth of the cells occurs (6–24 h), a concentrated substrate is supplemented to the culture. Occasionally, supplementation of emulsifiers (Tween, organic solvents) is required to solubilize substrates and/or products, e.g. steroid biotransformation. The substrate transformation to the product can be examined by chromatographic procedures. Biotransformation can be stopped when the formation is optimum.

6.3.2 Non-growing cells

Non-growing cells are considered for biotransformation reactions for a number of reasons, such as:

- A very high concentration of substrate can be used (with high substrate concentration, growing cells stop their growth).
- The conversion efficiency of substrate to product is high.
- Cells can be washed and used and thus there will be no contaminating substances.
- Biotransformation can be optimized by creating specific environmental conditions (pH, temperature, etc). Product isolation and recovery are easy.

6.3.3 Immobilized cells

Biotransformations can be performed constantly by utilizing immobilized cells. Further, the same cells can be used again and again. Several bioconversion reactions with single or multistage reactions are in fact performed using immobilized cells, e.g. large-scale production of L-alanine and malic acid.

6.3.4 Immobilized enzymes

Immobilized enzymes are frequently used in biotransformation, due to a number of advantages, such as:

- Cell-free enzyme systems in the form of immobilized enzymes.
- There is no transport barrier across the cell membrane for the substrate or product.
- The isolation and recovery of the product is simpler and easier.
- The desired products are not degraded.
- There is no occurrence of undesirable side reactions.

Several immobilized enzyme systems have been established for biotransformations, such as glucose isomerase and penicillin acylase.

6.4 Product recovery in biotransformations

During most biotransformation reactions, the anticipated end-products are extracellular. The product may either be in a soluble or a suspended state. When whole cells are utilized, they have to be isolated and frequently washed (with water or organic solvent) as required. The extracted product can be recovered, using common techniques such as precipitation by salts, extraction with solvents, adsorption to ion-exchangers, etc. The volatile products can be recovered by direct distillation from the medium. A number of biotransformations are reported in the literature. Of these, only a selected few are significant for large-scale purposes. The major drawbacks with a number of biotransformations are low yields, expensive processes or very limited markets. Recently, an increase in the rate of biotransformations and product recovery was observed in a membrane bioreactor by using direct electric current. The amount of production of benzoic acid was found to be escalated by 42% when the current was applied. This type of process has immense potential in increasing productivity from biotransformations by means of a simultaneous enhancement in metabolic activity and *in situ* product recovery [5]. It was also observed that recovery of 3-methylcatechol produced in a biphasic system (aliphatic alcohol/water bioreactor) was 10–20 fold more than in an aqueous medium. This demonstrates the effectiveness of double-phase systems for this particular biotransformation [6]. Similarly, the recovery and productivity of vicinal diol was increased by using a biphasic system in a hollow fiber membrane bioreactor [7]. In 1998, a combined bioreactor–separator system was developed for simultaneous biotransformation and recovery of the product by crystallization in an immobilized L-aspartate beta-decarboxylase reactor system. An increase in the biotransformation rate and recovery of the isoflavones genistein and daidzein from industrial antibiotic fermentations was observed in 2013 [9].

6.5 Application of biotransformation in the production of pharmaceutical products

6.5.1 Biotransformation of steroids

The structure cyclopentanoperhydrophenanthrene is present in all steroids. In general, steroids are hormones that have an extensive range of therapeutic functions, e.g. cortisone is extensively employed in the treatment of rheumatoid arthritis and skin diseases; progesterone and estrogen derivatives are used as contraceptives. Some derivatives of cortisone (e.g. prednisolone) are more effective in their therapeutic action. Steroid production at the commercial level is an important endeavor, which has required the attention of many researchers. Cortisone production once involved 37 reactions and the cost of this product was approximately $200/g in 1950. With the help of biotransformation reactions, this lengthy procedure of 37 reactions was reduced to 11. This also decreased the cost of the product to just $1/g in 1980. Thus developments in biotransformation allow the shortening of lengthy reactions which can ultimately impact the cost of the final product.

There are a number of reactions involved in the production of steroids, microbial transformation in particular mainly involves oxidation (introduction of hydroxyl groups, splitting of side chains, production of epoxides, etc), reduction (conversion of aldehydes or ketones to alcohols, hydration of double bonds), hydrolysis and ester formation. Steroid production exclusively by biotransformation reactions is not possible, thus in addition to microbial transformation, chemical reactions are required. Key steps involved in the biotransformation of steroids are shown in figure 6.1. There are a number of precursors available to initiate biotransformation reactions, such as stigma sterol extracted from soybeans or diosgenin isolated from the roots of the Mexican barbasco plant.

In steroid synthesis, stigma sterol can be chemically converted to progesterone, which is further subjected to biotransformation to ultimately produce 11 α-hydroxyprogesterone through microbial transformation (*Rhizopus nigricans*). Cortisol or hydrocortisone, derived from 11 α-hydroxyprogesterone by chemical reactions, is exposed to a microbe (*Corynebacterium simplex*) to produce prednisolone. In addition, product derived from cortisol can be subjected to biotransformation by a micro-organism (*Corynebacterium simplex*) to synthesize prednisone. Once diosgenin is used as the precursor compound, substance S can be synthesized by chemical reactions, which can be further converted to cortisol by biotransformation with the help of *Curvularia lunata.* Biotransformation of steroids is generally achieved by batch fermentation. In addition, several immobilization techniques are emerging in this area, which may further encourage more efficient biotransformation reactions by using immobilized cells or enzymes. This is more profitable as the biotransformation is more efficient with a high substrate concentration, short conversion time and good product recovery. As not all steroids are water soluble, the microbial transformation reactions are always performed in an organic solvent (water-immiscible) system. However, organic solvents are toxic to micro-organisms or enzymes. It is best to use a two-phase aqueous system for the biotransformation of steroids. Mibolerone is a synthetic anabolic and androgenic steroid utilized for the treatment of estrous in female dogs. Based on a recent report, microbial transformations of mibolerone with *Cunninghamella blakesleeana*, *Cunninghamella echinulata* and *Macrophomina phaseolina* led to production of 11β,17β-dihydroxy-7α,17α-dimethylestr-4-en-3-one of potential use [10]. Similarly, microbial transformation of drospirenone, an oral contraceptive drug, with *Cunninghamella elegans*

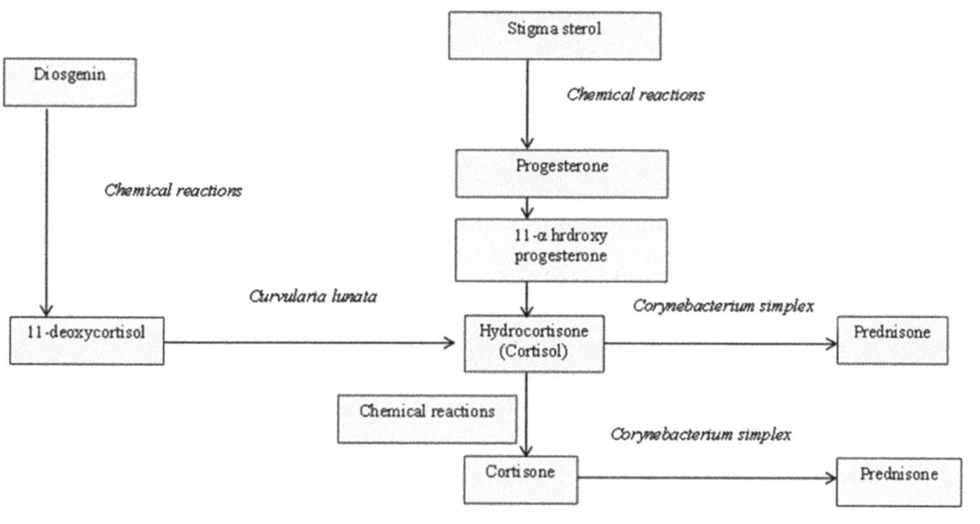

Figure 6.1. Biotransformation of commercially important steroids.

was also reported recently, which helps in understanding the metabolism of this important drug and provide new analogs of drospirenone [11].

6.5.1.1 Biotransformation of cholesterol

Recently, many research activities have focussed on the importance of the lymphatic route in reversing cholesterol transport, as well as the biliary and the non-biliary pathways for removal of cholesterol from the body, which will help in understanding and preventing more of the causes of ischemic coronary heart disease [12]. Studies are also conducted to understand the role of cholesterol in multiple sclerosis (a progressive, neurodegenerative disease of the central nervous system) by specific modification in the mRNA and protein expression of key molecules involved in the maintainance of cholesterol homeostasis in the rat spinal cord. It was observed that modification in the regulation of cholesterol metabolism at onset can result in the progression of disease, whereas during the course of the recovery period it may have beneficial effects, contributing to the regeneration of myelin sheath and restoration of neuronal function [13]. The importance of cholesterol for normal brain function was also realized and it is understood that astrocytes produce most of the brain cholesterol which is important for brain functioning. During this study it was observed that sterol regulatory element-binding protein 2 (a transcription factor decreased during diabetes which can ultimately lead to decrease in brain cholesterol synthesis) mediated cholesterol synthesis in astrocytes plays an important role in brain function and development. Alteration in the expression of this transcription factor can significantly affect brain cholesterol synthesis which draws a relationship between diabetes and altered brain function [14]. Similarly, the key importance of cholesterol in brain physiology and function, changes in cholesterol homeostasis and levels related to brain disorders and neurodegenerative diseases have been studied recently. It was observed that intracellular cholesterol pools in the compartments of mitochondria and endoplasmic reticulum play an important role in understanding the importance of cholesterol homeostasis disruption in neurodegeneration [15]. The above mentioned reports help in understanding the importance of cholesterol in our human body and its biotransformation. A number of commercially important steroids (e.g. androstendione, androstadiendione) can be synthesized directly from cholesterol by biotransformation (figure 6.2).

6.5.2 Biotransformation of antibiotics

One of the most important priorities of the pharmaceutical industry is the production of new antibiotics or modifications of existing ones for more effective treatment against particular diseases or infections. In addition, antibiotics with a broad spectrum, minimum toxicity, minimal allergic reactions and decreased resistance are highly preferred. The significant contribution of biotransformation allows the production of various new antibiotics. Based on recent reports it is well understood that modification in bacterial metabolism, e.g. those related to dormancy (metabolically inactive spores to survive harsh conditions) or biofilm formation, control bacterial susceptibility against antibiotics. This is called

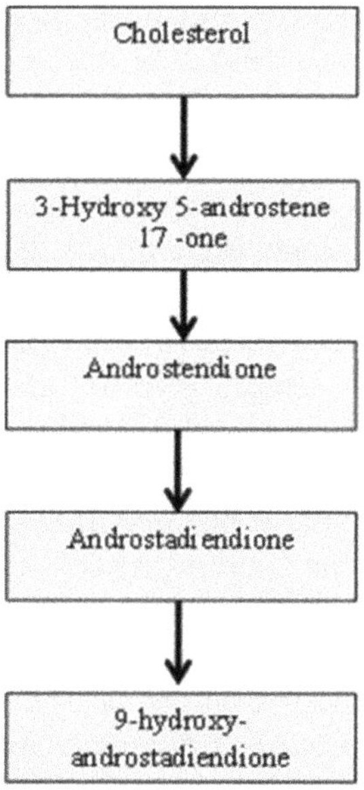

Figure 6.2. Biotransformation of cholesterol by mycobacteria to commercial products.

Table 6.2. Production of new antibiotics.

Organism	Antibiotics	Ref.
White rot fungi (*Irpex lacteus*, *Panus tigrinus*, *Dichomitus squalens*, *Trametes versicolor* and *Pleurotus ostreatus*)	Biodegradation of fluoroquinolones (norfloxacin, ofloxacin and ciprofloxacin)	[16]
Streptomyces davawensis and *Streptomyces cinnabarinus*	Antibiotics (riboflavin analogs) roseoflavin and 8-demethyl-8-aminoriboflavin	[17]
Ligninolytic fungi (*Irpex lacteus*, *Dichomitus squalens*)	Flumequine (fluoroquinolone antibiotic)	[18]
Marine-derived fungus *Paecilomyces* spp.	Tetracycline, minocycline, chlortetracycline, oxytetracycline, doxycycline	[19]
Cystobacter fuscus	Myxobacterial antibiotic (cystothiazole A)	[20]
Streptomyces venezuelae strain	Rosamicin antibiotic into 10,11-dihydrorosamicin	[21]
Xylaria longipes	Danofloxacin	[22]
Spores of *Streptomyces griseus*	Acylation of chloramphenicol	[23]
Deoxystreptamine-negative mutant of *Micromonospora purpurea*	2,4,6/3,5-pentahydroxycyclohexanone and 2,4/3, 5-tetrahydroxycyclohexanone	[24]

phenotypic resistance, presenting a connection between bacterial metabolism and antibiotic resistance [25].

Adoption of new techniques such as miniaturization, nanotechnology, micro-dosing, chemometrics and high-throughput analysis helps in the exploration of new molecules. Alternatively, it can be produced by 'green chemistry', a novel synthetic approach resulting in reduced production of waste and the smallest environmental impact [26]. Since the discovery of penicillin by Sir Alexander Fleming, more than 20 000 types of antibiotic molecules have been produced using micro-organisms, although only few are clinically useful. Reports also confirm a steep decline in the discovery of new molecules, in particular antibiotics [27]. During the last three decades, among the 28 new molecules discovered, only three semisynthetic molecules that have experienced chemical modifications have been approved for clinical use [28]. Even after advancement in synthetic chemistry, biotransformation is considered as the most cost-effective track to discover new molecules [29]. This requires a focus towards developing more approaches to produce a substantial number of molecules from micro-organisms. This gives scope for considerable further development biotransformation processes in the future [28]. However, there are still various challenges that need to be addressed, such as:

- Determining the relationship between structural changes and time and cost efficiency.
- Determining the relationship between the biological activity of products and the optimization of pharmacokinetic/pharmacodynamic properties and safety.
- The limited number of suppliers involved in large-scale drug production with good quality practices [30].

6.5.2.1 Direct biotransformation

Various reactions such as acylation and de-acylation, phosphorylation, adenylation and hydrolysis are involved in direct biotransformation, in particular during microbial transformation of antibiotics.

6.5.2.1.1 Biotransformation of penicillin G

The earliest extraction of penicillin G from *Penicillium chrysogenum* and its conversion into 6-aminopenicillanic acid and phenyl acetic acid was reported in 1992. In this study 6-aminopenicillanic acid was converted into ampicillin when a liquid membrane carrier system was electrocoalescence [31]. Recent study related to the production of 6-aminopenicillanic acid has revealed production of penicillin G acylase from immobilized whole cells of *Escherichia coli*. Later, *E. coli* was utilized for the biotransformation of penicillin V to 6-aminopenicillanic acid to express a highly active penicillin V acylase [32]. In association with chemical synthesis, microbial transformation is usually used for the commercial synthesis of semi-synthetic cephalosporins and penicillins. During this conversion, 6-amino-penicillanic acids derived from the enzymatic cleavage of penicillin by penicillin acylase is a very important reaction (figure 6.3). With the help of enzyme penicillinase (β-lactamase), penicillin G becomes inactivated by its conversion to benzylpenicilloic acid.

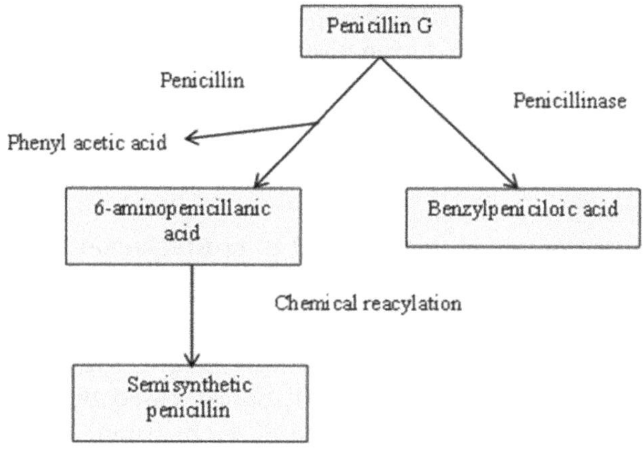

Figure 6.3. Biotransformation of penicillin G.

6.5.2.1.2 Biotransformation of narbomycin

During 1998, hydroxylation of narbomycin was achieved by the pikC (P450 hydroxylase) encoded cytochrome P450 in *Streptomyces venezuelae.* In this process hydroxylation of narbomycin to picromycin (by *Streptomyces* spp.) is performed with the help of P450 hydroxylase [33].

6.5.2.1.3 Biotransformation of macrolides

The macrolides are prodrugs and semisynthetic derivatives that are used in a variety of infections. The most important feature of macrolides is that they are frequently administered with other drugs, thus this may increase chances for pharmacokinetic interactions. The potential of macrolides to participate in the biotransformation of some other drugs has been extensively documented, typically with erythromycin and troleandomycin. Macrolides can induce their own hepatic biotransformation into nitrosoalkanes. Less active products are derived from de-acylation of macrolide antibiotics. These products can be used for the further synthesis of more efficient semisynthetic macrolides [34].

6.5.2.2 Indirect biotransformation

During antibiotic production, biotransformation reactions can often be controlled by the addition of certain inhibitors or modified substrates to the medium, or in other words the biosynthetic processes of antibiotics occur in a controlled fashion in indirect biotransformation.

6.5.2.2.1 Biotransformation of actinomycins

A good example of indirect biotransformation is biotransformation of actinomycins which involves the production of new actinomycins in the presence of 4-methylproline (a proline analog) in the medium by using the micro-organism *Streptomyces parvulus.* These actinomycins are more effective in their function as they have 4-methylproline in place of proline [35].

6.5.2.2.2 Biotransformation of ribostamycin

Ribostamycin, an intermediate during biosynthesis of neomycin, can be synthesized in large quantities by employing mutant strains of *Streptomyces fradiae.* Numerous other mutant strains of micro-organisms have been produced by rDNA technology for the synthesis of modified antibiotics of aminoglycosides and rifamycins.

6.5.3 Biotransformation of arachidonic acid to prostaglandins

Prostaglandins are significant in pharmaceutical and therapeutic application, e.g. PGE_1 as a contraceptive, PGG_1 in the treatment of CHF and PGG_2 for relieving labor pains. Arachidonic acid (unsaturated fatty acid) is the starting compound for the biosynthesis of prostaglandins. Several reports based on biotransformation of arachidonic acid to PGE_1, PGE_2, PGF_1 and PGF_2 by using fungi area are available.

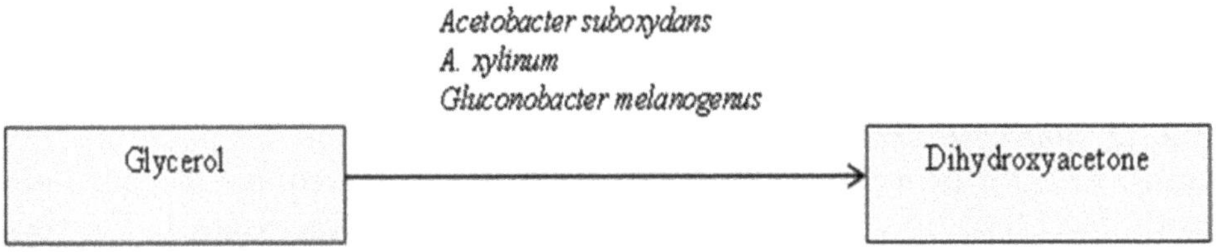

Figure 6.4. Biotransformation of glycerol to dihydroxyacetone.

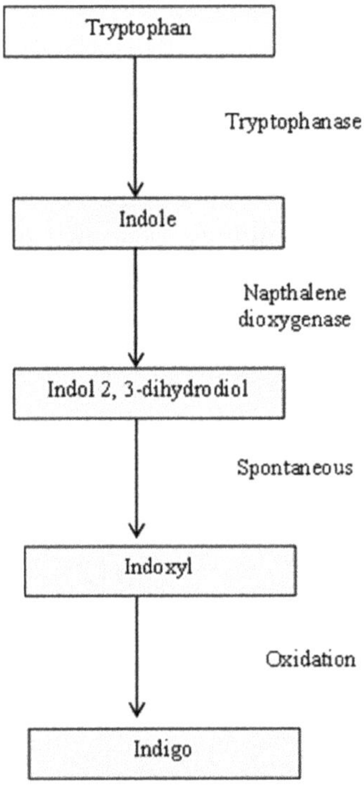

Figure 6.5. Microbial production of indigo.

It is assumed that prostaglandins with better efficacy will be synthesized by biotransformations in upcoming years.

6.5.4 Biotransformation for the production of ascorbic acid

Another common example of biotransformation includes the combination of chemical and microbial transformation processes for the commercial production of ascorbic acid (vitamin C).

6.5.5 Biotransformation of glycerol to dihydroxyacetone

Dihydroxyacetone is often employed in cosmetics and suntan lotions. By the process of biotransformation certain acetic acid bacteria can convert glycerol to dihydroxyacetone (figure 6.4). Sufficient oxygen supply, a temperature of 26–28°C and pH 6.0 are ideal for optimal biotransformation.

6.5.6 Biotransformation for the production of indigo

By means of microbial transformation, indigo can be produced. This can be achieved by cloning a single *Pseudomonas* gene that encodes naphthalene dioxygenase in the creation of *E. coli.* The related reactions of biotransformation for the synthesis of indigo are represented in figure 6.5.

References

[1] Lilly M D 1994 Advances in biotransformation processes *Chem. Eng. Sci.* **49** 151–9

[2] Leresche J E and Meyer H P 2006 Chemocatalysis and biocatalysis (biotransformation): some thoughts of a chemist and a biotechnologist *Org. Proc. Res. Dev.* **10** 572–80

[3] Collins A M and Kennedy M J 1999 Biotransformations and bioconversions in New Zealand: past endeavours and future potential *Austral Biotechnol.* **9** 86–94

[4] Woodley J M and Lilly M D 1994 Biotransformation reactor selection and operation *Applied Biocatalysis* ed J M S Cabral, D Best, L Boross and J Tramper (Chur: Harwood Academic)

[5] Mustacchi R, Knowles C J, Li H, Dalrymple I, Sunderland G, Skibar W and Jackman S A 2005 Enhanced biotransformations and product recovery in a membrane bioreactor through application of a direct electric current *Biotechnol. Bioeng.* **89** 18–23

[6] Rojas A, Duque E, Schmid A, Hurtado A, Ramos J-L and Segura A 2004 Biotransformation in double-phase systems: physiological responses of *Pseudomonas putida* DOT-T1E to a double phase made of aliphatic alcohols and biosynthesis of substituted catechols *Appl. Environ. Microbiol.* **70** 3637–43

[7] Gao P, Wu S, Praveen P, Loh K C and Li Z 2017 Enhancing productivity for cascade biotransformation of styrene to (S)-vicinal diol with biphasic system in hollow fiber membrane bioreactor *Appl. Microbiol. Biotechnol.* **101** 1857–68

[8] Takamatsu S and Ryu D D 1988 Recirculating bioreactor-separator system for simultaneous biotransformation and recovery of product: immobilized L-aspartate beta-decarboxylase reactor system *Biotechnol. Bioeng.* **32** 184–91

[9] Weber J M, Reeves A R and Seshadri R *et al* 2013 Biotransformation and recovery of the isoflavones genistein and daidzein from industrial antibiotic fermentations *Appl. Microbiol. Biotechnol.* **97** 6427–37

[10] Siddiqui M, Ahmad M S and Wahab A *et al* 2017 Biotransformation of a potent anabolic steroid, mibolerone, with *Cunninghamella blakesleeana*, *C. echinulata*, and *Macrophomina phaseolina*, and biological activity evaluation of its metabolites *PLoS One* **12** e0171476

[11] Baydoun E, Atia-Tul-Wahab, Iqbal S, Smith C and Choudhary M I 2017 Biotransformation of drospirenone, a contraceptive drug, with *Cunninghamella elegans Steroids* **126** 30–4

[12] Wang H H, Garruti G, Liu M, Portincasa P and Wang D Q 2017 Cholesterol and lipoprotein metabolism and atherosclerosis: recent advances in reverse cholesterol transport *Ann. Hepatol.* Suppl. 1: s3-105 s27–42

[13] Lavrnja I, Smiljanic K and Savic D *et al* 2017 Expression profiles of cholesterol metabolism-related genes are altered during development of experimental autoimmune encephalomyelitis in the rat spinal cord *Sci. Rep.* **7** 2702

[14] Ferris H A, Perry R J, Moreira G V, Shulman G I, Horton J D and Kahn C R 2017 Loss of astrocyte cholesterol synthesis disrupts neuronal function and alters whole-body metabolism *Proc. Natl Acad. Sci. USA* **114** 1189–94

[15] Arenas F, Garcia-Ruiz C and Fernandez-Checa J C 2017 Intracellular cholesterol trafficking and impact in neurodegeneration *Front. Mol. Neurosci.* **10** 382

[16] Čvančarová M, Moeder M, Filipová A and Cajthaml T 2015 Biotransformation of fluoroquinolone antibiotics by ligninolytic fungi—metabolites, enzymes and residual antibacterial activity *Chemosphere* **136** 311–20

[17] Matern A, Pedrolli D, Großhennig S, Johansson J and Mack M 2016 Uptake and metabolism of antibiotics roseoflavin and 8-demethyl-8-aminoriboflavin in riboflavin-auxotrophic listeria monocytogenes *J. Bacteriol.* **198** 3233–43

[18] Cvančarová M, Moeder M, Filipová A, Reemtsma T and Cajthaml T 2013 Biotransformation of the antibiotic agent flumequine by ligninolytic fungi and residual antibacterial activity of the transformation mixtures *Environ. Sci. Technol.* **47** 14128–36

[19] Shang Z, Salim A A, Khalil Z, Bernhardt P V and Capon R J 2016 Fungal biotransformation of tetracycline *J. Antibiot. Org. Chem.* **81** 6186–94

[20] Suzuki Y, Ojika M and Sakagami Y 2004 Biotransformation of cystothiazole A, a myxobacterial antibiotic, into novel derivatives by the mother producer, *Cystobacter fuscus Biosci. Biotechnol. Biochem.* **68** 390–6

[21] Huong N L, Hoang N H, Shrestha A, Sohng J K, Yoon Y J and Park J W 2014 Biotransformation of rosamicin antibiotic into 10,11-dihydrorosamicin with enhanced *in vitro* antibacterial activity against MRS *J. Microbiol. Biotechnol.* **24** 44–7

[22] Rusch M, Kauschat A, Spielmeyer A, Römpp A, Hausmann H, Zorn H and Hamscher G 2015 Biotransformation of the antibiotic Danofloxacin by *Xylaria longipes* leads to an efficient reduction of its antibacterial activity *J. Agric. Food Chem.* **63** 6897–904

[23] El-Kersh T A and Plourde J R 1976 Biotransformation of antibiotics. I. Acylation of chloramphenicol by spores of *Streptomyces griseus* isolated from the Egyptian soil *J. Antibiot.* **29** 292–302

[24] Daum S J, Rosi D and Goss W 1977 Production of antibiotics by biotransformation of 2,4,6/3,5-pentahydroxycyclohexanone and 2,4/3,5-tetrahydroxycyclohexanone by a deoxystreptamine-negative mutant of *Micromonospora purpurea J. Am. Chem. Soc.* **99** 283–4

[25] Martínez J L and Rojo F 2011 Metabolic regulation of antibiotic resistance *FEMS Microbiol. Rev.* **35** 768–89

[26] Tucker J L 2006 Green chemistry, a pharmaceutical perspective *Org. Process. Res. Dev.* **10** 315–19

[27] Muñoz Solano D, Hoyos P, Hernáiz M J, Alcántara A R and Sánchez-Montero J M 2012 Industrial biotransformation in the synthesis of building blocks leading to enantiopure drugs *Bioresour. Technol.* **115** 196–207

[28] Newman D J and Cragg G M 2012 Natural products as sources of new drugs over the 30 years from 1981 to 2010 *J. Nat. Prod.* **75** 311–35

[29] Zaks A and Dodds D R 1997 Application of biocatalysis and biotransformations to the synthesis of pharmaceuticals *Drug Discovery* 513–31

[30] Bauer A and Brönstrup M 2014 Industrial natural product chemistry for drug discovery and development *Nat. Prod. Rep.* **31** 35

[31] Barenschee T, Scheper T and Schügег K 1992 An integrated process for the production and biotransformation of penicillin *J. Biotechnol.* **26** 143–54

[32] Avinash V S, Chauhan P D, Gaikwad S and Pundle A 2017 Biotransformation of penicillin V to 6-aminopenicillanic acid using immobilized whole cells of *E. coli* expressing a highly active penicillin V acylase *Prep. Biochem. Biotechnol.* **47** 52–7

[33] Xue Y, Wilson D, Zhao L, Liu H W and Sherman D H 1998 Hydroxylation of macrolactones YC-17 and narbomycin is mediated by the pikC-encoded cytochrome P450 in *Streptomyces venezuelae Chem. Biol.* **5** 661–7

[34] Westphal J F 2000 Macrolide—induced clinically relevant drug interactions with cytochrome P-450A (CYP) 3A4: an update focused on clarithromycin, azithromycin and dirithromycin *Br. J. Clin. Pharmacol.* **50** 285–95

[35] Perlman D, Mauger A B and Weissbach H 1966 Microbial metabolism of actinomycins and other heterodetic antibiotic peptides *Biochem. Biophys. Res. Commun.* **24** 513–8

Chapter 7

Introduction to genomics

7.1 Introduction

In 1910 several researchers first worked together to identify and map genes in organisms of interest. From then, efforts progressively focused on specific organisms, e.g. maize, mice, *Drosophila*, bacteria and yeast.

Genetic mapping involves three important steps:

- Identification of spontaneous mutations, or mutants induced by chemical or physical agents or identifying the locus accountable for variation.
- Sequencing the region in cases and controls to define causal mutation(s).
- Investigating the molecular and cellular variations causes and changes in the functions of the genes discovered.

For spontaneous mutations, DNA sequence analysis has revelaed the types of mutations that are responsible for several hereditary diseases. Many disorders are due to deletions or duplications, including repeated sequences. Multiple sources, such as errors in DNA replication, lesions, and transposable genetics and elements are responsible for spontaneous mutations. When this type of mutation exists it can be used for linkage studies and planning for linkage maps. It was observed by Drake *et al* in 1998 that the rate of occurrence of spontaneous mutation per genome is extraordinarily identical within extensive groups of organisms, but varies extremely among groups [1]. Physical mappings of genes in organisms such as *Drosophila* were also created. This strategy was effective and is extensively used in genetics. However, one of the major shortcomings of this method is that at least one mutation for the respective gene in the genome is essential. Attaining mutations of each gene is very difficult and also labor intensive. Moreover, the genetic mutation may also impact phenotype. Mutations frequently have a harmful outcome, making it practically impossible to map the mutated gene. In the mid-1980s, researchers began using recombinant DNA technology for investigation. In this method, a number of clones, called a genomic library, are established. These are paired together into overlapping

doi:10.1088/978-0-7503-1302-5ch7

sets and brought together they provide genetic material and physical maps for the complete genome. The clones are initially sequenced and thereafter all the genes in the genome are identified from this sequence.

7.2 Characterizations in genomics

In 1987, the word genomics was coined by the geneticist Thomas H Roderick, at the Jackson Laboratory, Bar Harbor, ME [2]. During that period, the term was considered to mean mapping and sequencing to examine the arrangement and organization of genomes. Currently, genomics comprises the sequencing of genomes, determination of the complete set of proteins encoded by an organism, and the functioning of genes and metabolic pathways in an organism. Thus genomics not only deals with the determination of the genetic information present in an organism, but also with understanding the mechanism by which this information is used by the organism. The word genome was presented by H Winkler in 1920. He used the word genome for the complete set of chromosomal and extra chromosomal genes found in an organism, including viruses. This term is used in a similar sense in modern genetics [3].

Vast amounts of information are available in genomics. Understanding and management of this amount of information requires advanced computers and specific software systems. Bioinformatics is a developing area which involves development and application of computer hardware and software to the procurement, storage, analysis and imaging of biological information. Databanks, which are necessary for the storage and examination of genetic information, have now become important tools for geneticists. Genomics also involves the investigation of the products encoded by genetic material, including the number of genes that are expressed, their period of expression, the kind and level of any post-translational modification of the gene product, the role of the encoded protein, and its site in various cellular compartments.

The field of genomics is divided into the following two areas:

- Structural genomics.
- Functional genomics.

In structural genomics the complete sequence of genetic material or the whole set of proteins produced by an organism is determined by different techniques. This can be achieved using the following steps:

- Assembling of high resolution genetic and physical maps.
- Genome sequencing.
- Evaluation of the whole set of proteins produced in an organism. Analysis is required to determine the protein's three-dimensional structure.

Functional genomics deals with the study of the functioning of genes and metabolic pathways. This means the patterns of genetic expression in an organism.

7.3 Historical background

In 1984, the concept behind genome sequencing was first presented in the scientific community. Later, in 1986, a scheme was set for sequencing of the human genome.

The Human Genome Project was announced to the public on 1 October 1990 [4]. The aim of the genome projects is to collect information on the structure of DNA in human chromosomes and those of other organisms. The next objective is to develop novel skills to accomplish mapping and sequencing. European scientific communities explored a number of genome projects on yeast, bacteria, *Drosophila* and *Arabidopsis thaliana* in 1988, and in 1990 started a new two year program on the human genome [4]. In 1995, the genetic material of the gram-negative bacterium, *Haemophilus influenza* (which is naturally transformable), was the first to be sequenced [5]. In 1997, the complete genetic material of *E. coli* (a 4 639 221-base pair sequence) was sequenced [6]. Among eukaryotic organisms, genetic materials of yeast (*Saccharomyces cerevisiae*) and a worm (*Caenorhabditis elegans*) were the first to be sequenced in 1999. One year after this breakthrough, *Drosophila melanogaster* and *Arabidopsis thaliana* were sequenced. Finally, on 26 June 2001, a rough draft of the human genome was published.

Separate drafts were achieved by the public-funded Human Genome Sequencing Consortium and the private company Celera Genomics, established by Craig Venter. However, the draft sequence was published collaboratively after involvement by the US President.

7.4 Genome sequencing

Defining the arrangement of nucleic acid residues in living specimens is an essential part of an extensive range of research applications. A simplified procedure for the sequencing of the whole genome is depicted in figure 7.1. For almost 50 years,

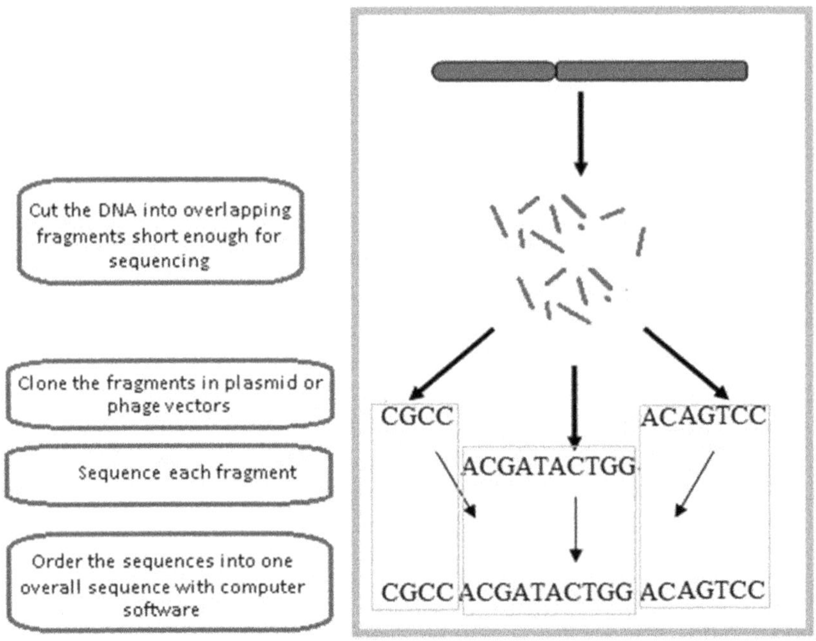

Figure 7.1. Sequencing of a complete genome.

investigators have explored various techniques to sequence DNA and RNA molecules, which involve sequencing short oligonucleotides to millions of bases, interpretation of the coding sequence of a single gene, and rapid and extensively available complete genome sequencing [7]. Genome sequencing is a highly complicated and challenging procedure. In one attempt, a portion of 500–600 bp (at best, 1000 bp) can be sequenced easily, however, most genomes are exceptionally large such as 4.2×10^6 bp for *E. coli* and 3.2×10^9 for humans. Thus for sequencing, the sequence of a genome has to be acquired in a particularly large number of small fragments. These small fragments then have to be assembled into a sequence for the genome. For sequencing, fragments are synthesized by splicing the genomic DNA into fragments at different locations. Consequently the actual site of the fragment in the genome has to be experimentally evaluated. In addition, the fragments subsequently produced typically overlap other fragments at their ends. The derived fragments from an organism's genetic material are cloned in a desirable vector. Ultimately this produces a genomic library of the organism.

For sequencing, cloning of the fragments is necessary to generate a large number of copies of each fragment required. The following two methods are available for the sequencing of genomes:

- Shotgun sequencing.
- Clone-by-clone sequencing.

In both of these approaches, the principal step is to produce a genomic library for a particular chromosome.

7.4.1 Clone-by-clone sequencing

The basis of shotgun and clone-by-clone genome sequencing is the same, however, the difference iss that in clone-by-clone sequencing researchers create a DNA library of the pieces of DNA clones obtained from the sequence that was used in the first place. Data management and the mounting of DNA contigs (overlapping sequences) is a lot easier computationally speaking. In shotgun sequencing the same end is achieved but without cloning and libraries. Researchers directly mount contigs from sequenced genome pieces, therefore better computers are required. Shotgun strictly means the sequencing based on the random shearing and consecutive construction of contigs.

During clone-by-clone sequencing the genome or considered genetic material is fragmented or broken into large pieces (150 kb). The actual position of these fragments in the chromosomes is mapped to support assembling them after sequencing. The fragments are first aligned into contigs (overlapping sequence data, i.e., a cluster of sequences which is derived after the completion of the sequencing process, which eventually provides the sequence of a single fragment). These contigs form a physical map of the genome which is eventually used to guide sequencing and assembly. It is also called directed sequencing of BAC contigs. During the initial procedure, BAC clones (functional fertility plasmids) are introduced to the contigs and these ae then further introduced into a bacterial cell for growth. The fragments of DNA under consideration are copied each time the bacteria multiply to synthesize multiple duplicate copies. A BAC clone (80–100 kb) has a long DNA fragment duplicated into it. This piece is then further employed to produce cosmid clones and plasmid clones. These synthesized clones have progressively smaller DNA fragments. Now each individual chunk is 500 bp stretched; a more suitable size for sequencing. These synthesized fragments are incorporated into a vector that has a well-known DNA sequence. Then DNA fragments are sequenced, starting with an identified sequence of the vector and extending out into the unidentified sequence of the DNA. Following this type of sequencing even small chunks of DNA are joined together by exploring available overlapping locations to assemble the bulky chunks that were initially incorporated into the BACs. This assembly is done by processors (computers) which identify locations of overlap and assemble the DNA sequence. Using the map created during the initial phase, the large pieces can be arranged again into chromosomes as part of the whole genome sequence. This method was introduced during the 1980s and 1990s to initially sequence the genetic material of the *C. elegans* (the nematode worm) and *S. cerevisiae* (yeast). During the sequencing of Human Genome Project (2001), clone-by-clone sequencing was favored. There are a number of advantages to clone-by-clone sequencing: 1) each fragment of DNA is taken from a well-known region of the genome, so it is comparatively easy to regulate whether there are any gaps in the sequence; 2) as a genome map is followed this method offers more reliable assembly, so researchers know where larger pieces/fragments belong in relation to each other; and 3) as each and every piece of DNA is different, numerous people can work on the genome simultaneously.

7.4.2 Human whole-genome shotgun sequencing

The determination of the whole sequence of genes in a genome is required to understand the genetic basis of an organism. The sequencing of large numbers of gene fragments and even complete genomes is only possible with the development of tools for DNA sequencing. Novel significant tools such as physical mapping, DNA sequencing and sequence analysis have been established. To increase the output, robotic processes for sample preparation and novel software for sequence analysis have been applied [8]. A simplified representation of human whole-genome shotgun sequencing is illustrated in figure 7.2.

In 1997, Weber and Myers published a study entitled 'Human whole-genome shotgun sequencing'. They contended, by means of computational models, that in the place of long reads, short reads would cover sufficient data to sequence the first human genome, which would not only be economical, but quicker than the Human Genome Project. This was so fiercely debated that Phillip Green published a review, describing why shotgun sequencing was not possible. Both studies were finally accepted and published side by side in the journal *Genome Research*.

Shotgun sequencing includes arbitrarily splicing of DNA sequences into numerous small fragments and then reconstructing the sequence by examining for regions of overlap. Because of their size and structural sophistication, large, mammalian genomes are the most problematic to clone, sequence and assemble [9, 10].

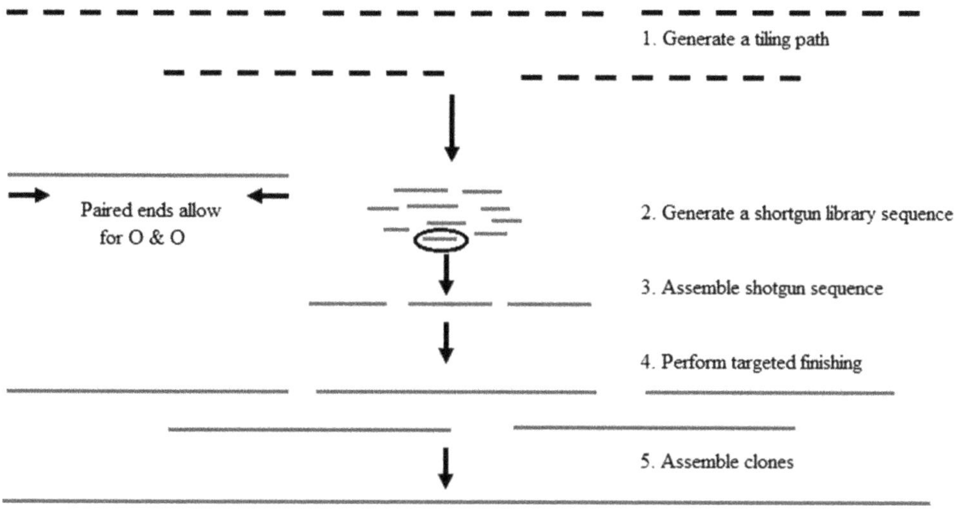

Figure 7.2. Human whole-genome shotgun sequencing.

For such genomes, the method known as clone-by-clone sequencing, although reliable and systematic, takes a very long time. With the development of cheaper sequencing techniques and more advance computer databases, geneticists can thus rely on whole-genome shotgun sequencing to deal with large, complex genomes. Fred Sanger and co-workers initially used shotgun sequencing to sequence small genomes, for example, the genomes of viruses and bacteria. Complete genetic material shotgun sequencing avoids the long mapping and cloning steps that make clone-by-clone sequencing so gradual. During this procedure the complete genome is fragmented into small pieces of DNA for sequencing. These pieces are often of variable sizes, extending from 2–20 kb (2000–20 000 bp) to 200–300 kb (200 000–300 000 bp). Then these pieces are sequenced to examine the order and arrangement of the DNA bases (A, C, G and T). Subsequently, the sequenced fragments are collected together by computer databases that determine where fragments overlap. In analogy, this technique is akin to chopping several copies of a book (which is here the genome or genetic material), blending up all the pieces and at that point reuniting the original text (genome/genetic material) by finding fragments/pieces of DNA with text that overlaps and piecing the book back together based on these. This technique was used by Venter (founder of Celera Genomics) to sequence the human genome. Venter desired to sequence the human genome quicker than the reported goal, and considered shotgun sequencing to be the best approach. For assembly, Venter initially used the reported clone-by-clone data from the Human Genome Project. Currently, with advancements in computational technology, entire genome shotgun sequencing is being employed to further advance the accuracy of present genome sequences, for example, the reference human genome. This procedure can be utilized to eliminate errors, fill in gaps or rectify parts of the sequence that were initially assembled inaccurately (by clone-by-clone sequencing). Consequently, the reference human genome is continuously being developed to guarantee that the genome sequence is of the utmost high standard. There are number of merits to shotgun sequencing, such as:

- By eliminating the mapping steps, entire genome shotgun sequencing (EGSS) is a much quicker process than clone-by-clone sequencing.
- EGSS utilizes a piece of the DNA that clone-by-clone sequencing needs.
- EGSS is mainly effective if there is any available reference sequence against an unidentified genome. It is much more convenient to assemble the genome sequence by assembling it to a present reference genome.
- Shotgun sequencing is much quicker and economical than the procedures employed for genetic mapping.

However, there are still various disadvantages to shotgun sequencing, such as:

- An enormous amount of computing power and advanced and sophisticated software are necessary to align shotgun sequences together.
- For sequencing the genomic material of a mammal (billions of bases long), 60 million individual DNA sequence reads are required.
- Since a genetic map is not used, the possibility of error in assembly is greater. However, comparatively these errors are usually easier to resolve than in other procedures and are minimized if a reference genome can be used.
- EGSS assembly is very challenging without an existing reference genome to match it to.
- EGSS can also result in errors which must be determined by other, more labor-intensive types of sequencing, such as clone-by-clone sequencing.

7.4.3 Compilation of genome resources

Four main objectives were set out for the Genome Sequence Compilation [11]:

- To determine evolutionarily conserved sequence motifs, mainly outside protein-coding genes, which are accountable for regulatory and other critical genomic functions.
- To explore new models of human disease and heritable phenotypes.
- To offer a starting point for the evaluation of the expansion, contraction, and adaptation of gene families in different evolutionary lineages.
- To offer a outline for the reconstruction of genome organization, content and dynamics that have happened through the mammalian radiations.

Genome sequencing projects required the production of high throughput tools that produce desirable information quickly. This can be achieved by the employment of advance computers to manage this vast store of information. These computational tools have given birth to a new area called bioinformatics. Manual manipulation of gene data is time consuming, but bioinformatics can deal with the storage, examination, analysis and utilization of the information about biological systems. For example, it involves processes such as accumulating genome sequences, documentation of genes, assigning functions to the identified genes, planning of databases, etc. Different genetic compilers are available today. A software compiler is a way to use genetic information and provides a design platform allowing the geneticist to manipulate and design everything from single genes to whole genomes. So as to guarantee that the nucleotide sequence of a genetic material is complete and error-free, the genetic material is sequenced more than once, such as by using the shotgun method. The bacterial genome (*Pseudornonas aeruginosa*) was sequenced seven times to make the sequence accurate and free from faults. Stover and co-workers (2010) reported the complete genome sequence of *Pseudomonas aeruginosa* PAO1. They suggest that the size and complexity of the *P. aeruginosa* genome reveals an evolutionary adaptation allowing it to flourish in various environments and limit the effects of a diversity of antimicrobial compounds [12]. However, the assembler software (a sequence assembler is innovative bioinformatics software for spontaneous DNA sequence assembly, contig editing, file format alteration, DNA sequence analysis and mutation recognition) identified 1604 regions that need further justification. These areas were reinvestigated and re-sequenced to successfully complete the genome sequence. The efficiency/accuracy of the shotgun technique was equated with the sequence derived from the clone-by-clone procedure of two extensively separated genomic regions of *P. aeruginosa.* These two considered areas together were 81 843 nucleotides long. The sequences derived by the two procedures were in perfect agreement. This assessment exposed the precision potential of the shotgun method of genome sequencing. This also demonstrates the safeguards to be considered during genome sequencing projects. This level of maintenance is not uncommon. Similar safety measures are employed in all genome projects. The Human Genome Project, an international scientific research project, sequenced the 3.2 billion bp of the human genome a total of 12 times. Celera Genomics (a private biotech company), using shotgun-cloning, used an approach of sequencing from both ends of DNA fragments. This project sequenced the human genome 35.6 times. While a rough draft of the human genome sequence is complete, some other objectives are yet to be accomplished. These include obtaining the remaining sequence and rectifying miscalculations (called proofreading the genome), filling in gaps and then sequencing the 7–10% of the genome that comprise heterochromatin. This is present in ample amounts inside the cells that are less active or not active, whereas euchromatin is predominant inside cells that are vigorous or active in the transcription of several of their genes. Since they comprise low priority sections of repetitive DNA sequences, heterochromatic regions of the genome were not initially sequenced. Moreover, intially it was thought that the heterochromatin does not contain genes. However, the *Drosophila* genome sequence revealed that their heterochromatic regions contain a very small number of genes (around 50). Based on this discovery, the heterochromatic regions of the human genome must be sequenced to confirm that all genes in the human genome have been recognized. As soon as the genome of an organism is sequenced, compiled and proofread, the subsequent process of genomics, namely annotation, begins.

7.5 Understanding bioinformatics and sequencing

Advance computers and high-speed networks are compulsory to evaluate genome projects. Genetic data signify a potential source for investigators and organizations interested in how genes contribute to our health and wellbeing [13]. Nearly half of the genes identified by the Human Genome Project have no known function. Scientists, by means of bioinformatics, can identify genes, establish their roles and develop gene-based strategies for preventing, diagnosing and treating disease [13]. Simplified presentation of bioinformatics and sequencing is presented in figure 7.3.

Genome sequencing and investigation is an area that has developed very quickly over the last 10–20 years, in particular after publication of the final draft of the human genome sequence in 2003. From procuring a complete genome sequence from a characteristic individual or strain of a small number of species, the genomics community has moved on to recording and authenticating genetic assortment within species, with a special focus on human beings, and by sequencing the genomes of a quickly growing number of species. The introduction of economical, very high throughput sequencing tools has also made it promising to sample transcriptomes (the summation of all mRNA fragments expressed from the genes of an organism), genomic areas bound by proteins, bacterial populations, or sometimes even whole ecosystems, by means of sequencing methods. This has germinated an innovative generation of software techniques to manage the very large numbers of short sequences, often called reads, which are produced by the new machines, and has also moved the area of computational genomics into the territory of 'big data', requiring large and sophisticated computers for the organization and investigation of these sequences.

The entities that are selected for genome projects are those typically used in genetic and other systematic studies. Thus they are usually called model organisms.

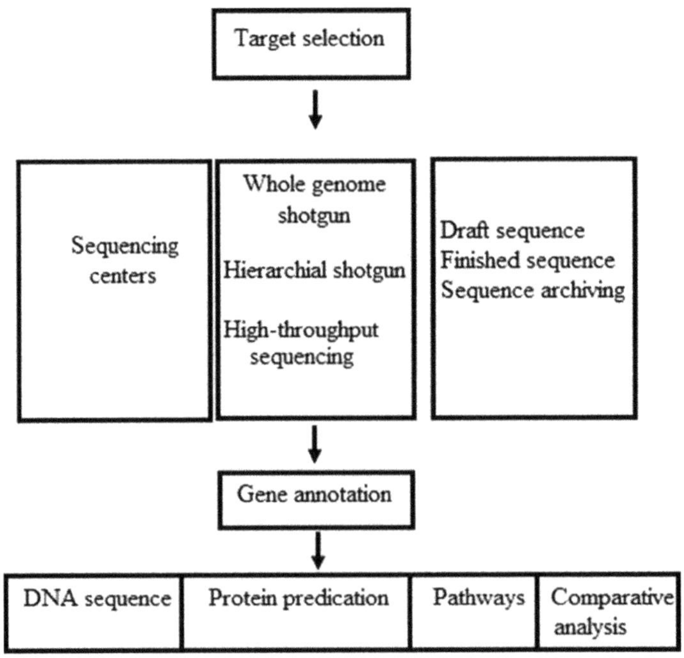

Figure 7.3. Bioinformatics and sequencing.

A model organism can be a prokaryotic or eukaryotic micro-organism, or an animal or plant, which is extensively investigated as it can be easily studied in the laboratory to understand biological processes. The Generic Model Organism Database project offers genetic researchers tools of open-source software machinery for imaging, interpreting, managing and procuring biological information. Model organisms include organisms such as *E. coli*, *Archaeoglobus fulgidus*, yeast (*Saccharorny cerevisiae*), *Bacillus subtilis, Arabidopsis thaliana,* the fruitfly (*Droso melanogaster*) and nematide worm (*Caenorhabditis elegans*). The Human Genome Project focused on sequencing the whole human genome.

E. coli is, by far, the most extensively studied micro-organism. Various tools that are available currently were established for the *E. coli* genome project. The sequencing of the *E. coli* genome was accomplished in 1997 [6]. The *E. coli* K-12 genome contains 4408 genes (4288 protein-coding genes annotated) with a size of around 4.64×10^6 bp. *B. subtislis* is a gram-positive bacterium that settles over leaf surfaces and is significant for both production of enzyme and food supply fermentation. 'GRAS' is an acronym for the phrase Generally Recognized as Safe. GRAS is well suited for enzymes, given the general availability of scientific data supporting enzyme safety, and the generally recognized (peer-reviewed) methodology and decision trees for evaluating the safety of microbial enzymes used in food processing and in animal feed, respectively. It is usually named as safe (GRAS) with genome of 4.21×10^6 bp (4212 genes). The reference database SubtiList was exclusively created for the genome of *Bacillus subtilis* 168, the paradigm of gram-positive endospore-forming bacteria [15]. Another example of a model organism is *Aracheoglobus fulgidus*, a strictly anaerobic archaebacterium with a genome size of 2.17 × 106 bp (2493 genes) [16] (1 563 423 bp, 1858 protein-coding, 52 RNA genes, according to the Genomic Encyclopedia of Bacteria and Archaea project) [16]. The yeast organism *Saccharomyces cerevisiae*, so called because of its ability to ferment saccharose (sugar), is the most key fungal species used in biotechnological developments. In 1996 *the S. cerevisiae* genome was the first completely sequenced from a eukaryote. The current version, called 'S288C 2010,' was determined from a single yeast colony by means of advance sequencing tools and functions as the anchor for further innovations in yeast genomic science [16]. In 1989, genome sequencing of *S. cerevisiae* was accomplished. The yeast genome is 12.8×10^6 bp in size and contains 6548 genes. PlantGDB (http://www.plantgdb.org/) is a databank of molecular sequence data for all plant species with important sequencing efforts. The database arranges EST sequences into contigs that represent tentative unique genes. PlantGDB offers genome browsing abilities that integrate all accessible EST and cDNA reported for existing gene models (for *Arabidopsis thaliana*, see the AtGDB site at http://www.plantgdb.org/AtGDB/) [15, 17].

The first plant for which the complete genome was sequenced and published was *Arabidopsis thaliana*. In 1907, it was first considered as a good organism for genetic investigations and has become a model plant for genetic investigations as it has the smallest genome of the higher plants. Furthermore, its genome contains only a small amount of repetitive DNA [18, 19]. The genomic sequencing project of the *A. thaliana* genome began in 1990 and was completed in 2000. According to the sequencing results, the size of the *A. thaliana* rabidopsis genome is 130×10^6 bp and a predicted 26 000 genes [18, 19]. The MIPS *Arabidopsis thaliana* Database (MAtDB; http://mips.gsf.de/proj/thal/db) started out as a repository for genome sequence data in the European Scientists Sequencing *Arabidopsis* (ESSA) project and the *Arabidopsis* Genome Initiative. Another model organism, *Caenorhabditis elegans* (a free-living nematode) was completely sequenced in 1999 and was the first multicellular animal whose genome was sequenced. The *C. elegans* genome has 97×10^6 bp and has over 20 000 estimated genes. WormBase (http://www.wormbase.org) is a web-based resource for the *C. elegans* genome and its biology. It is based on the existing ACeDB databank of the *C. elegans* genome and offers data curation services, and a considerably extended scope [20].

Drosophila melanogaster (the fruitfly) is often called the 'Queen of Genetics'. In 2000, the *Drosophila* genome was completely sequenced. It has 180×10^6 bp (16 000 genes). This breakthrough has offered a platform for further genomic research, as the number of genes present in the *Drosophila* genome is less than four times that of the bacterium *E. coli*. Since the human genome draft sequence was published in 2001, a number of observations have been made. Several significant features of the human genome are as follows:

- A minimum 50% of the genome is obtained from transposable elements.
- The genome sequences of individuals vary by less than 0.2% of their base pairs.
- It has approximately 35 000 genes.
- It encompasses over 3.2 billion base pairs.
- Only 5% of the genome encodes proteins.
- The genome contains gene-rich regions separated by gene poor regions, often called gene deserts.
- The largest gene is the gene encoding dystrophin (2.5×10^6 bp).

The majority of the variations in the human genome are present in the form of single-base differences in the sequence. A single-base difference is known as a single-nucleotide polymorphism (SNP), frequently called 'snips'. One SNP is present in roughly every 1000 bp of human genome. Approximately 85% of all variations in human DNA is due to SNPs. Advantages of genome projects include [21]:

- Better knowledge of human genetic diseases and their correlations should allow their management/cure.
- A number of methods have been established for genome sequencing projects.
- Information related to SNPs has become accessible; this may be beneficial in different ways.
- The complete genetic information existing in the genomes of various organisms can be determined.
- They have unlocked exciting opportunities for future research, e.g. functional genomics.
- They offer knowledge on why the individuals react differently to the same drugs (pharmacogenomics).
- They offer understanding of genome association and its evolution, and the most important mechanisms involved within that.

- Genome sequences allow the study of the many molecular interactions resulting in the normal development of organisms.
- Micro-organism pathogenicity can be better understood. This can offer protection from such diseases.
- The associations between genes can be studied with confidence.

7.6 Comparative genomics as a technique to understand evolution

The whole-genome sequence of an individual can be considered to be their final genetic map, in the sense that the heritable physical characteristics are encoded within the DNA and that the order of all the nucleotides along each chromosome is known. However, information on the DNA sequence does not tell us directly how these genetic data result in the observable traits and behaviors (phenotypes) that we want to understand. Identifying all the functional parts of genome sequences and using these data to improve the wellbeing of individuals and society are the focus of the next phase of the Human Genome Project [22]. Relative studies of genome sequences will be a major part of this effort. The investigation of variation and similarity in genomic structures and, in particular, their organization in various organisms is known as comparative genomics. The aims of comparative genomics are as follows:

- to recognize the route of evolution and
- to translate the DNA sequence data into proteins of known functions.

In order to equate the genomes of various organisms it is essential to understand the meaning of orthologues and paralogues. Orthologues are homologous genes present in different organisms and only encode proteins that have a similar function. Orthologues arise by direct vertical descent, and have deviated only by storing mutations. Comparatively, paralogues are homologous genes within similar organisms and always encode proteins that have interrelated but not similar functions. Paralogues have developed by gene duplication, followed by mutation accumulation, e.g. genes of the globin family.

7.6.1 The role of exon shuffling

Exon shuffling is a molecular tool for the development of new genes. It is a method by which two or more exons from diverse genes can be taken together ectopically (atypical gene expression in a cell, tissue, or at any developmental stage in which the gene is not typically expressed), or the same exon can be duplicated, to create a new exon intron structure. Exon shuffling has been considered as one of the main development powers determining both the genome and the proteome of eukaryotes. This is vital in the formation of multidomain proteins throughout animal evolution, obtaining a number of functional genetic novelties [23]. A simple presentation of exon shuffling is depicted in figure 7.4.

Various proteins contain distinct domains. Such proteins are known as mosaic proteins, e.g. serine proteases responsible for blood coagulation. The majority of

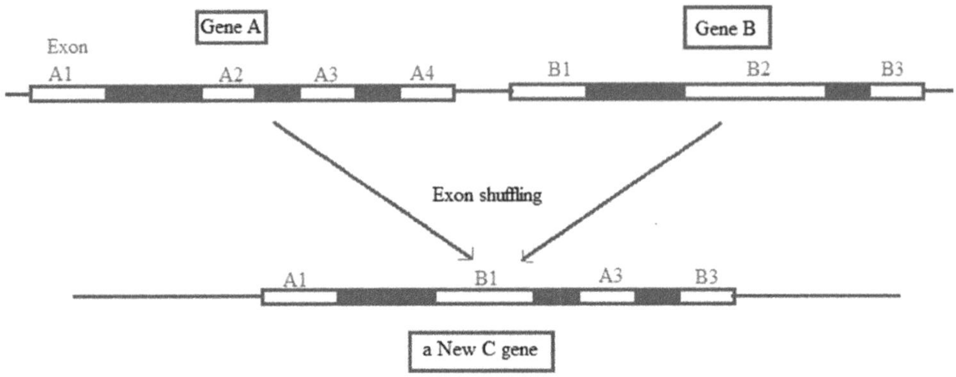

Figure 7.4. Exon shuffling. In this illustration the end gene is produced by shuffling exons from more than one gene.

mosaic proteins are extracellular and are mainly present in metazoa. In addition, these types of proteins are also present in unicellular organisms.

The study of the genes that encode mosaic proteins explores the strong association between domain establishment and intron–exon structure. Each domain is inclined to be encoded by one or a combination of exons, which are synthesized by recombination within the intervening sequences; this is known as exon shuffling. This process synthesizes new genes that encode proteins with improved function. As introns are much longer than exons, the chances of crossovers in introns are greater than in exons.

7.6.2 Horizontal or lateral gene transfer

Horizontal gene transfer was first reported in 1928, in an experiment by Frederick Griffith. During this experiment, Griffith demonstrated that virulence was able to transfer from virulent to nonvirulent strains of *Streptococcus pneumoniae*, establishing that genetic information can be horizontally transferred between bacteria via a mechanism called transformation [26]. A schematic representation of horizontal and vertical gene transfer is presented in figure 7.5.

Horizontal or lateral gene transfer involves genetic transfer/exchange between organisms of different evolutionary origin/lineages (sequences of cells, genes, organisms or populations, linked by a continuous line of origin from ancestor to descendent) [24]. It is usually assumed that such gene transfers/exchanges have happened many times throughout the course of evolution. The following two approaches are employed for the discovery of those genes that are likely to develop by horizontal gene transfer:

- Detection of genes having unusual base composition and bias in codon usage can also be used for the detection.
- Failure to find a similar gene in closely related species.

The accessibility of entire genome sequences has allowed the detection of such genes. It has been determined that hundreds (between 113 and 223) of human genes

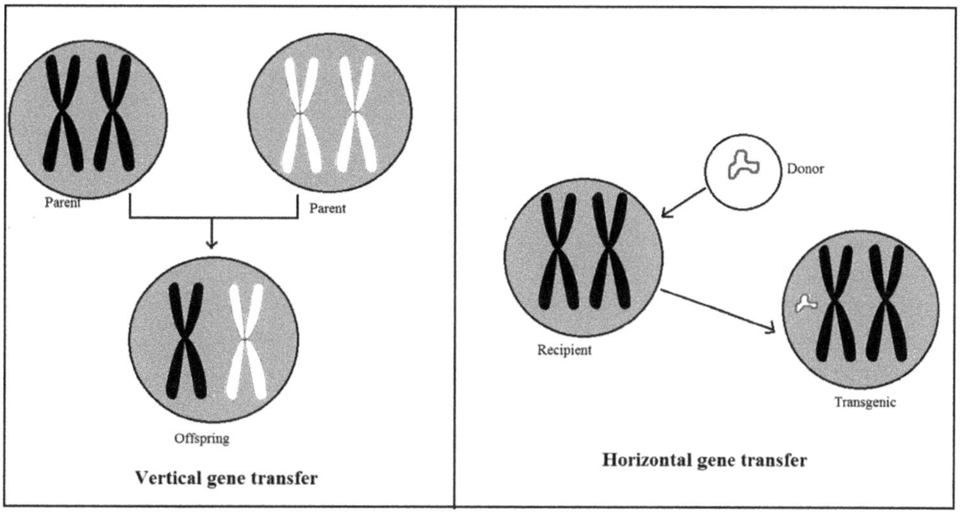

Figure 7.5. Horizontal and vertical gene transfer representation.

have been transferred directly from bacteria into the human genome rather than having developed from bacterial genes.

7.6.3 Genome similarity or homology

The sequence similarity searching tool is the most extensively used, and most reliable, approach for describing newly determined sequences. This tool can identify homologous proteins or genes by identifying a statistically significant similarity that reveals common ancestry [31].

Extensively used similarity searching databases are:

- BLAST [27].
- PSI-BLAST [27].
- SSEARCH [30].
- FASTA [28].
- HMMER3 [29].

These programs offer precise statistical evaluations, guaranteeing protein sequences that share considerable similarity also have similar structures. One of the amazing discoveries from the investigation of genome sequences of various organisms is as follows: organisms whose genomes are quite similar in appearance may look very different, for instance humans and mice.

These two organisms in fact have 97.5% of their DNA sequences in common that achieve the same genetic functions. It was also observed that humans and mice had a mutual ancestor some 100 million years ago. Since then, functional DNA has deviated only to a small extent, whereas the non-coding DNAs of the two species have deviated to a much greater extent. Likewise, it is estimated that human and chimpanzee genomes differ by only 1–3% of their DNA sequence. It may be noticed that the organisms above are more closely related to each other than organisms such as yeast (a fungus) and the nematode worm (an animal). The nematode worm is estimated to have 19 000 genes, of which over 2000 encode proteins that have functional equivalents as in yeast.

7.6.4 SNPs

For the last few decades, the utilization of molecular markers, highlighting polymorphism at the DNA level, has contributed a great deal in the development of animal genetics [32]. Among all the techniques, microsatellite DNA markers have been the most extensively utilized. A new marker tool, the SNP (see above) [32], is now in the picture and has recently gained high recognition. It is just a single base change in a DNA sequence, with a usual alternative of two likely nucleotides at a certain position [32].

SNPs, the most abundant type of genetic variation, are now the main raw material essential for most genetic investigations and databases. Variation comprising indels, copy number variants, microsatellites and epigenetic markers remain parameters to consider and can influence disease [33]. There are some reported 10 million SNPs in the human genome [34]. The basic concept and types of SNPs are depicted in figures 7.6 and 7.7, respectively.

As mentioned above, SNPs are the most common type of genetic variant among people. Each and every SNP signifies a variance in a single DNA building block, known as a nucleotide. For example, an SNP can substitute the nucleotide cytosine with the nucleotide thymine in a specific stretch of DNA. SNPs usually take place throughout a person's DNA. They occur once in every 300 nucleotides on average, i.e., there are approximately 10 million SNPs present in the human genome. As a

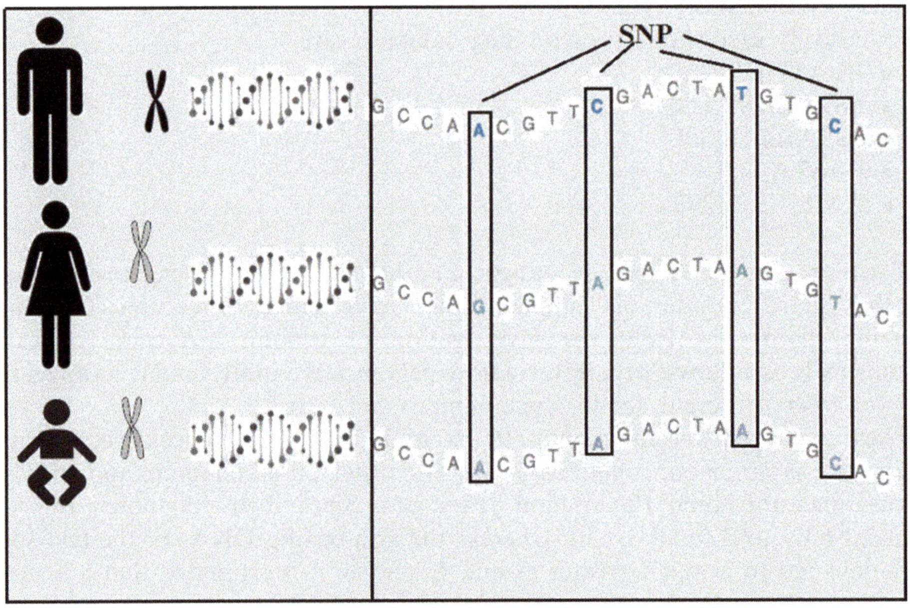

Figure 7.6. Representation of SNPs.

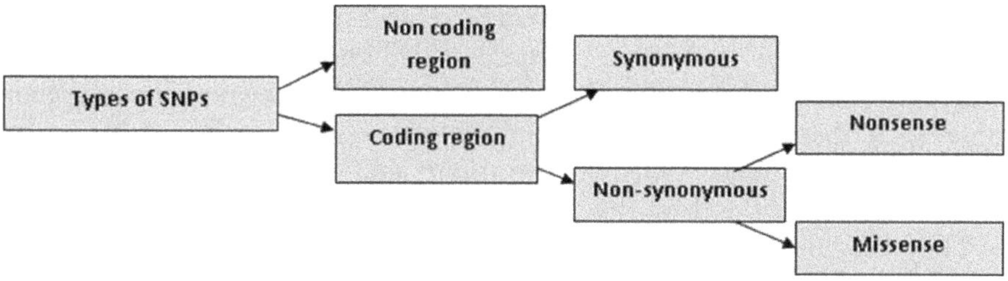

Figure 7.7. Types of SNPs.

rule, these deviations are present in the DNA between genes. They can behave as biological markers, facilitating researchers to locate genes that are related to disease. Once SNPs occur inside a gene or in a promoter region closer to a gene, they might show an additional direct role in disease by influencing the gene's function. The majority of SNPs have no influence on health or development. Certain genetic differences, however, have been shown to affect human health. Scientists have discovered SNPs that may help to predict an individual's reaction to certain drugs, vulnerability against external environmental factors such as toxins and the threat of developing specific illnesses. SNPs may also be utilized to track the causes of disease, in particular the inheritance of disease genes inside families. Upcoming investigations will work to recognize SNPs linked to complex diseases, e.g. heart disease, diabetes and cancer. Various academic institutions and industries are currently involved in the development of large-scale SNP detection projects, and these projects introduced data on thousands of publicly available SNPs. The most common example of these is the HGBASE database. It has been projected that 90% of sequence variation in humans is because of SNPs. Human genome is projected to contain 3–17 million SNPs. Out of these, 5% SNPs are likely to occur in genes. Thus each individual gene is likely to contain six SNPs. Therefore, SNPs offer a molecular marker that is present in the genome at a very high density. The SNPs can thus be utilized to map genes involved in human diseases. These genes are exceptionally difficult to map using other molecular markers as these latter markers are not distributed at an adequate density in the genome. Therefore, by using SNPs as markers, every gene in the human genome can be mapped. Additionally, it might be possible to identify all the genes involved in human diseases. An SNP might be present within the relevant gene or very near the gene which is tightly linked with the SNP. SNP1 and SNP2 are the two alleles of an SNP.

Some SNPs are situated in the recognition sequence of an endonuclease. They alter the recognition sequence (base sequence) and produce restriction fragment length polymorphism (RFLP). Some other SNPs are situated in the coding regions of genes. They are linked with definite phenotypes. The differences in the genetic sequences of individuals are understood to be responsible for the following:

- Disease susceptibility.
- Drug response.
- Normal development and aging.
- Reaction against environmental factors.

The field of study that is more concerned with the effect of genetic variation disease susceptibility and drug response is known as pharmacogenomics. Pharmacogenomics is the investigation of how genetic makeup regulates the response to a therapeutic intervention. Pharmacogenomics is likely to advance individualized treatments that will be safer and more effective [35]. Genetic variations may affect the drug reaction of an individual in the following ways:

- They may influence the metabolism of the drug itself.
- They may affect the action of the drug on its target molecule.

Pharmacogenomics aims to raise drug efficacy and safety by comparing drug properties to the genetic makeup of an individual. The response of the drug is expected to be influenced by various genes associated with drug metabolism, drug transport and creation of drug targets. Thus, various inputs have been made to employ SNPs to further map hundreds or thousands of genes that have an influence on the safety and efficiency of different drug treatments. These SNPs can be positioned on gene chips that can be employed to explore the genotype of those genes that are involved in responses to specific drugs. SNPs have the potential to describe the danger of an individual's susceptibility against numerous illnesses and reaction against various drugs. Biologically significant SNPs that are specifically related with the threat of disease need to be identified [35]. The identification and understanding of great numbers of these SNPs are essential before they can be used widely as genetic tools [35]. Depending upon these data, more effective and safe treatment can be considered for these individuals, such as variants in the gene encoding apolipoprotein E (apoE), which were found to be associated with variation in the reaction to a drug used for treatment. apoE supports lipid transport and binds to cell surface receptors to facilitate lipoprotein uptake. It is the main apolipoprotein produced in the brain [37]. The apoE protein is present in three different isoforms (E2, E3 and E4) that are the outcome of two non-synonymous SNPs (rs429358 and rs7412) present in exon 4 of the APOE gene. Gene apoE is involved in vulnerability to Alzheimer's disease, and a cholinesterase inhibitor is employed to mitigate the symptoms of illness [38].

7.6.5 Inferences from comparative genomics

In the case of prokaryotes, the following genomes and characteristics have been considered:

- *Bacillus megaterium* is important in the *Bacillus* phylogeny, as an evolutionarily important species and, in particular, in understanding genome evolution, dynamics and plasticity in the bacilli. *B. megaterium* is a nonpathogenic commercially existing host for the biotechnological production of numerous substances such as vitamin B12, penicillin acylase and amylases. The genome sizes of bacilli vary significantly, such as from 0.58×10^6 bp (*Mycoplasma genitalium*) to 30×10^6 bp (*Bacillus megaterium*) [39, 41]. The gene density is, however, comparatively constant at one gene per kilo base pairs.

- *Vibrio cholerae* O1 is reported as an important causative agent for infectious diseases [41]. A number of bacteria like *V. cholerae* have two or more circular chromosomes in their genomes. As a minimum, one of these chromosomes is obtained from a plasmid.
- In the case of larger genomes such as *E. coli*, almost 50% of the ORFs (open reading frames, the portion of a reading frame that has the prospect of being translated) are of unidentified biological function [42].
- Around 25% of all ORFs are exclusive and do not have substantial sequence resemblance to any other available protein sequence. Therefore, there are some new protein families yet to be revealed. The incidence of paralogues (pairs of genes that develop from similar ancestral genes) ranges from 26% for *M. genitalium* (the smallest self-replicating organism and an effective human pathogen related to a variety of genitourinary diseases) to 75% for *Pseudomonas aeruginosa*. The incidence of paralogues increases with genome mass [43].
- In close relation with bacteria, chromosomal transposals are more likely to arise nearby the origin or terminus of replication. Several bacterial genomes present in the database comprise phage DNA incorporated into the bacterial chromosome. It is not uncommon for bacteria to hold several prophages in their chromosomes, which then establish a substantial part of the total bacterial DNA. Several prophages and prophage remnants litter many genomes [44].
- Since severe complete genomes have been sequenced, gene order conservation between diverse organisms is evolving as a revealing feature of genome [45]. Gene order conservation has been employed for forecasting the function and functional interactions of proteins, as well as for investigating the developmental relationships between genomes [45]. The explanations for gene order maintenance are still not well understood, since the association of the prokaryote genome into operons and adjacent gene transfer cannot perhaps be considered for all the examples of conservation found [45]. In fact, there is very small preservation of gene order between phylogenetically distant genomes. A statistically important preservation of gene order is most likely indicative of the related genes being organized into operons. This characteristic has been employed to explore a number of new operons and to allocate functions to the genes based on the prediction of operon function.
- Genes that are out of use for an organism are often called pseudogenes. In general, inactivated gene copies are characterized by interference to their reading frames due to frameshifts and premature stop codons. Pseudogenes take place in prokaryotic genomes [46–49].

In eukaryotes, the following features of genomes have been observed:

- The detection of orthologous groups is beneficial for genome annotation, investigations on gene/protein evolution, comparative genomics and the determination of taxonomically restricted sequences. The procedures effectively utilized for prokaryotic genome examination have, however, proved difficult to apply to eukaryotes, since larger genomes can hold multiple paralogous genes and sequence information is often inadequate. A great number of the proteins encoded by the human genome have orthologues in other eukaryotic genomes. For example, 60% of human proteins have sequence resemblances to yeast, fly, nematode or plant proteins.

- Ancient transposon copies are found more in the human genome, whereas the genomes of *Arabidopsis*, *Drosophila* and *Caenorhabditis* have transposons of more recent origin [51, 52]. Bennett *et al* reported on comparisons between *Caenorhabditis* (~100 Mb) and *Drosophila* (~175 Mb) using flow cytometry. This work showed the genome size in *Arabidopsis* to be ~157 Mb and therefore ~25% larger than the *Arabidopsis* Genome Initiative Estimate of ~125 Mb [51, 52].
- Transcription factors control the expression of genes at the transcriptional level. *Arabidopsis* has not more than 20% transcription factors. These factors are zinc-coordinating proteins [53]. In contrast, in the fruitfly, yeast and nematode case, 51–64% of the transcription factors are of this type.
- *Arabidopsis* transcription factors are represented by several genes and by the assortment of gene families matched with those of *D. melanogaster* or *C. elegans*. Around 50% of the genes recognized so far have unknown function [53, 54].
- SINEs are retrotransposons that have appreciably significant reproductive success throughout the course of mammalian development, and have played an important role in shaping mammalian genomes [55]. Around 75% of the repetitive DNA in human genome is because of LINEs (long interspersed nuclear elements), these are sets of non-LTR retrotransposons [56] which are extensive in the genome of several eukaryotes and in SINE (short interspersed nuclear element) sequences. SINEs are sequences of non-coding DNA existing at high incidences in numerous eukaryotic genomes.
- As the genome size increases, gene density declines. It ranges from one gene per 2 kb in yeast to one gene per 10 kb in *Drosophila*, and one gene in 100 kb in humans.
- The beginning of the genomic era has unlocked the doors to the study of complete genome organization, not only in eukaryotes but also in prokaryotes, especially bacteria. Sufficient information on operon organization in *E. coli*, together with the completed chromosomal sequence of this bacterium, allowed the examination of the distances between genes and of functional interactions of adjacent genes in the same operon, as opposed to nearby genes in diverse transcription units. Bacterial genomes have distinguishing operons (a section of DNA that a repressor binds to); the *E. coli* genome has 600 operons [57].
- Approximately 15% of the ~20 000 *C. elegans* genes are present in operons, multigene clusters governed by a single promoter [58]. The *C. elegans* genome contains operons similar to those of bacteria, and they contain 25% of its genes.
- Eukaryotic genomes have relatively large amounts of repetitive DNA; this is a key reason for the variances in genome size.

- Transposable elements are found in the genomes of nearly all eukaryotes. The human genome has a much higher density of transposable elements (44.4% of the genome) than *A. thaliana* (10.5%), *D. melanogaster* (3.1%) or *C. elegans* (6.5%) [59].
- In plants such as maize and barley, genes are assembled in sections of DNA that can further be divided by long stretches of intergenic DNA. Earlier studies on the nuclear genomes of angiosperms showed that they are distinguished by a compositional compartmentalization and that the vast majority of genes derived from maize, rice, and barley are clustered in long DNA stretches that are separated by vast expanses of gene-empty DNA.
- Several human ailment networks are highly preserved in the fruit fly. Therefore, this insect can function as a model organism for the investigation of even complex human diseases, such as neurological disorders. *Drosophila melanogaster* is a well-studied and highly tractable genetic model organism to explore the molecular mechanisms of human diseases. Various biological, physiological and neurological characteristics are maintained between mammals and *D. melanogaster*, and approximately 75% of human disease-causing genes are assumed to have an efficient homolog in the fly.
- The incidence of sole genes (having no paralogues in the genome) is highest (>70%) in yeast and fruit flies, and the smallest in *Arabidopsis* (35%).
- Variances in intron and exon structure are recognized in eukaryotic genomes; the *Arabidopsis* genome varies considerably. It has been reported that there are three basic patterns of exon–intron variation in eukaryotic genomes, and these patterns were consistently present in almost all 13 genomes analyzed [60]. There is a better degree of 'alternative splicing' in humans than in the other three eukaryotes. This facilitates more proteins to be encoded by each gene.
- Various eukaryotic proteins are mosaic proteins, i.e., they are made of a number of different domains. Around 90% of the domains known in human proteins are present in *Drosophila* and *C. elegans* proteins. Therefore, vertebrate development has been central for the creation of the few new domains [61].
    1. Looking at both eukaryotes and prokaryotes, we have learned the following: certain bacterial species have more genes than lower eukaryotes.
    2. Certain genes are present in a wide variety of organisms. For instance, some of yeast genes have homologs among the human genes of unknown function.
    3. Unexpectedly, gene number does not essentially link with the complication of the organism or its evolutionary ladder.

Archaea is one of the three domains of life (alongside bacteria and eukaryotes). To date, 16 complete archaeal genomes have been sequenced, with a conserved core of 313 genes that are present in all sequenced archaeal genomes. The genomes of Archaea have similarities to eukaryotic genomes. For example, certain genes look like those of eukaryotes as they have histone genes and their DNA appears to be prearranged into chromatin [62].

7.6.6 Gene order comparisons (for phylogenetic inference)

Comprehensive information on gene maps or even complete nucleotide sequences for small genomes leads to the possibility of evolutionary interpretation based on the macrostructure of complete genomes. The mathematical modeling of development at the genomic level and the related inferential tools are, however, qualitatively different from the normal sequence comparison theory established to study evolution [63]. Equating gene order in dissimilar organisms is one of the techniques for developing molecular phylogeny. Once the gene instruction in a given region of the genome of two organisms is similar, they are called syntenic and the phenomenon is known as synteny. Gene order comparison in *S. cerevisiae* and *C. albicans* discloses numerous cases of inversions, around half of which are single gene inversions, in these two species separated by 140–330 million years [64, 65]. Comparatively, inversions in prokaryotes tend to include much larger regions of genomes and are focused on the origin and terminus of replication. Additionally, the degree of inversion is much higher in eukaryotes than in prokaryotes. Furthermore, the path of transcription appears to have little effect on gene location. A number of regions of the human genome are highly conserved in dogs, cattle and sheep. One of the main benefits of synteny is that information on gene location from a highly mapped organism can be employed to locate the analogous gene in a poorly mapped relative.

7.6.7 Phylogenetic footprinting (computational method)

Phylogenetic footprinting is a process for the detection of regulatory elements in a set of orthologous regulatory regions from various species [66, 67]. In other words, phylogenetic footprinting is a method employed to detect transcription factor binding sites (TFBS) inside a non-coding region of DNA of interest. This is done by equating it to orthologous sequences in different species. Scientists have discovered that non-coding pieces of DNA hold binding sites for regulatory proteins that direct the (spatiotemporal) expression of genes. A general representation of phylogenetic footprinting is shown in figure 7.8.

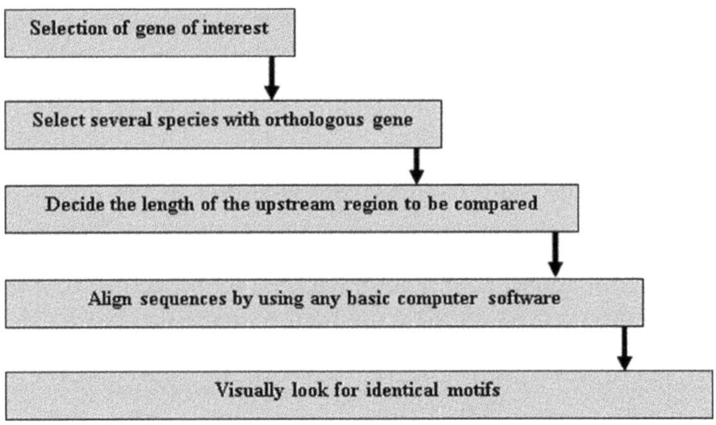

Figure 7.8. Phylogenetic footprinting.

These TFBS or regulatory motifs are difficult to detect, mainly because they are very small in length, and can display sequence variation. Two main concepts on which phylogenetic footprinting rely are:

- Transcription factor functions and their DNA binding preferences are well-conserved between diverse species.
- Significant non-coding DNA sequences that are important for directing gene expression will display differential selective pressure. A more gradual rate of change occurs in TFBS than in others.

7.6.8 Origins, evolution and phenotypic impact of new genes

The genetic code is shared, and the organization of the codons in the standard codon table is highly non-random [68]. The three main ideas on the origin and evolution of the code are the following:

- *Stereochemical theory*: In this theory codon assignments are dictated by physico-chemical affinity between amino acids and the cognate codons (anticodons);
- *Coevolution theory*: This postulates that the code structure coevolved with amino acid biosynthesis pathways.
- *Error minimization theory*: According to this theory selection to reduce the adverse effect of point mutations and translation errors was the principal factor of the code's evolution.

Thousands of genes present in eukaryotic genomes do not have complements in prokaryotes. Several 'new' genes are mostly regular prokaryotic genes that have been altered beyond recognition, such as domains involved in protein–protein interactions. Several 'new' eukaryotic domains are α-helical; these could have developed from the condensed coiled structures present in prokaryotes, which are now particularly abundant in eukaryotes. Certain other genes have arisen in completely unpredicted ways. For example, the hedgehog gene seems to have developed by gene fusion; during development an intein domain has fused with an extensively modified metalloprotease domain. However, the basis of several other genes remains unknown. A number of new genes might have risen from transposable elements. Among 13 799 human genes studied, 533 (4%) protein-coding regions comprised of transposable elements or sections of such elements were found. Such a transposable element could have incorporated into an exon. Otherwise, it may have initially first incorporated into an intron and then enrolled as an exon. This can become possible once the transposable element has possible splice sites. The incorporation of transposable elements into eukaryotic genomes could be one reason for the high incidence of alternative joining in human protein-coding genes. Therefore, incorporation of transposable elements into genes can produce new genes. For example, in their protein-coding regions, two mouse genes have one or numerous transposable elements, these genes have no human or rat orthologues.

7.6.9 The concept of minimum genome size

The 'minimal genome' method aims to estimate the smallest number of genetic elements adequate to create a modern-type free-living cellular organism. Or, in other words, the minimum genome size can be defined as the least number of genes essential to sustain life, and can at best only be predicted. A simplified scheme of the concept of minimum genome size is shown in figure 7.9.

For the following reasons, mycoplasma is considered a model organism for minimum genome size:

- It is a member of the mollicutes.
- It evolved by massive genome reduction.
- It lacks genomic redundancy.
- It is an obligate parasite.
- It has the smallest identified genome of any free-living organism capable of growing in axenic culture.
- It is wall-less.

Essential genes are considered important for its survival, e.g. genes that are encoding for protein that maintains the central metabolism, replicate DNA, translate genes into proteins, maintain a basic cellular structure, and mediate transport processes into and out of the cell. Some essential genes can tolerate mutations that are deleterious, but not wholly lethal, since they do not completely abolish the gene's function. Some of the essential genes, however, appear to perform non-essential functions, such as aging and cell death, while many of the non-essential genes play critical roles in cell survival. Essential genes can be difficult to study. If an essential gene is disrupted in the macronucleus, it is unlikely that complete replacement of the wild-type copies with the selectable marker will ever be achieved. There are a number of different transposon-based approaches available for defining essential genes. There are two ways to identify essential genes or regions of the bacterial chromosome which can be based on the location of the transposon insertion: the 'negative' and the 'positive' approaches. A problem in trying to identify essential genes is that a knockout of an essential gene is lethal. Therefore, the use of the negative approach will help with the identification of many regions that are not

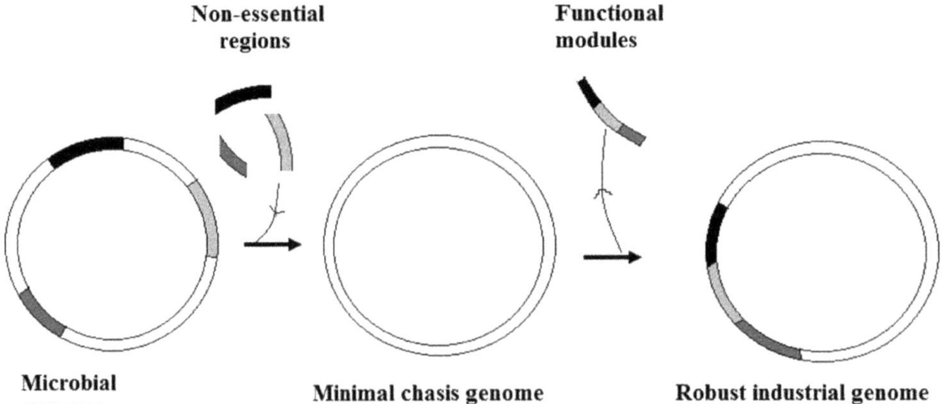

Figure 7.9. Concept of minimum genome size.

essential and enable one to say with some certainty that regions in which transposon insertions are not observed are likely to be essential.

Whole-genome sequences are now available for a great number of assorted species [70]. Gene content quantification, expansion of the gene family, orthologous gene conservation and gene displacement are promising for the assessment of the minimal set of proteins adequate for cellular life [70].

The genes found in different organisms with the smallest genomes and their roles are matched carefully. Moreover, investigational findings are collected by deactivating specific genes of these organisms and measuring the outcome on the organism's survival. Based on these investigations, it has been estimated that living creatures need at least 250–350 genes. This minimum gene number is essential for the organisms to exist as autonomous, self-replicating organisms. The unicellular eukaryote minimum genome size was found to be 2.9×10^6 bp (the parasite *Encephalitozoan cuniculi*). *E. cuniculi* infects different mammals, including humans, and can be a source of digestive and nervous clinical syndromes in HIV-infected or cyclosporine-treated people [71]. The genome of *E. cuniculi* has very little repetitive DNA other than that for rDNA, and is expected to contain approximately 2000 genes [71]. Therefore, transformation from prokaryotes to eukaryotes has augmented the gene number by a factor of 7–8, however, the genome size has not grown remarkably. The minimum gene number in multicellular eukaryotes, e.g. *Drosophila*, *Arabidopsis*, etc, appears to be 16 000; this significant increase would be essential for development and growth, and for sufficiently reacting to environmental or other external factors. Despite its toxicity, *Fugu rubripes*, the pufferfish is widely consumed in Japan and considered the most delicious of all fish. Interest in *F. rubripes* is not only restricted to taste and toxicity, however. The smallest genome size for a vertebrate is that for the *F. rubripes*; it has only 4×10^8 bp DNA, but this genome has 35 000 genes, which is similar to that of the human genome [72]. The pufferfish genome has tightly packed genes, lacks repetitive factors and has single short intergenic and intronic sequences. This drop in genome size is attained by decreasing the sequences not involved in genetic functions to the minimum; it is known as genome compaction. Otherwise, genome size may be decreased by gene loss; this has happened in several parasites.

7.6.10 Comparative genomics analysis of mitochondria and chloroplasts

Comparative genomics investigation has offered new understanding of the origin of organelles by endosymbiosis and exposed the vast evolutionary dynamics of organellar genomes [73]. Moreover, they have significantly assisted in elucidating phylogenetic relationships, particularly in algae and primary land plants with inadequate morphological and anatomical diversity [73]. The genomes of mitochondria are often circular but linear molecules also occur. A number of observations on mtDNA are as follows:

- Animal and fungal mitochondrial DNA (mtDNA) is very small (15–20 kb), compared to plant mtDNA (200–2000 kb) which is more complex than that of either animals or fungi [74].

- The greater genome size of plant mitochondria is mainly due to a greater amount of spacer DNA. For example, *Arabidopsis* mtDNA is almost 20 times larger than human mtDNA, but has less than twice the number of genes [74].
- The mtDNA of numerous species can be divided into two basic types: ancestral and derived DNA. Plant and animal mtDNAs are derived genomes. In the case of angiosperms, there has been extensive gene loss, however, the mtDNA has grown in size because of the duplication and capture of DNA from cpDNA and the nuclear genome.
- Mitochondria are thought to be derived from *Rickettsia prowazekii*, the agent responsible for epidemic, louse-borne typhus in humans [75]. The *Rickettsia* (alpha-proteobacteria) multiply in eukaryotic cells only. The organization, structure and gene content of this bacterium look exactly like the mtDNA of *Reclinomonas americana* [75].
- Seed plant mitochondrial genomes are exceptionally fluid in size, structure and sequence content, with the accumulation and activity of repetitive sequences underlying much of this variation. mtDNA genes have been transferred into the nucleus, as this transfer has stopped in animals, however, it continues to occur in the case of plants and protists [76]. For example, a gene from mitochondria, cox2, is still under the development of transfer in the case of legumes. In some legumes, cox2 is present in the mitochondria, whereas in other organisms it is present inside the nuclear genome, and in some it is present in both the nuclear and mitochondrial genomes [76].
- The three plant genomes gene types display differences in their evolutionary rates. Nuclear genes develop the fastest, followed by chloroplast genes and finally mitochondrial genes [77], in spite of the fact that the mitochondrial genome displays extensive reorganization of its structural organization [78]. This gradual rate of evolution has made mitochondrial gene sequences unattractive to plant phylogenetic studies at the suborder and subfamily levels, but has proven to be valuable in inferring ancient phylogenetic relationships [79] and in approximating the time of early diversification events in seed plants [80]. There has been cross-species acquisition of DNA or horizontal gene transfer of genes by plant mitochondrial genomes such as the homing group I intron, introns that encode site-specific endonucleases to further catalyze their effective distribution from introns including alleles to intron-lacking alleles of similar gene in genetic crosses. This intron is present in the cox gene of some angiosperm species [77].

7.7 Gene estimation and counting

Annonation is a procedure which detects genes, and explores their regulatory sequences and promising functions. Annonation designates the non-protein-coding genes (those that do not encode protein sequences), e.g. coding genes for rRNA, tRNA and nuclear RNAs, mobile genetic elements (DNA segments that encode enzymes and other proteins that control the DNA movement inside genomes, called intracellular mobility or, between bacterial cells, intercellular mobility) and repetitive sequence families existing in the genome. The role of annonation begins only after the genome sequence has been attained. Gene estimation is a significant issue for computational science. There are numerous logarithms that can be employed for gene estimation by utilizing identified genes as a training data set which can be further utilized to form a model or to test (or validate) the built model. Gene counting is problematic until the actual locations of genes in a genome are determined. The occurrence of overlapping genes and splice variations makes the job more challenging as it becomes more difficult to forecast which region of DNA should be considered as the same or as many different genes.

Numerous genes in eukaryotes have an arrangement of exons (coding regions) trailed by introns (non-coding regions). Because of this, genes are not systematized as continuous ORFs (open reading frames). An ORF has a series of codons that require an amino acid sequence.

Also, eukaryotic genes are often widely spaced, thus there is a greater probability of finding false genes. However, novel methods of ORF scanning software for eukaryotic genes allow more effective scanning. It has been observed that approximately 99.8% of the 3.2 billion base pairs of two humans are identical and merely 0.2% are different. Of every 500 nucleotides, only one nucleotide varies between two individuals. It means that a difference in a few sites in the DNA sequence can result in severe disorders and dissimilar features in human beings.

7.7.1 Genome similarity, SNPs and comparative genomics

As discussed above, approximately 99.8% of every human genome is identical to any other human genome and 0.2% different, which indicates that two individuals vary only in 6 million locations out of 3.2 billion locations (figures 7.10 and 7.11). Two human beings will be similar if these locations have little influence. As already discussed, the human genome is closely related to that of chimpanzees (more than 98%). Therefore, variance in some locations in DNA results in a unique individual. One nucleotide among every 500–1000 nucleotides always differs between two

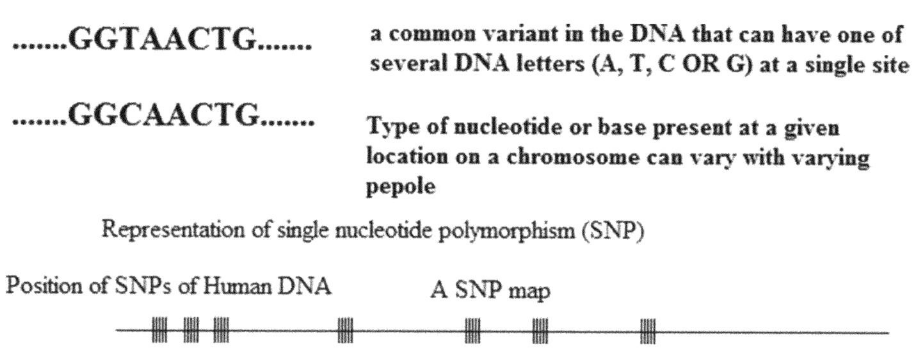

Figure 7.10. A SNP map.

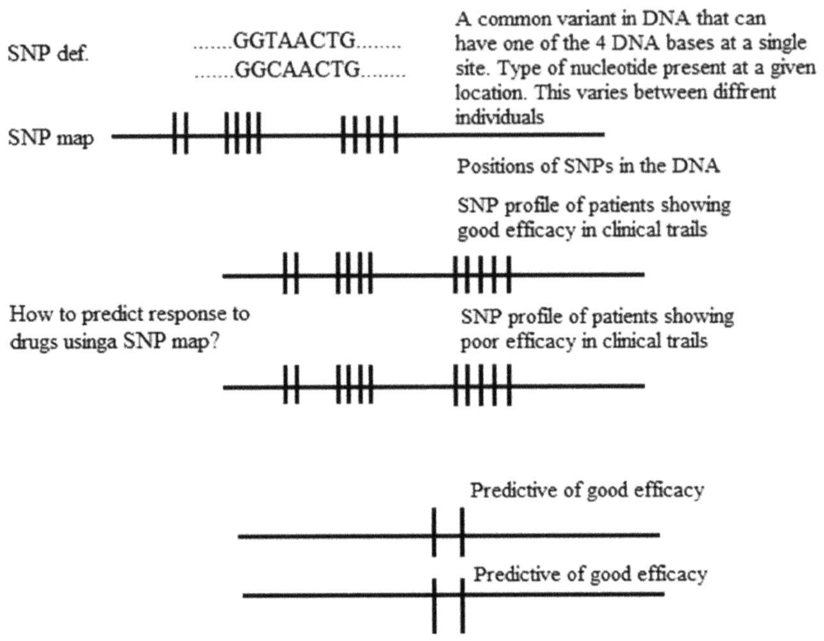

Figure 7.11. An SNP map to determine how patients are likely to respond to a particular drug.

individuals. The most considerable variations in individual genomes are SNPs, which have been detected in both coding and non-coding regions of genome. SNPs as discussed above are fundamentally varied in a single base, i.e., A, C, G or T resulting in DNA variations which further result in the presence of different bases at such locations. With varying individuals the type of nucleotide or base existing at a given position on a chromosome can also vary. It has been observed that 90% of sequence variation among humans is because of the presence of SNPs. Thus SNPs offer a molecular marker that is present in high density. During DNA finger-printing in mainly non-coding parts of the genome, such genetic deviations among diverse human beings are used. These genetic variations are also accountable for the severity of disease and the reaction of the body to treatment.

7.8 Genomes: genome evolution

Genome trees are a way to contain the astounding amount of phylogenetic data that is found in genomes. Diverse formalisms have been presented to construct genome trees on the basis of various features of the genome [83–85]. On the basis of these characteristics, we divide genome trees into five classes:

- alignment-free trees based on statistical properties of the genome;
- gene content trees based on the presence and absence of genes;
- phylogenomics-based genome trees;
- trees based on average sequence similarity; and
- trees based on chromosomal gene order.

Despite their current advancement, genome tree approaches have already had some influence on the phylogenetic arrangement of bacterial species. However, their main impact so far has been on our understanding of the nature of genome evolution and the role of horizontal gene transfer therein [83–85]. Relative genome analyses disclose that most functional domains of human genes have homologs in commonly divergent species. These shared functional domains, however, are differentially shuffled among evolutionary lineages to create an increasing number of domain architectures. Combined with duplication and adaptive evolution, domain shuffling is accountable for the great phenotypic complexity of higher eukaryotes [83–85]. Mobile elements within genomes have determined genome evolution by different means. Mainly in plants and mammals, retrotransposons have been shown to establish a large fraction of the genome and have shaped both genes and the entire genome. Although the host can frequently govern their numbers, massive expansions of retrotransposons have been accepted during evolution. Currently, mobile elements are becoming valuable tools for learning more about genome evolution and gene function [83–85].

Based on paleological investigation, it was demonstrated that approximately 3.5 billion years ago, cells identical to bacteria existed on our planet. Based on this observation we can only accept that the genomes of these organisms or the organisms that evolved soon afterward contain double-stranded DNA molecules. According to previous findings (dated to about 1.4 billion years ago) the first eukaryotic fossils were similar to single-celled algae, which means that both prokaryotes and eukaryotes experienced variations in size, shape and complexity, thus their genomes also transformed vigorously. These variations were motivated by mutation, transposition, gene transfer and recombination, along with gene deletion and duplication. Through investigating and equating genomes of organisms that exist currently, we can gain understandings into how these mechanisms have shaped genomes.

7.8.1 Microbial genome reduction in bacteria

Once bacterial pathogens make the evolution from free-living life cycles to a permanent relationship with a host, they experience a significant loss of genes and DNA [86]. Complete genome sequences offer information on how significant genome reduction affects the developmental directions and metabolic capabilities of obligate pathogens and symbionts [86]. Reconstruction of the gene content and order of the last common ancestor of human pathogens can be achieved, for example in the case of *Mycobacterium leprae* and *Mycobacterium tuberculosis* [87, 88]. An understanding into which genes are the most indispensable for existence comes from equating the genomes of two related species of *Mycobacterium*: *M. tuberculosis*, which is responsible for tuberculosis [87] and *M. leprae,* which is the causative agent of leprosy [88]. As per reports, *M. tuberculosis* has 3959 protein-coding genes and a 4.41 Mb genome. In contrast, *M. leprae* has 1604 protein-coding genes and a 3.26 Mb genome. Therefore, during development or evolution, *M. leprae* has misplaced or lost approximately 2000 genes (more than 50% of its ancestral gene set). Several reasons can account for this, such as codon elimination or deletion and deterioration, through mutation and incorporation of transposable elements, which have disabled many genes. Consequently, metabolic functions vital for cell development have been abolished, and the bacterium develops very slowly; it multiplies once every 14 days. Moreover, genetic mutations for recombination and DNA repair are incapable of averting additional loss to the *M. leprae* genome. This blend of mutations may have comdemmed *M. leprae* to extinction, as epidemiologists have projected that the organism is at the limit of being able to sustain itself by infecting new individuals.

7.8.2 Role of duplications in the origin and evolution of the eukaryotic genome

Assessments of prokaryotic and eukaryotic genomes are providing indications of the lineages of the eukaryotic genome and by what means eukaryotes arose from prokaryotes. Several lines of inquiry, comprising amino acid analysis of proteins, gene sequences and metabolic pathways, determined that the eukaryotic genomes are actually mosaics that have notable contributions from both the Archaea and the Eubacteria. For example, the eukaryotic nuclear genome has few, if any, operons, and its genes encompass introns, both of which are features of the Archaea [89].

The mitochondrial genome of a eukaryotic looks very much like that of alpha-proteobacteria. To explain these observations, it has been suggested that eukaryotes arose as a result of a symbiotic relationship, or fusion, between an anaerobic archaebacterial host and an alpha proteobacterium (such as *Rickettsia*), which developed into the mitochondrion [90]. The monophyletic character of all eukaryotes suggests that this incident occurred effectively only once in the history of the Earth.

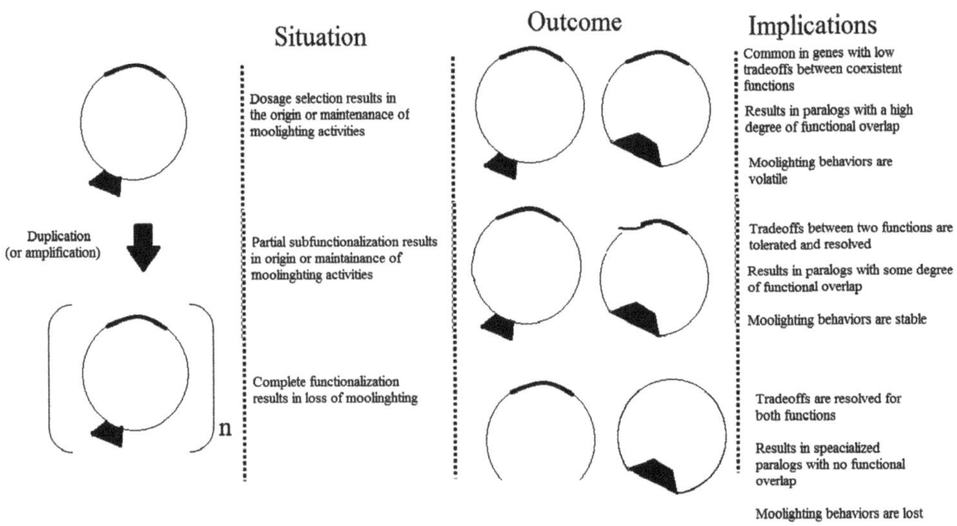

Figure 7.12. Role of duplications in the origin and evolution of the eukaryotic genome.

The duplication–divergence concept was suggested many decades ago [92, 93], i.e., that new genes evolve from pre-existing ones via gene duplication and subsequent divergence of the extra copy to acquire a new function (figure 7.12). Based on this concept Susumu Ohno wrote the book *Evolution by Gene Duplication* [94–102]. Since then, comparative genomics, genetics and biochemistry have clearly demonstrated that duplication–divergence mechanisms are a key contributor to the evolution of new genes [103]. In contrast, the *de novo* origination model of new gene evolution is more recent and less supported. *De novo* origination of protein-coding genes takes place when genes arise from a nonfunctional DNA sequence that was formerly not a gene. It would seem highly questionable whether functional proteins could arise spontaneously from non-coding DNA, as the DNA must both be transcriptionally active and comprise a translatable open reading frame. Also, translation of any random ORF devoid of genes is anticipated to offer unimportant polypeptides rather than proteins with specific functions. Certainly it has been claimed that *de novo* origination of new genes is tremendously unlikely, however, irrespective of these reports, the arrival of large-scale sequencing and comparative genomics has delivered increasing proof that new genes have developed and continue to originate from non-coding sequences [104–109].

Gene duplication plays a significant role in the development of eukaryotic genomes, particularly their size and complexity. Susumo Ohno has predicted that complete genome duplications are a powerful development mechanism [91]. Examination of nucleotide sequence information from genome projects encourages the impression that a considerable portion of the difference in gene number that differentiates prokaryotes from eukaryotes arose from genome expansions. Currently, it is understood that a significant expansion in eukaryote genome size arose from a genome duplication event that accompanied the arrival of vertebrates in the fossil record. Genome duplications have also occurred at other times during eukaryotic development. A current investigation of the yeast genome displays traces of its earliest development by genomic duplication. The yeast genome has been determined to hold at least 55 duplicated regions encompassing 376 genes, which covers 5% of the genome. Of these 55 regions, 50 are in a similar comparative location on diverse chromosomes. For example, chromosomes XI and XIII comprise duplicated blocks of genes in which gene order and positioning relative to the centromere has been well-maintained. Additional chromosomes cover internal duplications, such as the region positioned on either side of the centromere on chromosome XII. Phylogenetic examination designates that this genome duplication incident took place around 100 million years ago. Examination of the human genome also displays proof of an early large-scale duplication, tracked by reorganizations and gene damage in certain duplicated regions. These duplications range from minor sections covering only few genes to great stretches that cover nearly a complete chromosome. The larger duplications date to the origin of the vertebrates, around 500 million years ago. Overall, there are 1077 blocks of duplicated regions in the human genome, covering 10 000 genes (around one-third of the genome). Such duplication includes chromosomes 18 and 20. Different potential areas covered under proteomics are presented in figure 7.13.

7.8.3 Gene duplications increase genetic diversity and complexity

Genome duplication is both a historic and continuing process in yeast, plants and animals. It was recently established that the yeast genus *Saccharomyces* experienced an ancient whole-genome duplication [112–119], and genome duplication is pervasive in plants [112–119]. Numerous incidents of genome duplication have occurred during the divergence of angiosperms, including ancient polyploidization processes and the assembly of self-governing duplications in different ancestries [112–119]. Although not as common as in plants, many genome duplications have also happened in independent lineages in metazoans. During chordate development, whole-genome duplications have occurred coincident with the origin of vertebrates gnathostomes, and teleosts [112–119].

Gene and genome duplications offer a basis of genetic material for mutation, drift and selection to act upon, making new evolutionary changes more likely. Thus, several researchers have claimed that genome duplication is a dominant factor in the evolution of complexity and diversity. However, a strong connection between a genome duplication incident and increased complexity and diversity is not conclusive, and there are variations in the patterns of diversity raised to support this claim. Remarkably, many studies of genome duplication processes in vertebrates show they are preceeded by numerous extinct ancestries, resulting in preduplication gaps in extant taxa.

Genomic nucleotide sequence information indicates that multigene families are found in the human genome, if not all genomes. In addition to genome-wide duplications, small chunks of genes and single genes can be duplicated by numerous mechanisms, including:

- unequal crossing over and
- replication errors.

Throughout replication of a template molecule, a slippage can cause the introduction of a short segment into the freshly produced strand. When produced, participants of multigene families may remain associated on a single chromosome or can scatter to other parts of the genome. Numerous mechanisms determine this procedure, such as inversions, translocations and transposition by mobile elements.

The ancestry and relationships among members of gene families have been recently established by several groups of molecular phylogeneticists. Among the most studied examples is the globin gene superfamily. Some 800 million years ago, duplication of a family gene encoding an oxygen transport protein occurred in this family. This has formed two sister genes, one of which developed into the modern day myoglobin gene. Myoglobin is an oxygen-carrying muscular protein, i.e., it is found in muscles. Around 500 million years ago, the ancestral globin gene duplicated to produce the prototypes of the α- and β-globin subfamilies. The α- and β-globin genes encode for the proteins that are present in hemoglobin, the oxygen-carrying molecule in red blood cells. Further duplications within the α- and β-globin genes occurred in the last 200 million years. Subsequent events dispersed members of this superfamily, and each is now on a separate chromosome. Similar patterns of development are reported in other gene families, including the trypsin-chymotrypsin family of proteases, the homeotic selector genes of animals and the rhodopsin family of visual pigments.

7.9 Algae bioinformatics

An alga is considered a unique source of various bioactive compounds and having significant biological activities [120–136]. Algae bioinformatics, as the name suggests is the application of information technology to decipher more algae with the aid of computational tools and software. This emerging field, just like the other streams of bioinformatics employs computational or dry lab techniques for applications in various areas such as gene prediction, comparative genomics, genome analysis, and functional genomics and so on. Algae bioinformatics cannot be performed without computer science, biology and genetics with a good-sized dollop of mathematics, statistics and other medical specialties thrown into the mix. There are many tools available for algae bioinformatics research, which are briefly discussed at the very end of this article.

7.9.1 Scope of algae bioinformatics

Algae bioinformatics is a valuable resource for geneticists, phycologists and other scientists who sequence algal genomes as a part of their wet lab research. Therefore, there is an immense requirement for more reliable, sophisticated, computerized methods for examining this information bringing in the requirement of an algae bioinformaticist.

7.9.2 What is involved in algae bioinformatics

Algae bioinformatics involves the development of novel algorithms and statistics with which relationships among the different algal species can be measured. Examination and elucidation of different types of data and their related information is done. This data study involves DNA, RNA and protein sequences and structures. Development of various techniques allows the efficient access and management of different types of algal samples.

7.9.3 Role of algae bioinformatics

As discussed earlier, algae bioinformatics will only involve dry-lab research and it requires the use of RNA, DNA, protein sequence data. A series of wet-lab work is done so as to obtain the sequencing data and is subjected to bioinformatics analysis.

7.9.4 Steps involved in obtaining the data for analysis using bioinformatics

- Design of primers;
- Extraction of DNA, RNA;
- PCR amplification;
- Denaturizing gradient gel electrophoresis;
- Sequencing.

Nucleic acids are usually derived from the algal samples which are then utilized to synthesize and design primers and then PCR amplification is done. As the products of the PCR reactions are equivalent in their size, a denaturing gradient gel electrophoresis is done to identify variation between the sequences of the products. After this step algal bioinformatics work is performed. This can be achieved by making phylogenetic trees, BLAST searches for similar sequences, and annotation of the new sequences.

References

[1] Drake J W, Charlesworth B, Charlesworth D and Crow J F 1998 Rates of spontaneous mutation *Genetics* **148** 1667–86

[2] Kuska B and Roderick T H 1998 Beer, Bethesda, and biology: how 'genomics' came into being *J. Natl Cancer* **5** 93

[3] Lederberg J and McCray A T 2001 'Ome Sweet' omics—a geneological treasure of words *Scientist* **15** 8

[4] Watson J D and Cook-Deegan R M 1991 Origins of the Human Genome Project *FASEB J.* **5** 8-11

[5] Smith H O, Tomb J F, Dougherty B A, Fleischmann R D and Venter J C 1995 Frequency and distribution of DNA uptake signal sequences in the *Haemophilus influenzae* Rd genome *Science* **269** 538-40

[6] Blattner F R *et al* 1997 The complete genome sequence of *Escherichia coli* K-12 *Science* **277** 1453-62

[7] Heather J M and Chain B 2016 The sequence of sequencers: the history of sequencing DNA *Genomics* **107** 1–8

[8] Sterky F and Lundeberg J 2000 Sequence analysis of genes and genomes *J. Biotechnol.* **76** 1-31

[9] Weber J L and Myers E W 1997 Human whole-genome shotgun sequencing *Genome Res.* **7** 401-9

[10] Venter J C, Adams M D, Sutton G G, Kerlavage A R, Smith H O and Hunkapiller M 1998 Shotgun sequencing of the human genome *Science* **280** 1540-2

[11] Lewin H A, Larkin D M, Pontius J and O'Brien S J 2009 Every genome sequence needs a good map *Genome Res.* **19** 1925–28

[12] Stover C K *et al* 2000 Complete genome sequence of *Pseudomonas aeruginosa* PAO1, an opportunistic pathogen *Nature* **406** 959-64

[13] Jongeneel C V 2014 Bioinformatics in genome sequencing projects *Encyclopedia of Life Sciences* (New York: Wiley)

[14] Moszer I, Jones L M, Moreira S, Fabry C and Danchin A 2002 SubtiList: the reference database for the *Bacillus subtilis* genome *Nucleic Acids Res.* **30** 62–5

[15] Dong Q, Schlueter S D and Brendel V 2004 PlantGDB, plant genome database and analysis tools *Nucleic Acids Res.* **32** D354–9

[16] von Jan M *et al* 2010 Complete genome sequence of *Archaeoglobus profundus* type strain (AV18T) *Stand. Genomic Sci.* **2** 327–46

[17] Weigel D and Mott R 2009 The 1001 Genomes Project for *Arabidopsis thaliana Genome Biol.* **10** 107

[18] Goodman H M, Ecker J R and Dean C 1995 The genome of *Arabidopsis thaliana Proc. Natl Acad. Sci. USA* **92** 10831–5

[19] Schoof H *et al* 2002 MIPS *Arabidopsis thaliana* Database (MAtDB): an integrated biological knowledge resource based on the first complete plant genome *Nucleic Acids Res.* **30** 91–3
[20] Stein L, Sternberg P, Durbin R, Thierry-Mieg J and Spieth J 2001 WormBase: network access to the genome and biology of *Caenorhabditis elegans Nucleic Acids Res.* **29** 82–6
[21] Khromykha A and Solomon B D 2015 The benefits of whole-genome sequencing now and in the future *Mol. Syndromol.* **6** 108–9
[22] Collins F S, Green E D, Guttmacher A E and Guyer M S 2003 A vision for the future of genomics research *Nature* **422** 835–47
[23] França G S, Cancherini D V and de Souza S J 2012 Evolutionary history of exon shuffling *Genetica* **140** 249-57
[24] Ravenhall M, Škunca N, Lassalle F and Dessimoz C 2015 Inferring horizontal gene transfer *PLoS Comput. Biol.* **11** e1004095
[25] Hiramatsu K, Cui L, Kuroda M and Ito T 2001 The emergence and evolution of methicillin-resistant *Staphylococcus aureus Trends Microbiol.* **9** 486–93
[26] Griffith F 1928 The significance of Pneumococcal types *J. Hyg.* **27** 113–59
[27] Altschul S F, Madden T L, Schaffer A A, Zhang J, Zhang Z, Miller W and Lipman D J 1997 Gapped BLAST and PSI-BLAST: a new generation of protein database search programs *Nucleic Acids Res.* **25** 3389–402
[28] Pearson W R and Lipman D J 1988 Improved tools for biological sequence comparison *Proc. Natl Acad. Sci. USA* **85** 2444–8
[29] Johnson L S, Eddy S R and Portugaly E 2010 Hidden Markov model speed heuristic and iterative HMM search procedure *BMC Bioinform.* **11** 431
[30] Smith T F and Waterman M S 1981 Identification of common molecular subsequences *J. Mol. Biol.* **147** 195-7
[31] Pearson W R 2014 An introduction to sequence similarity ("homology") searching *Curr. Protoc. Bioinform.* June (chapter 3) unit 3.1
[32] Vignal A, Milan D, SanCristobal M and Eggen A 2002 A review on SNP and other types of molecular markers and their use in animal genetics *Genet. Sel. Evol.* **34** 275–305
[33] Johnson A D 2009 SNP bioinformatics: a comprehensive review of resources *Circ. Cardiovasc. Genet.* **2** 530–6
[34] Kruglyak L and Nickerson D A 2001 Variation is the spice of life *Nat. Genet.* **27** 234-6
[35] Alwi Z B 2005 The use of SNPs in pharmacogenomics studies *Malays. J. Med. Sci.* **12** 4–12
[36] Bekris L M *et al* 2008 Multiple SNPs within and surrounding the apolipoprotein E gene influence cerebrospinal fluid apolipoprotein E protein levels *J. Alzheimers Dis.* **13** 255–66
[37] Mahley R W, Weisgraber K H and Huang Y 2006 Apolipoprotein E4: a causative factor and therapeutic target in neuropathology, including Alzheimer's disease *Proc. Natl Acad. Sci. USA* **103** 5644
[38] Huang Y, Weisgraber K H, Mucke L and Mahley R W 2003 Apolipoprotein E: diversity of cellular origins, structural and biophysical properties, and effects in Alzheimer's disease *J. Mol. Neurosci.* **23** 189
[39] Kiss F M, Schmitz D, Zapp J, Dier T K, Volmer D A and Bernhardt R 2015 Comparison of CYP106A1 and CYP106A2 from *Bacillus megaterium* identification of a novel 11-oxidase activity *Appl. Microbiol. Biotechnol.* **99** 8495-514
[40] Eppinger M *et al* 2011 Genome sequences of the biotechnologically important *Bacillus megaterium* strains QM B1551 and DSM319 *J. Bacteriol.* **193** 4199–213

[41] Snow J 1849 *On The Mode Of Communication Of Cholera* (London: John Churchill)
[42] Kolisnychenko V, Plunkett G, Herring C D, Fehér T, Pósfai J, Blattner F R and Pósfai G 2002 Engineering a reduced *Escherichia coli Genome Genome Res.* **12** 640–7
[43] Saikolappan S, Sasindran S J, Yu H D, Baseman J B and Dhandayuthapani S 2009 The mycoplasma genitalium MG_454 gene product resists killing by organic hydroperoxides *J. Bacteriol.* **191** 6675–82
[44] Canchaya C, Proux C, Fournous G, Bruttin A and Brüssow H 2003 Prophage genomics *Microbiol. Mol. Biol. Rev.* **67** 238–76
[45] Tamames J 2001 Evolution of gene order conservation in prokaryotes *Genome Biol.* **2** research0020.1
[46] Liu Y, Harrison P M, Kunin V and Gerstein M 2004 Comprehensive analysis of pseudogenes in prokaryotes: widespread gene decay and failure of putative horizontally transferred genes *Genome Biol.* **5** R64
[47] Vanin E F 1985 Processed pseudogenes: characteristics and evolution *Annu. Rev. Genet.* **19** 253–72
[48] Mighell A J, Smith N R, Robinson P A and Markham A F 2000 Vertebrate pseudogenes *FEBS Lett.* **468** 109–14
[49] Harrison P M and Gerstein M 2002 Studying genomes through the aeons: protein families, pseudogenes and proteome evolution *J. Mol. Biol.* **318** 1155–74
[50] Li L, Stoeckert Jr C J and Roos D S 2003 OrthoMCL: identification of ortholog groups for eukaryotic genomes *Genome Res.* **13** 2178–89
[51] Bennett M D, Leitch I J, Price H J and Johnston J S 2003 Comparisons with *Caenorhabditis* (~100 Mb) and *Drosophila* (~175 Mb) using flow cytometry show genome size in *Arabidopsis* to be ~157 Mb and thus ~25 % larger than the *Arabidopsis* genome initiative estimate of ~125 Mb *Ann. Bot.* **91** 547–57
[52] Arabidopsis Genome Initiative 2000 Analysis of the genome sequence of the flowering plant *Arabidopsis thaliana Nature* **408** 796-815
[53] Mitsuda N and Ohme-Takagi M 2009 Functional analysis of transcription factors in *Arabidopsis Plant Cell Physiol.* **50** 1232–48
[54] Riechmann J L, Heard J, Martin G, Reuber L, Jiang C and Keddie J *et al* 2000 *Arabidopsis* transcription factors: genome-wide comparative analysis among eukaryotes *Science* **290** 2105–21
[55] Wang W and Kirkness E F 2005 Short interspersed elements (SINEs) are a major source of canine genomic diversity *Genome Res.* **15** 1798–1808
[56] Eickbush T H 1992 Transposing without ends: the non-LTR retrotransposable elements *New Biol.* **4** 430-40
[57] Salgado H, Gabriel M H, Smith T F and Collado-Vides J 2000 Operons in *Escherichia coli*: genomic analyses and predictions *Proc. Natl Acad. Sci. USA* **97** 6652–7
[58] Blumenthal T, Davis P and Garrido-Lecca A 2015 Operon and non-operon gene clusters in the *C. elegans* genome *WormBook* 1–20
[59] Kaminker J S *et al* 2002 The transposable elements of the *Drosophila melanogaster* euchromatin: a genomics perspective *Genome Biol.* **3** research0084.1–84.2
[60] Zhu L, Zhang Y, Zhang W, Yang S, Chen J-Q and Tian D 2009 Patterns of exon-intron architecture variation of genes in eukaryotic genomes *BMC Genomics* **10** 47

[61] Barakat A, Matassi G and Bernardi G 1998 Distribution of genes in the genome of *Arabidopsis thaliana* and its implications for the genome organization of plants *Proc. Natl Acad. Sci. USA* **95** 10044–9

[62] Makarova K S and Koonin E V 2003 Comparative genomics of Archaea: how much have we learned in six years, and what's next? *Genome Biol.* **4** 115

[63] Sankoff D, Leduc G, Antoine N, Paquin B, Lang B F and Cedergren R 1992 Gene order comparisons for phylogenetic inference: evolution of the mitochondrial genome *Proc. Natl Acad. Sci. USA* **89** 6575-9

[64] Berbee M L and Taylor J W 1993 *The Fungal Holomorph: Mitotic, Meiotic and Pleomorphic Speciation in Fungal Systematics* ed D R Reynolds and J W Taylor (Wallingford: CAB International), pp 67–78

[65] Pesole G, Lotti M, Alberghina L and Saccone C 1995 *Genetics* **141** 903–7

[66] Blanchette M and Tompa M 2002 Discovery of regulatory elements by a computational method for phylogenetic footprinting *Genome Res.* **12** 739–48

[67] McCue L A, Thompson W, Carmack C S, Ryan M P, Liu J S, Derbyshire V and Lawrence C E 2001 Phylogenetic footprinting of transcription factor binding sites in proteobacterial genomes *Nucleic Acids Res.* **29** 774–72

[68] Koonin E V and Novozhilov A S 2009 Origin and evolution of the genetic code: the universal enigma *IUBMB Life* **61** 99–111

[69] Itaya M 1995 An estimation of minimal genome size required for life *FEBS Lett.* **362** 257-60

[70] Mushegian A 1999 The minimal genome concept *Curr. Opin. Genet. Dev.* **9** 709-14

[71] Katinka M D *et al* 2001 Genome sequence and gene compaction of the eukaryote parasite *Encephalitozoon cuniculi Nature* **414** 450-3

[72] Taylor M S and Semple Colin A M 2002 Sushi gets serious: the draft genome sequence of the pufferfish *Fugu rubripes Genome Biol.* **3** reviews1025.1

[73] Bock R and Knoop V *Genomics of Chloroplasts and Mitochondria*vol 1 (Dordrecht: Springer)

[74] Lowe A, Harris S and Ashton P 2009 *Ecological Genetics: Design, Analysis, and Application* (New York: Wiley), p 10

[75] Andersson S G E *et al* 1998 The genome sequence of *Rickettsia prowazekii* and the origin of mitochondria *Nature* **396** 133-40

[76] Alverson A J, Zhuo S, Rice D W, Sloan D B and Palmer J D 2011 The mitochondrial genome of the legume *Vigna radiata* and the analysis of recombination across short mitochondrial repeats *PLoS One* **6** e16404

[77] Laroche J, Peng L, Maggia L and Bousquet J 1997 Molecular evolution of angiosperm mitochondrial introns and exons *Proc. Natl Acad. Sci. USA* **94** 5722–7

[78] Wolfe K H, Li W-H and Sharp P M 1987 *Proc. Natl Acad. Sci. USA* **84** 9054–8

[79] Palmer J D 1985 *Molecular Evolutionary Genetics* ed R J MacIntyre (New York: Plenum), pp 131–240

[80] Hiesel R, von Haesler A and Brennicke A 1994 *Proc. Natl Acad. Sci. USA* **91** 634–8

[81] Laroche J, Li P and Bousquet J 1995 *Mol. Biol. Evol.* **12** 1151–6

[82] Pertea M and Salzberg S L 2010 Between a chicken and a grape: estimating the number of human genes *Genome Biol.* **11** 206

[83] Babushok D V, Ostertag E M and Kazazian Jr H H 2007 Current topics in genome evolution: molecular mechanisms of new gene formation *Cell Mol. Life Sci.* **64** 542-54

[84] Cordaux R and Batzer M A 2009 The impact of retrotransposons on human genome evolution *Nat. Rev. Genet.* **10** 691–703

[85] Snel B, Huynen M A and Dutilh B E 2005 Genome trees and the nature of genome evolution *Annu. Rev. Microbiol.* **59** 191-209

[86] Moran N A 2002 Microbial minimalism: genome reduction in bacterial pathogens *Cell* **108** 583-6

[87] Rondini S *et al* 2007 Ongoing genome reduction in *Mycobacterium ulcerans Emerg. Infect. Dis.* **13** 1008–15

[88] Gómez-Valero L, Rocha EP, Latorre A and Silva F J 2007 Reconstructing the ancestor of *Mycobacterium leprae*: the dynamics of gene loss and genome reduction *Genome Res.* **17** 1178-85

[89] Levasseur A and Pontarotti P 2011 The role of duplications in the evolution of genomes highlights the need for evolutionary-based approaches in comparative genomics *Biol. Direct* **6** 11

[90] Michael W 2012 Gray mitochondrial evolution *Cold Spring Harb. Perspect. Biol.* **4** a011403

[91] Ohno S 1970 *Evolution by Gene Duplication* (Berlin: Springer)

[92] Haldane J B S 1932 *The Causes of Evolution* (London: Longmans)

[93] Fisher R 1935 The sheltering of lethals *Am. Nat.* **69** 446–55

[94] Jensen R 1976 Enzyme recruitment in evolution of new function *Annu. Rev. Microbiol.* **30** 409–25

[95] Piatigorsky J 1991 The recruitment of crystallins: new functions precede gene duplication *Science* **252** 1078–9

[96] Hughes A L 1994 The evolution of functionally novel proteins after gene duplication *Proc. R. Soc. London* B **256** 119–24

[97] Force A, Lynch M, Pickett F B, Amores A, Yan Y and Postlethwait J 1999 Preservation of duplicate genes by complementary, degenerative mutations *Genetics* **151** 1531–45

[98] Lynch M and Conery J S 2000 The evolutionary fate and consequences of duplicate genes *Science* **290** 1151–5

[99] Lynch M and Force A 2000 The probability of duplicate gene preservation by subfunctionalization *Genetics* **154** 459–73

[100] Kondrashov F A, Rogozin I B, Wolf Y I and Koonin E V 2002 Selection in the evolution of gene duplications *Genome Biol.* **3** 8

[101] Bergthorsson U, Andersson D I and Roth J R 2007 Ohno's dilemma: evolution of new genes under continuous selection *Proc. Natl Acad. Sci.* **104** 17004–9

[102] Näsvall J, Sun L, Roth J R and Andersson D I 2012 Real-time evolution of new genes by innovation, amplification, and divergence *Science* **338** 384–7

[103] Dittmar K and Liberles D A 2010 *Evolution after Gene Duplication* (Hoboken, NJ: Wiley)

[104] Cai J, Zhao R, Jiang H and Wang W 2008 *De novo* origination of a new protein-coding gene in *Saccharomyces cerevisiae Genetics* **179** 487–96

[105] Knowles D G and McLysaght A 2009 Recent *de novo* origin of human protein-coding genes *Genome Res.* **19** 1752–9

[106] Tautz D and Domazet-Lošo T 2011 The evolutionary origin of orphan genes *Nat. Rev. Genet.* **12** 692–702

[107] Wu D-D, Irwin D M and Zhang Y-P P 2011 *De novo* origin of human protein-coding genes *PLoS Genet* **7** e1002379

[108] Carvunis A R *et al* 2012 Proto-genes and *de novo* gene birth *Nature* **487** 370–4

[109] Wu D and Zhang Y 2013 Evolution and function of *de novo* originated genes *Mol. Phylogenet. Evol.* **67** 541–5
[110] Kellis M, Birren B W and Lander E S 2004 Proof and evolutionary analysis of ancient genome duplication in the yeast *Saccharomyces cerevisiae Nature* **428** 617–24
[111] Soltis D E and Soltis P S 1999 Polyploidy: recurrent formation and genome evolution *Trends Ecol. Evol.* **14** 348–52
[112] Ku H M, Vision T, Liu J P and Tanksley S D 2000 Comparing sequenced segments of the tomato and *Arabidopsis* genomes: large-scale duplication followed by selective gene loss creates a network of synteny *Proc. Natl Acad. Sci. USA* **97** 9121–6
[113] Vision T J, Brown D G and Tanksley S D 2000 The origins of genomic duplications in *Arabidopsis Science* **290** 2114–2117
[114] Bowers J E, Chapman B A, Rong J K and Paterson A H 2003 Unravelling angiosperm genome evolution by phylogenetic analysis of chromosomal duplication events *Nature* **422** 433–8
[115] Blanc G and Wolfe K H 2004 Widespread paleopolyploidy in model plant species inferred from age distributions of duplicate genes *Plant Cell* **16** 1667–78
[116] Holland P W, Garcia-Fernandez J, Williams N A and Sidow A 1994 Gene duplications and the origins of vertebrate development *Development* 125–33
[117] Meyer A and Schartl M 1999 Gene and genome duplications in vertebrates: the one-to-four (-to-eight in fish) rule and the evolution of novel gene functions *Curr. Opin. Cell Biol.* **11** 699–704
[118] Postlethwait J H, Woods I G, Ngo-Hazelett P, Yan Y L, Kelly P D, Chu F, Hill-Force H H A and Talbot W S 2000 Zebrafish comparative genomics and the origins of vertebrate chromosomes *Genome Res.* **10** 1890–902
[119] Prohaska S J and Stadler P F 2004 The duplication of the Hox gene clusters in teleost fishes *Theory Biosci.* **123** 89–110
[120] Bhatia S, Rathee P, Sharma K, Chaugule B B, Kar N and Bera T 2013 Immunomodulation effect of sulphated polysaccharide (porphyran) from *Porphyra vietnamensis Int. J. Biolog. Macromol.* **57** 50–6
[121] Bhatia S, Kumar V, Sharma K, Nagpal K and Bera T 2014 Significance of algal polymer in designing Amphotericin B nanoparticles *Sci. World J.* **2014** 1
[122] Bhatia S, Sharma K, Nagpal K and Bera T 2015 Investigation of the factors influencing the molecular weight of porphyran and its associated antifungal activity *Bioact Carbohydr. Diet. Fiber* **5** 153–68
[123] Bhatia S, Sharma K, Nagpal K, Sharma S and Bera T 2015 Evaluation of pharmacognostical, phytochemical and anti-microbial properties of *Porphyra vietnamensis Int. J. Green Pharm.* **2** 1–5
[124] Bhatia S, Sharma K and Bera T 2015 Structural characterization and pharmaceutical properties of porphyran *Asian J. Pharm.* **9** 1–3
[125] Bhatia S, Sharma K, Sharma A, Nagpal K and Bera T 2015 Anti-inflammatory, analgesic and antiulcer properties of *Porphyra vietnamensis Avicenna J. Phytomed.* **5** 1–7
[126] Bhatia S *et al* 2009 Novel algal polysaccharides from marine source: porphyran *Pharmacog. Rev.* **2** 271–6
[127] Bhatia S *et al* 2011 Mycosporine and mycosporine-like amino acids: A paramount tool against ultra violet irradiation *Pharmacogn Rev.* **5** 138–46

[128] Bhatia S *et al* 2010 Broad-spectrum sun-protective action of Porphyra-334 derived from *Porphyra vietnamensis. Pharmacog. Res.* **2** 45–9
[129] Bhatia S, Namdeo A G and Nanda S 2015 Factors affecting the gelling and emulsifying property of natural polymer *Sys. Pharm. Rev.* **9** 93–101
[130] Bhatia S, Sharma K, Sharma A, Namdeo A G and Chaugule B B 2011 Anti-oxidant potential of Indian Porphyra *Pharmacologyonline* **1** 248–57
[131] Bhatia S *et al* 2015 *Modern Applications of Plant Biotechnology in Pharmaceutical Sciences* (Amsterdam: Elsevier), pp 277–87
[132] Bhatia S 2016 *Nanotechnology in Drug Delivery: Fundamentals, Design, and Applications* (Boca Raton, FL: CRC Press), pp 121–7
[133] Bhatia S and Goli D 2016 *Leishmaniasis: Biology, Control and New Approaches for Its Treatment* (Boca Raton, FL: CRC Press), pp 177–9
[134] Bhatia S 2016 *Natural Polymer Drug Delivery Systems: Nanoparticles, Plants, and Algae* (Berlin: Springer), pp 117–27
[135] Bhatia S 2016 *Systems for Drug Delivery: Safety, Animal, and Microbial Polysaccharides* (Berlin: Springer), pp 122–7
[136] Bhatia S 2018 *Introduction to Pharmaceutical Biotechnology* Volume 1 (Bristol: IOP Publishing), pp 178–87

IOP Publishing

Introduction to Pharmaceutical Biotechnology, Volume 2

Enzymes, proteins and bioinformatics

Saurabh Bhatia

Chapter 8

Basics of proteomics

8.1 Introduction

In 1995, the word 'proteomics' was coined and was defined as the characterization of the large-scale complete protein complement of a cell line, tissue or organism [1–3]. The main objective of proteomics is not only to recognize all the proteins in a cell but also to make a three-dimensional map of the cell representing where proteins are located. The first protein studies that can be considered proteomics began in 1975, when O'Farrell [4], Klose [5] and Scheele [6] began mapping proteins from *E. coli*, mice and guinea pigs, respectively, using two-dimensional gel electrophoresis (2DE). Although several proteins could be separated and visualized, they could not be identified. Despite these restrictions, soon thereafter a large-scale investigation of all human proteins was planned. The aim of this scheme, called the Human Protein Index, was to use 2DE and other approaches to sequence all human proteins [7]. However, a lack of funding and technical limitations prevented this project from continuing.

The proteome is defined as the whole set of proteins expressed in a cell's entire lifetime. In brief, it can also be described as the set of proteins expressed in a cell at a particular time. Proteomics is the investigation of the proteome; it uses tools varying from genetic analysis to mass spectrometry. Among all the genomes sequenced to date, several recently discovered genes have no known role, and others have only apparent functions determined by similarity to known genes. For instance, in two intensively studied organisms (*E. coli* and *S. cerevisiae*), more than half of the genes encoded by their genomes have no known function. Such genes are frequently known as orphans or ORFans as they do not fit into any known gene family. The promising fields and basic procedures involved in proteomics are presented in figures 8.1–8.3.

doi:10.1088/978-0-7503-1302-5ch8

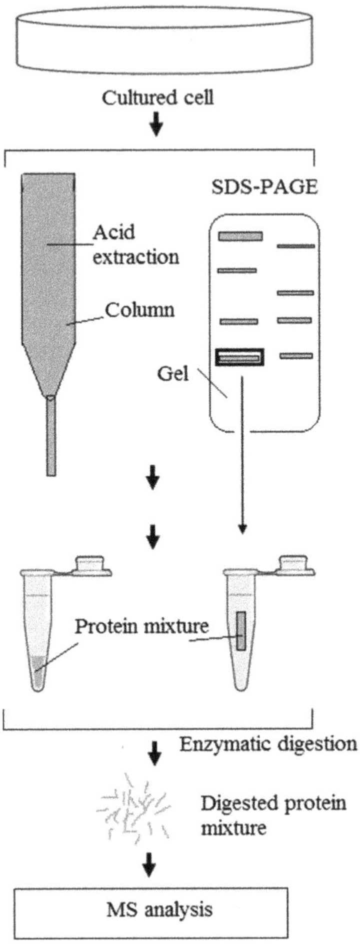

Figure 8.1. Basic protocol for proteomics.

8.2 Types of proteomics

8.2.1 Structural proteomics

In the postgenomic era, structural proteomics is one of the promising areas of investigation, and explains the structure–function relationships of uncharacterized gene products based on the three-dimensional protein structure (figure 8.4) [8]. It suggests the biochemical and cellular roles of unannotated proteins and thus classifies possible drug design and protein engineering targets [8]. Recently, several innovative groups in structural proteomics research have attained proof of structural proteomic theory by forecasting the three-dimensional structures of theoretical proteins that correctly recognized the biological functions of those proteins.

Structural proteomics can show the three-dimensional structure and nature of protein complexes present in a specific cell/organelle [8]. The final goal of structural proteomics is to form an organization of structural data that will assist in predicting

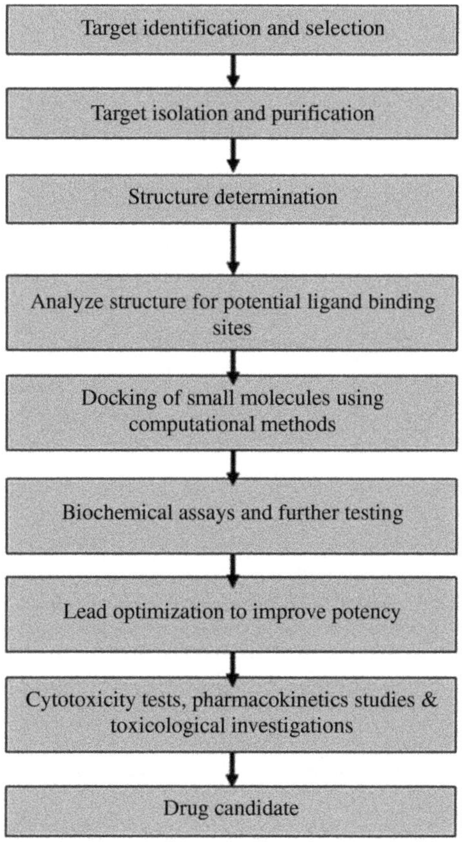

Figure 8.2. Basic procedure involved in proteomics.

the possible structure and possible function of virtually any protein from information of its coding sequence. This proteomics branch can also help in accumulating data about protein–protein interactions and about the construction of cells, to describe how the expression of certain proteins results in a cell's unique features. Recent mass spectrometry tools have offered a useful platform that can be combined with several procedures to examine protein structure and dynamics. Moreover nuclear magnetic resonance (NMR)-based structural proteomics coupled with x-ray crystallography can provide a comprehensive structural database to predict the basic biological functions of hypothetical proteins identified by genome projects.

8.2.2 Functional proteomics (strategy)

As the number of genome sequencing projects is increasing, there is a parallel exponential growth in the number of protein sequences whose function is still unidentified (figure 8.5) [9]. Functional proteomics is a developing research area in the field of proteomics whose methods are geared towards two main objectives: the

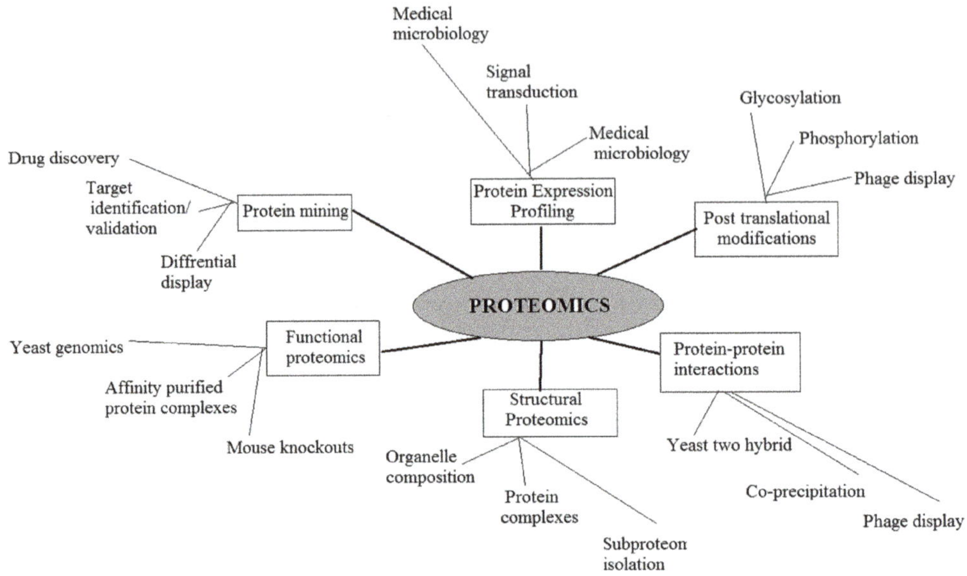

Figure 8.3. Different areas in proteomics.

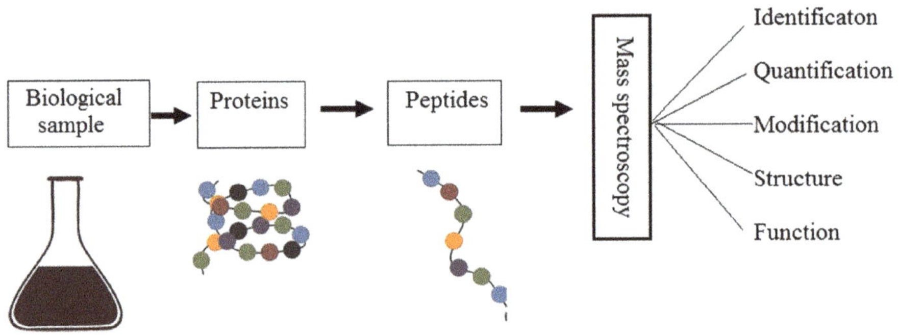

Figure 8.4. Basic procedure involved in structural proteomics.

interpretation of the biological function of unidentified proteins and the description of cellular mechanisms at the molecular level [10].

Functional proteomics uses proteomics tools to examine the features of the molecular protein networks involved in a living cell. Documentation and investigation of molecular protein networks involved in the nuclear pore complex in yeast is one of the recent accomplishments of functional proteomics. This accomplishment allows understanding of the translocation of molecules from the nucleus to the cytoplasm and *vice versa* [9].

8.2.3 Expression proteomics

Expression proteomics is the investigation of dysregulated proteins as a function of stimulation or condition (disease, time, drug, etc). In other words it is the

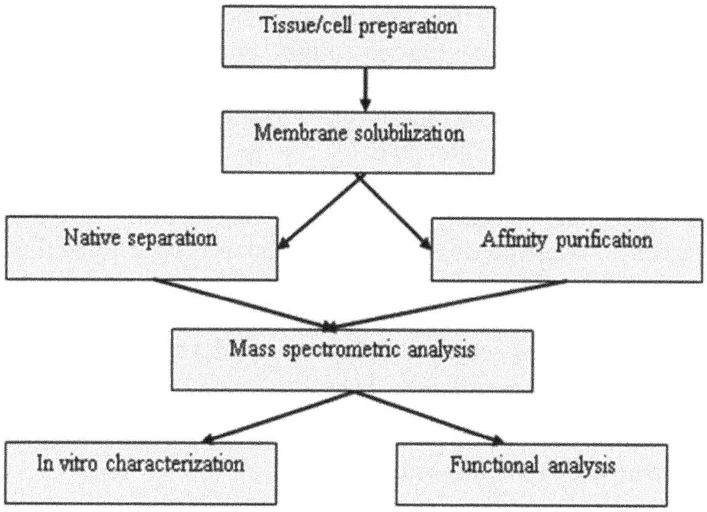

Figure 8.5. Functional proteomic analysis.

quantitative investigation of protein expression between samples differing in a number of variables. It can also be defined as the examination of protein expression at a larger scale. The arrangement of expression of the entire proteome or a portion of it (a subproteome) between samples can be ascertained with the help of this method. Expression proteomics is quite beneficial in finding disease-specific proteins. For example, over-expression or under-expression of proteins in cancerous cells and normal cells derived from a cancer patient and a normal individual, respectively, can be examined using several tools such as 2DE, mass spectrometry, microarrays, etc. This approach can recognize the growth of cancers and allows improvement of drugs in the management of cancer.

8.3 Basic techniques involved in proteomics

8.3.1 Sequence alignment (algorithms)

Fast emerging new sequencing tools offer information on an unparalleled scale. A principal challenge to the examination of these data is sequence alignment, wherein sequence reads must be matched to a reference [10]. An extensive variety of alignment algorithms and software have been developed over the past few years [10]. These techniques involve the study of the sequences of the DNA, RNA or protein to detect sites of resemblance that may be a result of functional, structural or evolutionary relationships between the sequences. Nucleotide or amino acid residues which are aligned sequences are characteristically presented as rows within a matrix. The residues gaps are incorporated between these, so that identical or similar features are associated in successive columns. The fastest way to allocate a new gene or cDNA is to perform pairwise assessment (a procedure of relating objects in pairs to judge which object is favored) with cDNA and EST sequences, and recognize gene sequences procured in databanks using alignment techniques such as BLAST (Basic Local Alignment Search Tool) or FASTA (a DNA and protein sequence alignment software package). However, this method is really only beneficial when there is >30% sequence uniqueness between the new gene and the databank sequence. If a similar gene/protein is found, it proposes the possible role of the product of new gene. However, additional experiments will be essential to determine the real *in vivo* function of the new gene product. To find resemblances below 30% sequence uniqueness, numerous sequence assessments are made by means of an algorithm such as PSI-BLAST (Position-Specific Iterated Blast). However, such a search may provide false positive findings. PSI-BLAST develops a position-specific scoring matrix or outline from the various sequence alignments detected above a given score threshold by protein–protein BLAST [11].

8.3.2 Protein structure (annotation resources)

An important reason three-dimensional protein structures are interpreted with supportive or derived data is to determine the molecular background of the protein function. In this endeavor, protein structure annotation databanks curate important evidence and explanations, based on community-accepted standards, for the ~100 000 three-dimensional investigational protein structures that allow further understanding of the structure–function relationship [12].

A method to control the function of an orphan gene is to inspect the three-dimensional structural arrangements of the protein, that the gene responsible for this protein is expected to encode; such a protein is called a theoretical protein. This three-dimensional structural arrangement is matched with the protein structures stored in databases. One of the bases for this method is the fact that the three-dimensional structures of proteins are better maintained than are their primary structures. For this reason protein function depends on the three-dimensional structure instead of the primary structure of the protein. It has been projected that 20–30% of orphan genes could be allocated function by defining the three-dimensional structures of the proteins encoded by them, and equating these to protein structures in databanks. For example, the three-dimensional structural configurations of hemoglobin and myoglobin are very comparable, considering that they both are oxygen carriers. However, a BLAST search will fail to detect resemblance between myoglobin and α- or β-globin (both are hemoglobin polypeptides). Thus it has been projected that if a theoretical protein has a similar structure to a known protein, there is a 66% chance that it has a role related to that of the known protein. The structure of a theoretical protein can be evaluated in the following three ways:

- If this target sequence displays >25% resemblance to a sequence whose structure is already known (called a template sequence), relative modeling can be employed to forecast the structure of the target protein sequences.
- When the uniqueness is <25%, the sequence can be matched to known structures to examine the level to which the experimental data match the values anticipated by theory.

- The target sequence structure can also be examined from the primary data without reference to known structures. Protein structures are experimentally evaluated using x-ray crystallography and NMR spectroscopy. Both of these methods are slow, time consuming and require greater quantities (milligrams) of the purified protein. However, current developments in both procedures will allow their alteration for high throughput analyses.

8.3.3 Protein structural investigation

Figure 8.6 provides a schematic representation of the process involved in protein structural investigation.

Structural investigation can often disclose the general function of a protein when sequence and structural assessments fail to suggest function. For example, protein surface scanning may disclose splits that may signify ligand-binding sites. A protein with a large split or cleft could be cautiously assigned the name 'enzyme'. A greater resolution in the description can be conceivable by comparing the shape of the split/cleft to a library of small molecular shapes using drug design software, e.g. DOCK and HOOK. These programs are capable of identifying possible ligands with binding sites present on protein surfaces. Scanning of the protein surface also facilitates identification of the domains most likely involved in interaction with other proteins. Knowing gene function involves more than recognizing the gene products. After they are produced, several gene products are altered by the breakdown of end groups (e.g. signal sequences, propeptides or initiator methionine residues), by the accumulation of chemical groups (such as methyl-, acetyl-, phosphoryl) or sometimes by adding linkages to sugars and lipids. In addition to the wide variety produced by the alternate splicing of mRNA, nearly a hundred mechanisms of post-translational modification are recognized. Therefore, the human genome, which may have 35 000–40 000 protein-coding genes, may offer over 350 000 different gene products. The main aim

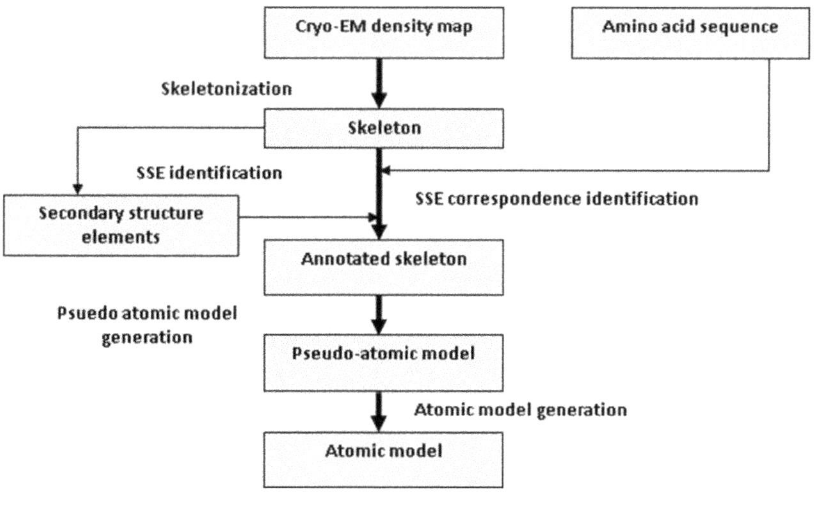

Figure 8.6. Procedure involved in protein structural investigation.

of proteomics is to provide data on each protein encoded in a genome, on its role, structure, cellular localization, post-translational modifications, associations (shared domains, evolutionary history) with other proteins and variants.

8.3.4 Two-dimensional gel electrophoresis in proteomics

Gel-based proteomics is one of the most multipurpose approaches for fractionating protein complexes [14]. Among these approaches, two-dimensional polyacrylamide gel electrophoresis (2DE) provides a key orthogonal method [14]. It is commonly used to simultaneously fractionate, detect and quantify proteins when coupled with mass spectrometric identification or other immunological tests [14].

The basic tools in proteomics include separation and identification of proteins that are isolated from cells. To achieve this, 2DE can be utilized. 2DE involves placement of the protein extract on a polyacrylamide gel after which an electric charge is applied across the gel to separate proteins based on their molecular weight (figure 8.7). Once this is complete, the gel is turned 90 degrees, and through a second phase of electrophoresis, the proteins are separated in a second dimension based on their molecular mass. Once the gels are stained, the proteins are exposed as spots; different gels display 200–10 000 spots. To detect different proteins, spots are cut from the gel and are allowed to digest with enzymes, e.g. trypsin, to offer a characteristic set of fragments. These fragments are further studied by mass spectrometry, which is known as peptide mass fingerprinting. To recognize the proteins, the peptide's mass is matched with the masses predicted from data in genetic or protein databanks.

8.3.5 Domain fusion method (or Rosetta stone method)

The examination of amino acid sequences from different individuals often provides results in which two or more proteins encoded for discretely in a genome also act as fusions, either in a similar genome or that of some other organism [15]. These fusion proteins, called Rosetta stone sequences, support the linkage of dissimilar proteins, and suggest the probability of functional interactions between the linked

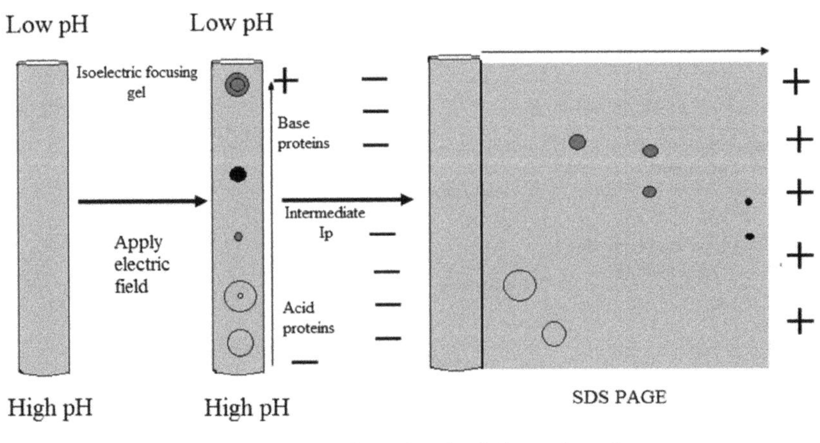

Figure 8.7. Two-dimensional gel electrophoresis.

objects, creating local and global relationships within the proteome [15]. The domain fusion method searches for functionally active proteins that are distinct in a number of organisms, but are fused into a single protein, a Rosetta stone protein, in some other organisms. The basic principle of the Rosetta stone technique is as follows. The protein A sequence from an organism is utilized to search for a homolog in a different organism. This examination can recognize a single domain protein such as A. However, in a number of organisms, it may recognize a Rosetta stone protein, such as A–B. As domains A and B are portions of a single protein, they are likely to be functionally related. Now the sequence of domain B is used to search for its homolog in the first organism. The homolog of the domain identified in this organism will be functionally related to protein A. Thus if the function of protein A were known, this strategy will reveal the function of protein/gene B, which was formerly an orphan gene.

8.4 Complete proteome of *Mycoplasma genitalium*

M. genitalium is the smallest member of the Mollicutes, has a genome size of 580 kb and the potential to express 480 gene products, and is thus an outstanding model to assess [16]:

- The minimum metabolism necessary for a free-living cell.
- Proteomic tools and the information derived by proteome analysis.

Wasinger and co-workers utilized proteomics to offer a portrait of what type of genes are expressed in the bacterium *M. genitalium* during exponential and stationary growth periods. *M. genitalium* is one of the simplest independent bacteria with a reduced genome. Using 2DE, the researchers explored 427 protein spots in exponentially growing cells. Of these, 201 were examined and recognized. The rest of the spots present in fragments derived from larger proteins, variants of similar proteins (isoforms) and post-translationally modified forms. The recognized proteins included enzymes involved in DNA replication, transcription, translation, delivery of materials across the cell membrane and energy metabolism. The study, however, could not cover 158 known proteins (33% of the proteome) and 17 unknown proteins. It was analyzed that there was a 42% reduction in the number of proteins synthesized during the transition from the exponential growth phase to the stationary phase. Moreover, a number of new proteins appeared, and additional proteins experienced intense variations because of nutrient exhaustion, enhanced acidity of the growth medium and other adaptations against environmental changes. In *M. genitalium,* a mere 33% of the proteome is expressed during maximal growth, and the remaining 67% of the proteome are the proteins to be expressed under different environmental conditions. Wasinger's examination of the *M. genitalium* proteome assisted in establishing the least number of expressed genes essential for independent existence and the modifications in gene expression that accompany the changeover from the exponential growth phase to the stationary phase. It was also noticed that the extensive range of information offered by proteome analysis cannot be derived only by genome sequencing.

8.5 Architecture and design of the nuclear pore complex

Three-dimensional examination of the nuclear pore complex discloses the fundamental, highly symmetric framework of this supramolecular assembly, how it is attached in the nuclear membrane and how it is constructed from many distinct, interconnected subunits [17]. The organization of the subunits within the membrane pore makes a large central channel, by which active nucleocytoplasmic transport is known to occur, and eight smaller peripheral channels that are probable routes for passive diffusion of ions and small molecules [17]. The nuclear pore complex (NPC) is made up of proteins embedded in the nuclear membrane. This complex joins the cytoplasmic and nuclear compartments of the cell and permits transport of materials between the nucleus and the cytoplasm. As this transportation contains mRNA intended for the cytoplasm, nuclear pores are vital control points for regulating gene expression. Their structural and functional sophistication has been an obstacle in understanding their molecular organization. Genomics in association with proteomics has offered scientists techniques to investigate the three-dimensional construction of nuclear pores and to recognize their proteins. Depending on its size, the yeast nuclear pore complex contains around 200 different proteins. By using a rat genome databank as a resource, scientists were able to recognize potential nuclear pore complex proteins, also called nucleoporins, by means of sequence resemblance to already identified nuclear pore complex proteins from other organisms. In another example, proteomics was employed to outline the molecular construction of a nuclear pore complex. Accordingly, pore complexes were separated and purified, and then nucleoporins were detected by means of mass spectrometry on peptide digests of proteins purified from the NPCs. Yeast's NPC only includes around 30 different proteins, but with each protein present in multiple copies. The nucleoporins are prearranged into 16 subunits (eight on the pore's nuclear side and eight on its cytoplasmic side). Moreover, there are filament-base structures on both sides of the pore, formed into a bag on the nuclear side. Using multiple tools such as mutant analysis, microscopy and proteomics, the actual regions/sites of NPC proteins in the three-dimensional arrangement of the pore have been mapped. The majority of nucleoporins are proportionally distributed on the cytoplasmic and nuclear sides of the pore. Five proteins are present only on one side or the other, and seven are distributed more on one side than on the other. The function of nucleoporins in nucleocytoplasmic transportation or the regulation of transport can be studied by examining protein–protein interactions within the pore, as well as interactions of pore proteins with transport proteins, signal molecules and other cellular components.

8.6 Functional genomics and systems biology

Functional genomics is an efficient method for determining the roles of the novel genes revealed by complete genome sequences (figure 8.8) [18]. Such a method should take a hierarchical approach, as this will both bound the number of trials to be done and allow a closer and closer estimate of the function of any individual gene to be attained [18]. Furthermore, hierarchical studies have, in their early stages, remarkable

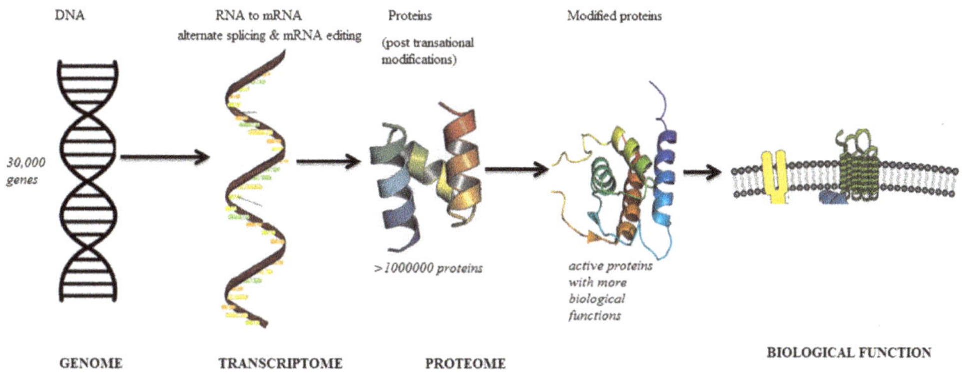

Figure 8.8. Schematic representation of functional genomics.

integrative power and functional genomics aims at a comprehensive and integrative view of the workings of living cells [18]. Functional genomics is an area where the function of a gene product is determined. This includes answers to the questions:

- How are genes expressed?
- How are the gene sequence and structure related to the end product, and its relationship with gene versus an individual from the same generation?
- Exactly how is a product associated with a sequence and structure, and how is it associated with the end products of other genes of similar organisms?
- What are the implications of the environment over a gene and how does this interaction help in understanding the rate of genetic variation? Gene–environment interaction helps in understanding genetic influences modeled as latent [40].

These questions can be answered by studying the following:

- Determining at what time and exactly at which place specific genes are expressed on a genomic scale (called expression profiling). This permits the identification of those particular genes that are over-expressed or under-expressed. This concept can be used to understand different biological processes in health and disease. This concept can also be utilized to study the expression in a particular tissue/organ at the molecular level. This type of gene-based profiling, and more specifically recent RNA sequence based profiling, can be used as an important tool for drawing correlations between gene activity and various physiological or developmental states. It also helps in creating collections of gene expression data across various cell types, development times, varied species and against different stimuli [39].
- The replacement of a particular natural gene with mutated gene under *in vitro* conditions to access its *in vivo* function (the knockout method) [41].
- The genetic relationship between proteins and other molecules (as first suggested by Archibald Garrod in 1902, who predicted that genes control the function of proteins during his study on patients suffering from alkaptonuria).

Functional genomics aims towards answering such questions thoroughly for most of the genes present in a genome, in comparison to classical strategies that report one gene at a time. Among potential branches of functional genomics, the field of transcriptomics helps in predicting and understanding the exact functions of genes and their respective products. High-throughput screening of gene transcripts in biological systems, through the development of the necessary bioinformatics analytical tools, has allowed scale-up of experiments classically performed with single genes to identify genetic variation at large scales. Functional genomics can be theoretically further categorized into gene-driven (dependent on genomic data to identify, clone and express genes in a genome) and phenotype-driven approaches (dependent on phenotypes from random mutation screens or naturally occurring variants to recognize and characterize genes for the phenotype, without having any previous information of the basic molecular mechanism or function). The most important aspect of functional genomics is that it explores the functional arrangement of all genomes, in particualr parts that are external to any coding gene sequences which play an important role in gene expression. The concept of the epigenome covers all the molecules and proteins which can affect DNA function, usually by turning genes on or off. Studies related with this field usually cover factors responsible for controlling development and differentiation against external changes. During this process epigenetic tags are made in the form of DNA methylation and covalent modifications of histones to determine the arrangements of cell- and tissue-specific gene expression, as described in figure 8.9. Epigenetic readers also help in determining gene expression. In certain cases, small RNA molecules are used to guarantee sequence specificity.

Overall, these studies of epigenetic modifications explain the forms of epigenetic 'flavors' (i.e., the existence of chromatin in different states). This means the presence of chromatin in heterochromatin (inactive, repressed) and euchromatin (potentially active), decides the tags and reads which may further decide flavors. In addition functional genomic studies also cover the complex organization of RNA molecules in the genome, i.e., well characterized coding RNA and non-coding RNAs. Functional genomic investigations include highly complex genome and proteome information which requires the development of potential computational and hardware tools. This discipline gave birth to an independent discipline called bioinformatics. Recent developments in bioinformatics have allowed collaboration between different disciplines, resulting in further new disciplines ending with the suffix '-omics', e.g. proteomics, genomics and metabolomics. All these disciplines have the common objective of computational analysis of sophisticated and vast information at each and every biological level; starting from the molecular level, i.e., genes and molecules, to cells, tissues and organs, and finally to the study of whole-body systems.

8.6.1 Gene expression profiling

As mentioned above, functional genomics is defined as the branch of genomics that predicts the functions of genes and gene products by high-throughput screening of

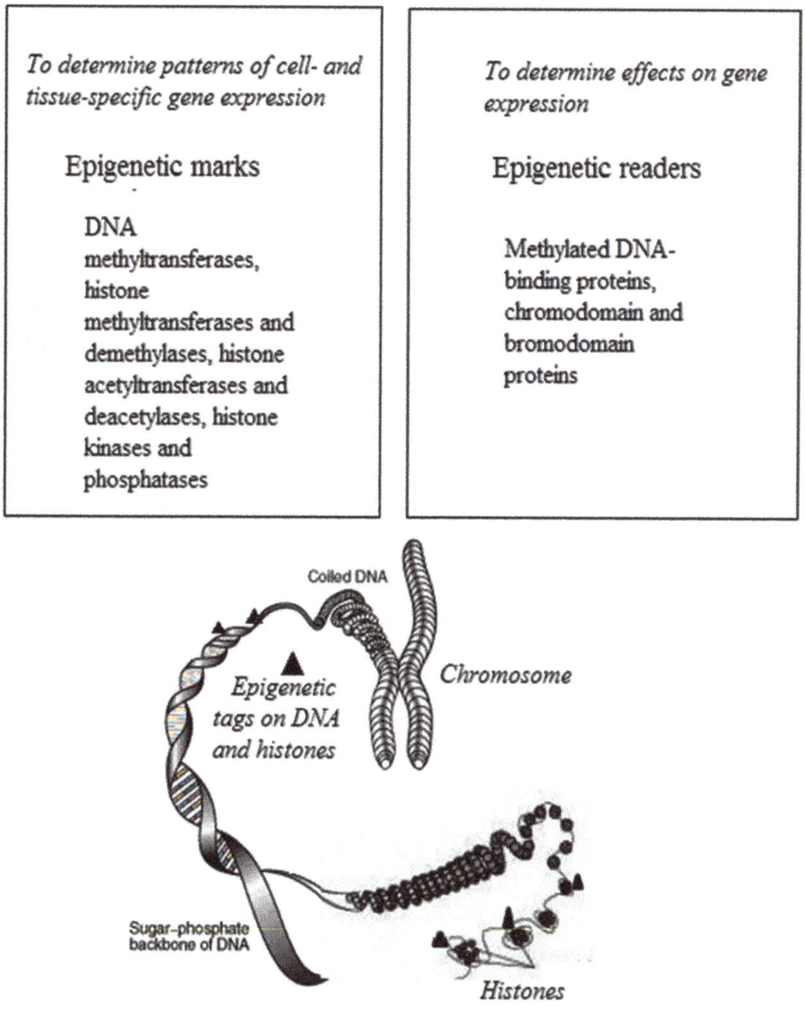

Figure 8.9. Epigenetic tags and reads.

gene transcripts in a biological system. Functional genomics has numerous applications in clinical medicine [19]. Functional genomics is a broad approach that allows the concurrent examination of mRNA levels for the complete human transcriptome from as few as 1000 cells. Currently, functional genomics is employed to categorize the development of disease and survival in response to traumatic and burn injuries, sepsis and visceral ischemia, and reperfusion injury, as well as to designate patterns of gene expression in reaction to erratic microbial pathogens. Due to the number of new bioinformatics techniques emerging, functional genomics is providing a foundation to reveal the fundamental complexity of the biological response to a range of inflammatory diseases and is offering new methods for their exploration [19]. Functional genomics has now been developed as a standard tool in inflammation research to undo basic biological processes.

Expression profiling is a discipline that deals with the extent of expression of multiple genes at the same time, i.e., it attempts to determine the complete picture of cellular function at a molecular level. This can be achieved using certain tools, e.g. DNA microarrays, which help in measuring the relative expression of multiple genes and their correlation with already identified target genes. The microarray is an advanced tool to determine genetic expression by evaluating the actual amount of mRNA present in the sample. In addition to exploring the role of existing genes and their relevant association, microarrays also include genes with currently unknown functions. This can provide the prospect of new gene discovery, which in particular plays an important role in functional genomics, e.g. microarray based analysis of previously unknown genes selectively expressed in T-helper 2 type lymphocytes can offer new targets for asthma [42]. In addition, by using comprehensive arrays it is now possible to carry out deletion mapping to recognize selective areas of chromosomes whose genes are not present in a large microarray.

Functional genomics covers global expression profiling to study genes under epigenetic and transcriptional regulation to further study several forms of coding and non-coding RNA molecules. It is a branch that deals with the simultaneous examination of the expression pattern, usually of all genes present in the genome at the RNA level or at the protein level. In contrast to global expression profiling, traditional approaches are more focused on a single gene, which can restrict access to these complex interactions. Systematic investigation of gene expression profiling helps in interpreting the transcriptomes of consecutive developmental stages. This information is required to understand the developmental mechanism which can shed light on conservation and diversification at the molecular level. Thus functional genomics is directed towards studying global expression profiling at the RNA level (by nucleic acid arrays or direct sequence analysis) or at the protein level (by 2DE followed by mass spectrometry or protein arrays).

Global expression profiling is a valuable approach which can help in the identification of those particular genes that play an important role in development, and also help in understanding the complex behavior of a gene in a particular environment against different levels of stress or any other stimuli which can cause the onset of a disease, etc. Moreover, it also offers a unique method for characterizing cellular phenotypes and exploring novel drug targets which can help in developing potential drugs.

8.6.2 Transcriptome, proteome and genomes

The transcriptome is the initial product of genome expression. Basically, it is an assortment of RNA molecules obtained from those protein-coding genes whose biological information is essential for the cell at a specific time [20]. These RNA molecules further direct the production of the final product of genome expression, the proteome, the cell's repertoire of proteins, which stipulates the nature of the biochemical reactions that the cell is able to carry out [20]. The transcriptome is created by the process known as transcription, in which individual genes are copied into RNA molecules.

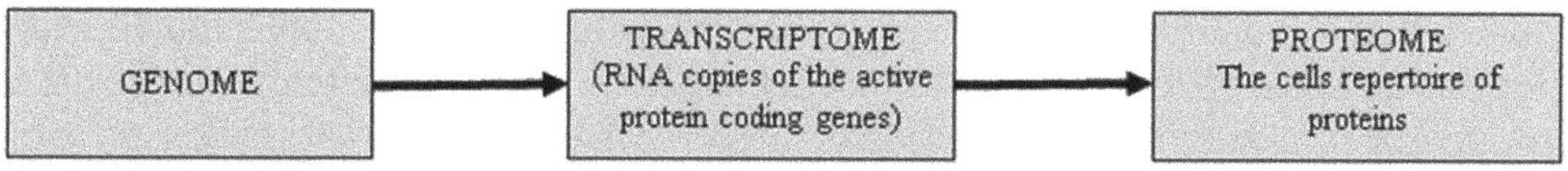

Figure 8.10. General overview of transcriptomes.

Transcriptomic studies are part of integrative genomics, which includes the set of all RNA molecules from protein-coding (mRNA) to non-coding RNA, of a complete organism or a particular cell. As the name suggests, the transcriptome encompasses the complete set of transcripts (the copies of DNA or RNA) in a particular cell or tissue. Transcripts are complementary strands of DNA or RNA synthesized by the genome (figure 8.10). Thus the whole objective of transcriptome studies is to recognize genes differentially expressed under different environmental conditions. This may help in understanding genes or their associated pathways. To understand this more deeply we have to consider a eukaryotic system, in which genetic expression of a single gene results in the synthesis of more than one type of developed mRNA. This process of alternative splicing or differential splicing results in a single gene coding for multiple proteins, i.e., mRNA is directed to synthesize different proteins that may have different functions or properties. Alternative splicing involves elimination of introns (nucleotide sequences) from the gene (complementary RNA/ DNA) and fixing exons together in the right direction to yield the final functional mRNA product. The information which is not of any use, i.e., non-coding sections of an RNA transcript, are removed and the coding sections (the code for proteins) are joined together to form the functional product in the form of mRNA. Alternative splicing can take place in different ways, but still results in well-defined patterns. This information helps in understanding the whole human genome project. For example, when a single human gene has undergone alternative splicing to produce functional RNA to further produce three different proteins, you can assume how many proteins the projected 35 000 human genes could create (the estimated number is 105 000 different proteins, i.e., different in their properties and functions). This diversity of proteins present inside the body helps in controlling the whole metabolism, structure, etc. A popular example of alternative splicing is seen in the case of the *Drosophila* gene Dscam [21]. Dscam is a (immunoglobulin superfamily) protein essential for the development of neuronal connections in *Drosophila* [21]. By means of alternative splicing, Dscam can potentially develop 19 008 different extracellular domains associated with one of two alternative transmembrane segments, resulting in 38 016 isoforms [21]. All these isoforms share the same domain structure, however, they encompass variable amino acid sequences within three Ig domains in the extracellular region [21]. A gene identified in this study produced approximately 40 000 different mRNAs. These functional mRNAs can be further translated in different receptor proteins. After the determination of the complete human genome sequence in 2001, it is now understood that only a small portion of the human transcriptome is translated into proteins. The rest of the residual transcripts have unidentified functions. It has been reported that a noticeable increase in transcriptional complexity is connected with organization of the transcriptional units in the genome. There are genomic regions that are highly enriched for transcripts which may be subjected to shared epigenetics regulatory control over larger regions. This complexity is due to the production of a manifold of functional mRNAs from each RNA transcript, which makes them more vulnerable to mutation. The multiple bioconversion steps, which involve the passage of large quantities of information from one molecule to another, and further the multiple copies made in the form of a functional message (mRNA) to synthesize different proteins, involve cascades of molecular events where each time elements are expose against different conditions, there is a chance of being mutated. The extent and nature of transcriptome varies from tissue to tissue because no tissue will express all of the genes. The genetic expression of the transcriptome always varies, which is why each tissue has its own transcriptome to synthesize unique proteins. This process is very selective in nature as genes expressed in a unique tissue will vary from those present in another tissue. Consequently, based on this variation it is obvious to discuss different transcriptomes, such as the human brain transcriptome, mouse liver transcriptome, etc.

8.6.3 DNA arrays: a potential genomic tool

DNA microarray techniques are primarily used to determine the transcriptional levels of RNA transcripts derived from thousands of genes within a genome in a single experiment (figure 8.11). By using this approach we can relate physiological

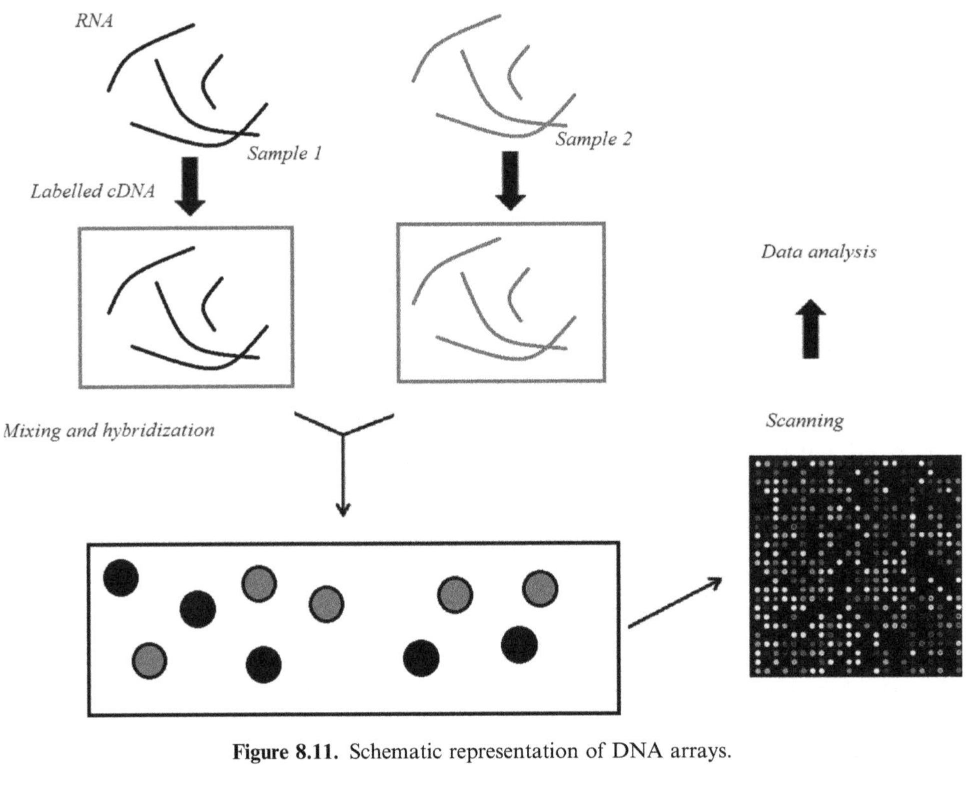

Figure 8.11. Schematic representation of DNA arrays.

cell states to gene expression patterns for examining tumors, disease progression, cellular response to stimuli and drug target identification. Currently DNA microarrays are not limited to gene expression, as they are also being used to detect the single nucleotide polymorphisms (SNPs) of the human genome (the Hap Map project) [22], aberrations in methylation patterns [23], alterations in gene copy-number [24], alternative RNA splicing [25] and pathogen detection [26, 27].

DNA microarray or biochip techniques include synthesis of DNA sequences in two- or three-dimensional format. These sequences are synthesized such that the DNA sequences are covalently or non-covalently linked to the surface. In use, a DNA array allows the hybridization of targets (labeled nucleic acids) to the probes present on the array to evaluate the relative amount of nucleic acid present in a given solution, as shown in figure 8.12.

Generally, during this process small sequences of nucleic acids are immobilized or fixed on an appropriate solid support such as a glass chip, silicon chip or nylon membrane.

DNA arrays are available in three types:

- Spotted arrays on glass.
- *In situ* synthesized arrays.
- Self-assembled arrays.

Spotted arrays on glass. Spotted DNA arrays allow high-density DNA arrays (10^6–10^9 double-stranded DNA molecules with dense spots up to 5000 spots per cm^2) to fix over glass substrates [43]. Polylysine coated glass microscope slides are used for immobilizing DNA and for spotting a robotic spotter can be utilized.

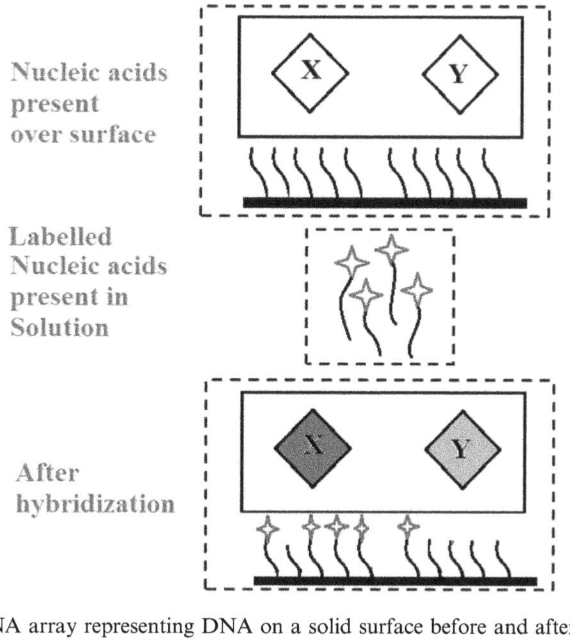

Figure 8.12. DNA array representing DNA on a solid surface before and after hybridization.

This robotic spotter spots various glass slide arrays with DNA from microtiter dishes. These small fragments of DNA are derived from genomic libraries, in particular from cDNA clones or by PCR amplification. This robotic spotter contains slotted pins which are quite similar to fountain pens in design (figure 8.13). During this process a single dip is made in the DNA solution, which can then be used over numerous slides. This robotic facility also allows one to fluorescently label the samples for fluorescence based detection, which offers greater sensitivity and a broad dynamic range, and can provide different colored labels, which can further help in their detection after hybridization. Flourescent labels are also cheaper than radioactive or chemilluminescent labels. Spotting is done in such a manner that each individual spot or dot signifies a separate gene, i.e., the spots are not characterized by repetition. Spotting can be studied using confocal scanning [43].

In situ synthesized arrays. Fodor first suggested, in 1991, that single-stranded oligonucleotides are synthesized in the presence of light on a solid substrate (the process is called photolithography) [44, 51]. Initially, ten amino acid peptides were synthesized using di-nucleotides, and later 256 different octa-nucleotides were synthesized [45]. During 1995, Affymetrix array technology was utilized to determine the variation in the reverse transcriptase and protease genes of the highly polymorphic HIV-1 genome [46] and also to determine mutation in the human mitochondrial genome [47]. Affymetrix technology has been used to produce an extensive set of DNA arrays for use in expression analysis [48, 49], genetic characterization [46] and gene sequencing [50]. *In situ* synthesized arrays can be synthesized by two approaches:

- Inkjet oligosynthesis methods.
- Photolithographic methods (e.g. Affymetrix).

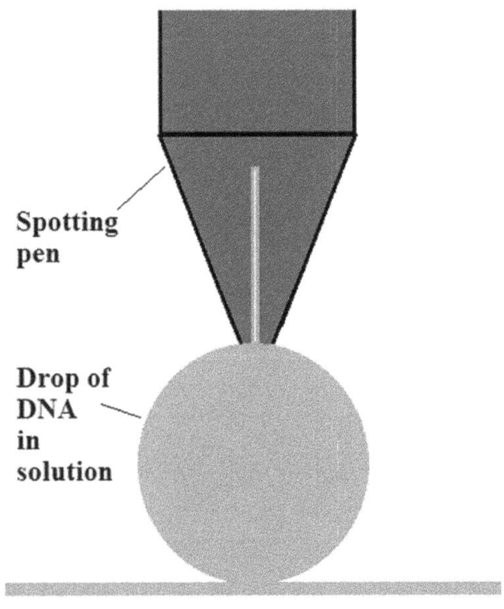

Figure 8.13. Spotting arrays. A glass fountain-pen-like structure (usually multiple pens are present) is allowed to dip into solutions containing DNA and this is then deposited on glass slides.

Self-assembled arrays. Another method to construct arrays was initially suggested by David Walt at Tufts University [52–55]. This technique was later licensed to Illumina (a sequencing and array technology company in San Diego, CA). In this method small DNA strands are synthesized on small polystryrene beads. Ultimately these DNA coated beads are deposited on the ends of a fiber optic array (figure 8.14). The fibers are engraved to provide a well which is larger than a single polystryrene bead. The beads are optically encoded with different fluorophores so as to identify which oligonucleotide was in which position on the array. This method is called an optical sensor array [52–55].

Small single-stranded strands of oligonucleotide (20–25 base pairs) are synthesized and fixed (printed) over a solid matrix with the light-directed printing technique known as photolithography, to form a printed oligonucleotide chip. Oligonucleotides are directly synthesized over the matrix. To reduce the chance of false positive outcomes, each fragment sequence is described by multiple oligonucleotides (20 non-overlapping oligonucleotides). A high density of oligonucleotides over the matrix (64 000 per cm^2) offers more accurate results for investigating gene expression patterns. One of the most interesting applications of a microarray is to establish the association between the gene expression patterns of two cell types (e.g. cancerous tissue and normal tissue). Microarrays have potential applications in transcriptomics, because the whole concept is based on the synthesis of complementary strands (cDNA) or probes from mRNA. For the development of probes, targeted cells must first be identified to isolated mRNA, e.g. to study the genetic variation between normal and cancerous cells mRNA from both is isolated. However, since RNA is not very stable and can be quickly degraded, using reverse transcriptase mRNA is converted into complementary DNA (cDNA). By using the RNA molecule, an enzyme known as reverse transcriptase copies strand to synthesize a new complementary template DNA. This newly formed cDNA is labeled with suitable fluorescent dyes (fluors, for short). These labeled cDNA are typically called probes. Labeling of target molecules (cDNA or cRNA) is s crucial step in a microarray as it helps to determine the amount of mRNA indirectly through the labeled molecules. In the case of investigating the genetic variation between normal and cancerous cells, cDNA derived from both types of cells is labeled with a different fluor that will yield fluorescence of different colors. Now, the amount of cDNA from each sample signifies the amount of mRNA present in that sample, i.e., the amount of cDNA in the probe will be comparable to that of the

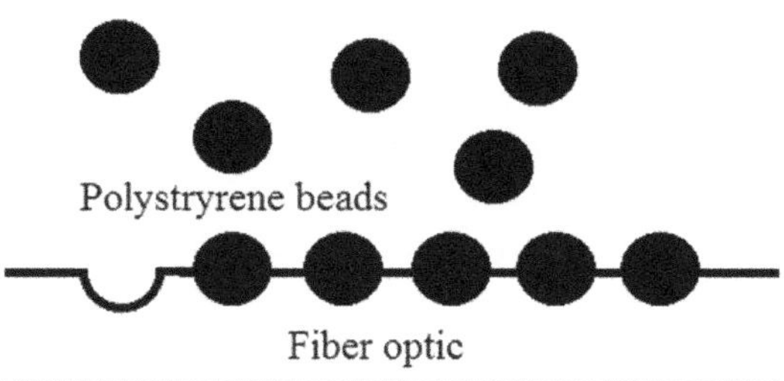

Figure 8.14. Self-assembled arrays.

mRNA in the cell. These cDNA are allowed to hybridize with cDNA probes. Each labeled cDNA molecule will hybridize with the DNA molecules of that dot of the microarray to which it is complementary. Since at this stage some strands are labeled and some are not, the strands utilized in hybridization can be identified by their respective fluorescence pattern. So, because of the fluor existing on the cDNA probe, these can be identified by their colored spots, however, those spots that do not have complementary cDNA in the probe will not fluoresce and will appear as a blank. Fluorescence intensity also helps in the identification of regions of blot paper where hybrid strands are present. This will indirectly determine the relative amount of mRNA present in the cell. To compare the genetic expression between cancerous and normal cells, cDNA probes of both are used concurrently for hybridization under same DNA array conditions. In such cases two different fluors are employed. The information based on fluorescence imaging (spots) of the two fluors can be accessed using suitable software. This software helps in establishing the relationship between the intensities of the images and the amount of mRNA. The examination of a microarray such as that described above will offer data on the following:

- The genes that are expressed in both normal and cancerous cells.
- The genes that are expressed in normal cells but are not expressed in cancerous cells.
- Those genes that are expressed only in cancerous cells. This information would allow the identification of cancerous cells, and also to devise suitable drugs to selectively target such cells.

8.6.4 Gene function determination from sequence information

Scientists have synthesized a vast number of genome sequences from different organisms. The available databases, such as GenBank at the NCBI, store many of these sequences. These databases can be potentially utilized for studying comparative biology. Not only do the databases store the genome sequences, but also information about the function (if it is known) of the genes. By making comparisons between known genes in the database, GenBank can be used to identify unknown genes. In database, one program exclusively used for this purpose is BLAST (Basic Local Alignment Search Tool). BLAST is a sequence similarity searching algorithms, which is based on the principle that if two sequences are similar then they are possibly homologous (that is, they share a common evolutionary ancestor). By means of this database, one can understand the function of an unknown gene by discovering similar sequences of known genes and proteins. BLAST is an advanced program, in that it searches at the nucleotide level, as well as making assessments at the amino acid level, offering much greater sensitivity. Thus, if one is mainly interested in the DNA sequence itself, it is better to hunt for genes using proteins. In addition to whole proteins, similarity searches can find protein motifs. A motif is a characteristic arrangement of amino acids, stored across many proteins, which offers a particular function to the protein.

The primary objective of classical functional genomics is to regulate the function of specific genes. Initial gene cloning is followed by mutation under *in vitro* conditions and finally reincorporation of the mutated gene into the host organism to examine its expression or any other effect. This approach has its limitations, as the process is complicated, slow and has a low success rate. The applications of a combination of the experimental approach to functional genomics and high-throughput sequencing analysis are as follows:

- To identify the gene accountable for specific biological phenotypes and diseases.
- To help in determining the function of gene.
- To facilitate the rapid identification of genes.
- To enable the discovery of genes that participate in other biological processes.
- To mutate every gene in the genome and accumulate the mutant strain to create genome-wide mutant libraries.

The last point, establishing genome-wide mutant libraries in several organisms such as bacteria, yeast, plants and mammals, is called mutational genomics. A mutational genomics based library can be established using one of the following two approaches.

The systematic approach. To establish a library, the systematic mutation of individual genes in the genome at one time should be studied, which allows the development of a bank of specific mutant strains. This can only be attained if the complete genome sequence is known. In this method, homologous recombination (which functions in the repair of DNA double-stranded breaks and inter-strand crosslinks) is utilized to accommodate a selectable marker gene (which helps in the selection of the targeted gene) inside the gene of interest. Usually, the insertion of the gene of interest in a selectable marker disturbs the selectable marker function, which can be further checked against the expression of normal selectable marker gene. However, in contrast, here homologous recombination allows the selectable marker to integrate into the gene of interest, which allows the dislocation of the targeted gene. The whole process is called gene knockout. By using suitable sequences from or around the target gene the selectable marker gene is flanked to attain homologous recombination. So it is essential to recognize the sequences of the gene of interest. After the employment of suitable vectors, integration performed by homologous recombination may possibly occur up to 90% in prokaryotes. However, the opposite is the case for plants and animals, the rate of integration by homologous recombination is only 0.001% of the entire integration process. Homologous recombination or integration in the case of embryonic stem cells derived from mice has a rate up to 0.1%, which is much lower than that of yeast, which has 6200 genes and a rate of homologous integration of up to 90%. Knockout investigations have allowed knockouts of almost 85% of yeast genes. The mutants produced by this approach are studied to determine the functions of the missing genes. It is probable that a similar mutagenesis procedure might be considered for *Drosophila* in the near future.

The random approach. In contrast to the systematic approach, in random integration the gene is randomly mutated and this can be functional for any species. Random insertional mutagenesis can be achieved by the addition or deletion of nucleotides from a target gene sequence. This type of insertion or deletion results in overall change in the sequences of the gene of interest, primarily a new region of assortment that cannot be achieved by only point mutation. Insertional mutagenesis is often studied during gene transfer by the *Agrobacterium* strain. In this process a gene of interest is inserted into a transposable element called T-DNA. This T-DNA of *Agrobacterium*, whenever it is introduced into the genome, results in the disruption and loss of gene function. It clearly suggests the scope of T-DNA in the transfer of a gene of interest. The gene of interest is labeled before its insertion into T-DNA, which can facilitate its isolation from the place where it is positioned. Thus this approach is often called signature-tagged mutagenesis. The incorporated DNA sequence after labeling is called a tag. Inserted labeled DNA sequences in the form of tags can be further employed as probes in hybridization or to synthesize PCR primers. These primers can be utilized to isolate the DNA sequences located on either side of the tag or labeled gene of interest. They can further be connected to the DNA sequence with a genomic cDNA clone and used to identify its function, etc. The tool called BLAST can be utilized for this in databases. In the case that this random insertion results in a mutant phenotype, the gene can be assigned a tentative function. Moreover, the incorporated sequence or gene of interest can be altered into a gene trap vector (the process known as gene trapping is a high-throughput approach) by introducing insertional mutations to ultimately offer and generate the data regarding the gene it interferes with. Gene trap mutations are present in equal numbers to the insertional targeted mutations, as long as the trapped gene is known. During gene trapping, the incorporated element covers a selectable marker gene such as lac Z (responsible for β-glucosidase synthesis) or gus A (responsible for β-glucuronidase synthesis) situated in a splice acceptor site. The selectable marker gene is, thus, triggered only if this element is introduced within the transcribed region of a gene. This approach facilitates a convenient assortment for incorporation inside genes, and is valuable in all organisms.

In this strategy, a number of tools are employed to prevent the expression of genes. These tools produce variations in phenotype, which means the organisms with a similar appearance are mutants; however, they have the typical variant form of a given gene. High throughput tools to explore protein and gene function *in situ* are still necessary to exploit these emerging advances in gene and protein discovery and in order to validate these identified targets. High-throughput target-selected gene inactivation in zebrafish was reported in 2011 [56]. High throughput functional inactivation of genes can be attained by various approaches such as virus-induced gene silencing for plants and RNA interference. In the case of animals, short double-stranded RNAs are being considered as a potential approach for gene silencing.

8.6.5 Protein interactions

The identification of genetic fusion across genomes can be used for the prediction of functional interrelationships of proteins, including physical interactions or complex formation. These predictions are obtained by the detection of resemblance between pairs of 'component' proteins and 'composite' proteins [28]. The most important criterion is how the gene functions, which can reveal the actual performance of the encoded proteins. This performance can be understood as multiple interactions between different proteins, and among proteins and molecules. For example, the drugs usually employed to treat disease act by modifying protein interactions in a positive manner. The process of investigating protein interactions is as follows. Let us say, for example, the function of protein A is unknown. Protein A is found to resemble proteins B and C, which contribute to RNA splicing. This type of relationship between proteins would suggest that protein A is involved in RNA splicing. Before the genomics age, communication between proteins was investigated on an individual basis by means of a range of approaches, including suppressor mutations and co-immunoprecipitation (to study protein–protein interactions). These tools are very useful, however, they are not suitable for high throughput investigation. Moreover, they do not offer an easy way to link the proteins to their respective genes.

Different types of protein–protein interactions are involved in all cellular processes. For systems biology, the mapping of these interaction networks to explain the organization of the proteome into functional units is essential [29–32]. Several approaches have been developed for screening protein interactions. One of the most traditional approaches is copurification (also called affinity purification or co-immunoprecipitation) of protein complexes, which requires *in vitro* handling of protein extracts. In addition, this approach also has the limitation of restricted sensitivity and bias towards high affinity interactions [29–32]. When a partner has been identified, detection by mass spectrometry (MS) is usually straightforward, although expensive. Cloning of the resultant cDNA may be time consuming, but clone repositories, e.g. RIKEN or IMACE, can provide a suitable alternative. Currently, surface plasmon resonance has been reported for screening protein–protein interactions. In this procedure purified cellular extracts are introduced onto a sensor chip covered with an immobilized binding partner. The tool arrangement combines the capture of the binding partner with a quantitative readout of the binding event, such that the putative partners can be eluted and detected by using MS [29–32]. An additional method for interaction screening uses 'cDNA-expression' libraries [29–32], such as phage display or yeast two-hybrid (Y2H) methods, for which genomic scale, highly parallel and automated processes are necessary. However, only a few detection approaches for protein–protein interactions can be effortlessly adapted for high-throughput approaches. These include, in particular, the Y2H method and affinity purification coupled with MS (AP/MS).

The Y2H method is a protein–protein interaction assay to determine if two proteins are interacting. In this procedure, a 'bait' protein is attached to half of a protein such as GAL4 that is attached to a reporter gene promoter. To settle on the promoter and activate transcription of the reporter gene, the 'prey' protein interacts with the 'bait', which will bring the other half of the GAL4 protein together (figure 8.15). This approach is discussed further below.

Protein interactions are investigated by means of high-throughput tools. For efficient screening of many proteins some library-based protein interaction mapping is required to assess their interactions by *in vitro* or *in vivo* assays. Such mapping helps in establishing relationships between proteins and genes, or the complementary DNAs that encode them. To assess their biological potential *in vitro* examination is essential, which can be based on standard expression library. Based on expression cloning, an expression library is composed of clones, where each clone presents its own protein. This library helps in screening for suitable features and clones of interest, which can be recovered for further examination. Thus an expression library contains a variety of clones that carry cDNA derived from an organism which can be further incorporated in an expression vector to produce their protein products (figure 8.16).

As mentioned in figure 8.16, the last step of protein screening, i.e., identification of desirable proteins, can be achieved by immunological assays (antigen–antibody interaction). Other tools include the protein detection tools western blot and ELISA and protein characterization by mass spectrometry, and the function of the proteins can be assessed by functional protein analysis with HaloTag Technology. The identification of proteins is a crucial step which requires its respective antibody (the antibody functions as a probe) for its effective screening. Moreover, various proteins such as transcription factors can also be employed to recognize their interacting partners. For example, for the transcription factor c-Jun, also known as AP-1, is encoded by the JUN gene.

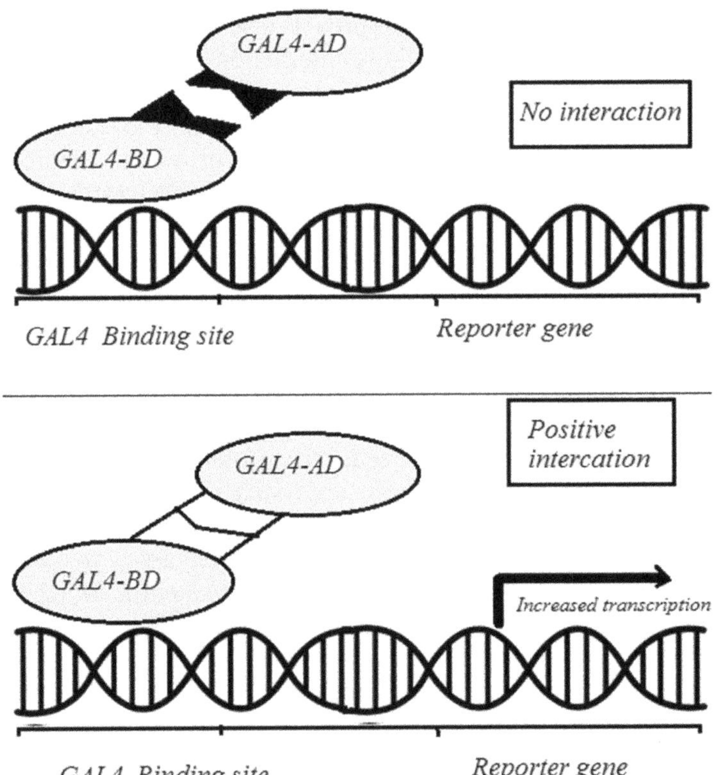

Figure 8.15. The Y2H hybrid method.

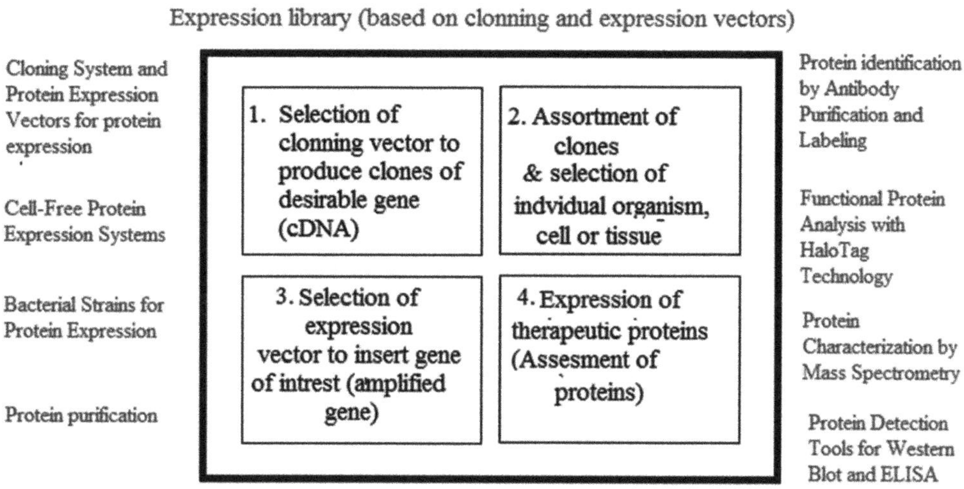

Figure 8.16. An expression library.

Another high-throughput approach is the phage display method, which offers a means to identify target-binding proteins from a library of millions of various proteins without the requirement of screening each molecule independently. Phage display is an approach to produce and screen a high number of novel therapeutic proteins and polypeptides by introducing a gene of interest into a gene responsible for the expression of a surface protein of a bacteriophage. This allows the expression of a display of proteins of our own choice over the surface coating of the phage. One of the most important advantages of phage display is the manipulation and testing of biological activity. As mentioned above, in this method the proteins considered for screening are allowed to express as fusion proteins (e.g. virus coat proteins) in such a manner that they are accessible to the surface of the bacteriophage particle (figure 8.17). This probe protein is further immobilized over a matrix or support available in the form of a microtitre plate or membrane. Since the phage genetic material is responsible for the synthesis of surface proteins, this approach is based on the idea that phage phenotype and genotype are physically associated with each other. Certainly, the gene encoding the displayed molecule is packed as a single-strained DNA. Proteins that are synthesized in the form of displayed proteins are expressed in fusion with phage coat protein. Several genes are expressed in a bacteriophage library in the form of fusion proteins in the bacteriophage coat protein, for their display over the surface of the virus particle. Now at this point the proteins displayed over the surface resemble the genetic sequence within the phage. Expression libraries containing phage displays have been created for screening with the probes (figure 8.17). This is done for the selection of particles that can efficiently interact with the probe.

After screening of the phage display with the probe, the surface particles that can efficiently interact with the probes are selected and further utilized to re-infect colonies of *E. coli*. Phage particles obtained after this are again selected to meet the requirements and this is repeated again until the desired expression of proteins is

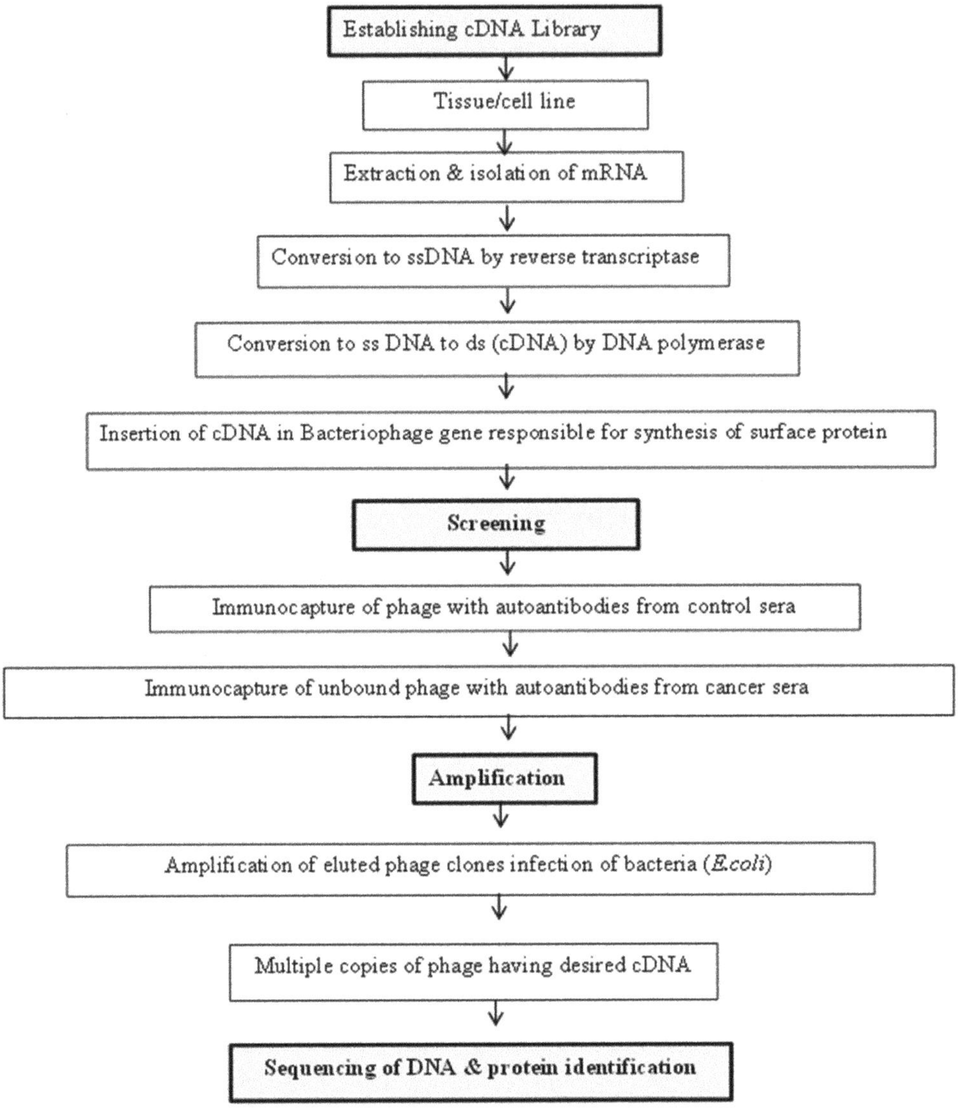

Figure 8.17. Important steps in creating a phage display based genomic library.

achieved (or not, as the case may be). During the last step the recovered phages are selected to isolate the interactive protein and the gene is responsible for their expression. Genomic libraries compiled using the phage display approach are found to be of great complexity. Thus more efficient screening tools in the array setup are required to allow the selection of suitable proteins with different probes. However, this strategy is yet to be used for genome-scale interaction investigations.

Yeast two-hybrid (Y2H) screening (introduced above) is a well established *in vivo* genetic approach for protein interaction studies, and is currently the most sensitive high-throughput method for interactomics research. This approach has offered the basis for a variety of interaction networks formed by different components, primarily proteins. One of the major applications of the hybrid system is to understand transient and dynamic protein–protein interactions. By using the yeast cell system we can test whether interaction between two different proteins takes place or not. The yeast cell system contains DNA which further contains transcription activator binding sites, along with the gene of interest which allows the expression of the desirable reporter protein. So to produce reporter protein, the transcription activator should be activated. This can only be achieved when the target protein makes a complex with the transcription activator protein, which will further bind with the transcription activator site and transfer a signal to the cell to start the transcription of the desired gene, which will lead to the production of desired reporter protein, which ultimately confirms interaction between two proteins. In this case both proteins, i.e., the transcription activator protein and the target protein, contain binding sites in the form of the transcriptional activation domain and DNA-binding domain (figure 8.18), which are also known as functionally self-governing domains. These domains interact with each other by covalent or noncovalent association. In addition, they both ('bait' and 'prey') contain their own interactive proteins called the target protein and binding partner (figure 8.18). Their

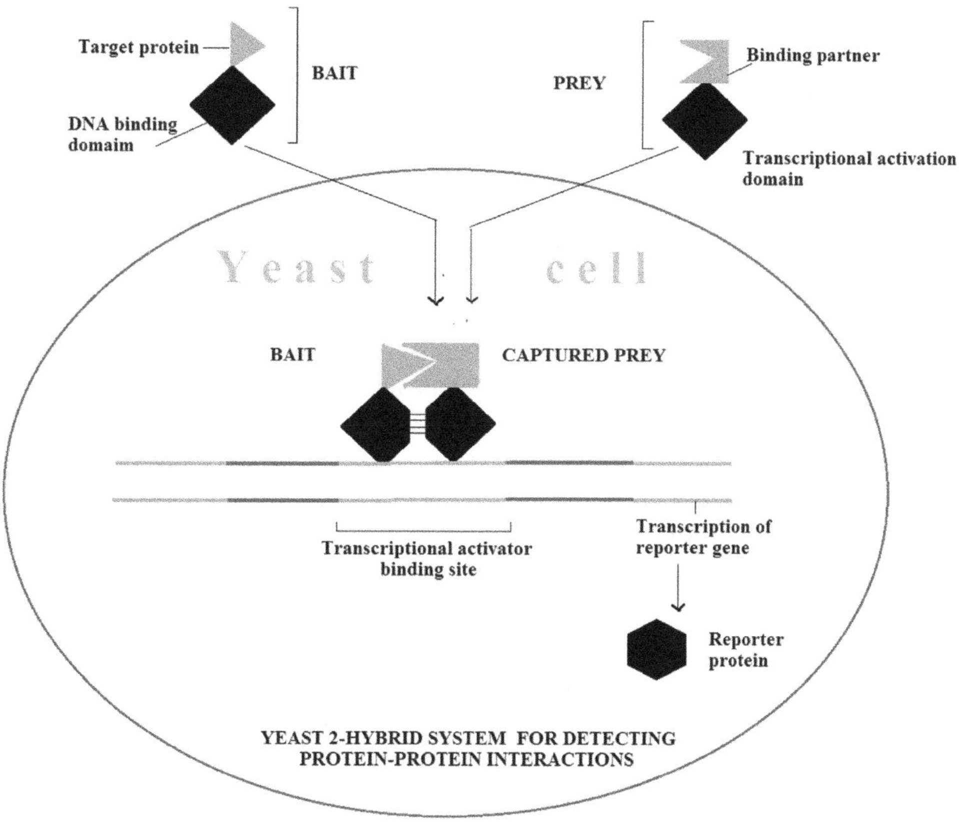

Figure 8.18. The Y2H system for detecting protein–protein interactions.

interaction allows the formation of a complex which can bind with the transcriptional activator binding site to yield a reporter protein.

So, if this complex does not bind with the transcription binding site it will not produce the reporter protein, and the response for protein production will be suppressed or halted. Thus if there is no interaction between two proteins then no complex will form in the form of a dimer, and consequently the transcription activator gene will not be activated to activate the gene for protein production. To understand this more deeply we have to understand one concept, that an active transcription factor is only produced if independently expressed DNA-binding and activation domains are allowed to interact (figure 8.18). The approach for the Y2H system is as follows: the target protein is allowed to express as a fusion protein along with the DNA-binding domain (known as the 'bait') and diverse proteins are allowed to express as binding proteins with the activation domain (known as 'prey'). The ultimate goal of Y2H system is to produce a protein by a reporter gene, which is only transcribed when it is activated by the complex formed by the 'bait' with the 'prey'. During the genome-wide mapping of protein interactions, complete libraries of 'baits' and 'preys' are produced. These libraries are assessed for protein interactions using the following:

- the matrix method.
- the random library approach.

Y2H, however, tends to give a high incidence of false positive and false negative outcomes. Another approach is known as GST (glutathione-S-transferase)-pull down. GST is a bacterial enzyme with a high affinity for its substrate glutathione. In particular, the 'bait' protein is expressed as fusion with GST. The proteins that interact with the 'bait' can be copurified, i.e., dragged down from a cell lysate by passing the lysate through a glutathione–sepharose column. This approach is functional on a genomic scale. Protein interaction data from different sources are integrated in databases. Numerous bioinformatics based techniques have been advanced to derive data from such databanks. One of the main challenges is to discover a simple approach to present protein interaction information in an easily available format and in a consistent structure.

8.7 Synthetic genomics

One of the first aims of synthetic genomics is to design new organisms and biological systems to satisfy human needs [33]. In synthetic biology, the aims is to model and create biological components, functions and organisms that do not actually exist in nature or to redesign already existing biological systems to execute new functions. On the other hand, synthetic genomics involves tools for the production of chemically synthesized whole genomes or larger parts of genomes, allowing the simultaneous engineering of numerous changes to the genetic material of organisms (figure 8.18) [33].

Synthetic genomes (artificial gene production to produce new DNA or complete live forms) can be defined as a genome assembled from smaller precursor molecules derived by either PCR amplification or chemical synthesis (figure 8.19). Viral genomes were the first to be produced. In 2002 a virus was reconstructed from its chemically synthesized genomes. Itaya and co-workers (2008) synthesized the mouse mitochondrial and rice chloroplast genomes [34–36]. During the same period Gibson and co-workers produced the complete genome (0.58 million base pairs) of *Mycoplasma genitalium*, which has the smallest genome among the known culturable free-living organisms [37]. The synthetic genome function was assessed by incorporating it into *M. genitalium* cells and transferring their native genomes. Again Gibson and co-workers utilized the *M. genitalium* genome sequence to chemically produce a great number of oligonucleotides, demonstrating the complete genome [37]. Exact chemical synthesis was achieveable for fragments of only up to 100 bases. By means of *in vitro* ligation, these small fragments were suitably arranged to produce 101 minimally overlapping DNA molecules 5–7 kb in length (overlapping 'cassettes'). During the following step, these DNA molecules were ligated successively to produce larger and larger genome segments. Assembly of the genome segments was then performed together, initially in *E. coli* and later in *S. corevisiae,* to derive the entire circular genome of *M. genitalium* (582 970 bp) [37]. To guarantee that the created genome had precisely the same base sequence as the original *M. genitalium*, the sequence used for chemical synthesis, a rough sequence confirmation of the precursor molecules at every stage was performed.

The tomato fruit and leaf are promising for investigating the evolution of gene regulation and the underlying development of different organs [36]. Itaya and colleagues identified over 350 and 700 small RNAs from the tomato fruit and leaf,

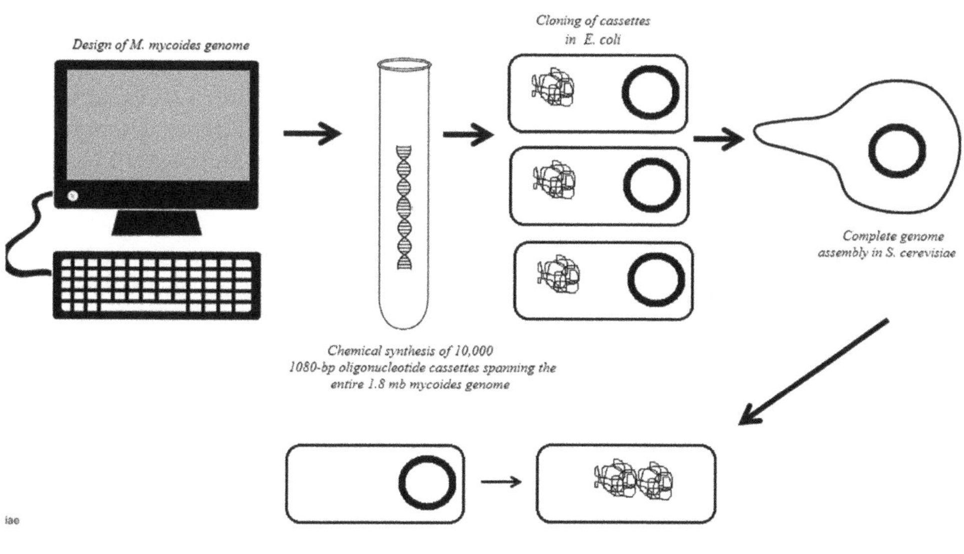

Figure 8.19. Creating a synthetic genome in a bacterial strain.

respectively, and further established a website (http://ted.bti.cornell.edu/digital/sRNA/) for public, searchable access to all of the small RNA sequences, their expression patterns in respective tissues, and their matching genes or predicted target genes [36].

With the help of PCR, Itaya and colleagues generated DNA fragments comprising the entire mouse mitochondrial and rice chloroplast genomes [36, 38]. To design PCR primers that encouraged the amplification of overlapping DNA fragments for the entire genome, they used the genome sequence information. By means of their serial integration into the genome of *B. subtilis,* these precursor DNA fragments were assembled. The genome was removed from the *B. subtilis* genome and purified when the entire set of precursors was incorporated into the *B. subtilis* genome. Synthetic genomes offer several applications. They can allow the regeneration of already vanished organisms, if their genome sequences are known. They will allow documentation of the set of genes vital for cellular life by examining a diversity of minimal genomes. New discoveries in biology may be enabled by designing, synthesizing and examining genomes of different specifications. Synthetic genomes may also allow the complete restructuring of industrial microbes, producing inexpensive, better and even new products. However, numerous technical improvements/developments are required to attain these goals. It should be noted that so far only existing genomes have been reconstructed—designing a functional genome is yet to be accomplished.

References

[1] Anderson N G and Anderson N L 1996 Twenty years of two-dimensional electrophoresis: past, present and future *Electrophoresis* **17** 443–53

[2] Wasinger V C, Cordwell S J, Cerpa-Poljak A J, Yan X, Gooley A A, Wilkins M R, Duncan M W, Harris R, Williams K L and Humphery-Smith I 1995 Progress with gene-product mapping of the Mollicutes *Mycoplasma genitalium* *Electrophoresis* **16** 1090–4

[3] Wilkins M R, Sanchez J C, Gooley A A, Appel R D, Humphery-Smith I, Hochstrasser D F and Williams K L 1995 Progress with proteome projects: why all proteins expressed by a genome should be identified and how to do it *Biotechnol. Genet. Eng. Rev.* **13** 19–50

[4] O'Farrell P H 1975 High resolution two-dimensional electrophoresis of proteins *J. Biol. Chem.* **250** 4007-21

[5] Klose J 1975 Protein mapping by combined isoelectric focusing and electrophoresis of mouse tissues. A novel approach to testing for induced point mutations in mammals *Humangenetik* **26** 231–43

[6] Scheele G A 1975 Two-dimensional gel analysis of soluble proteins. Characterization of guinea pig exocrine pancreatic proteins *J. Biol. Chem.* **250** 5375–85

[7] Anderson N L *et al* 1980 *Electrophoresis '79* (New York: Walter de Gruyter)

[8] Shin J, Lee W and Lee W 2008 Structural proteomics by NMR spectroscopy *Expert Rev. Proteomics* **5** 589–601

[9] Monti M, Orrù S, Pagnozzi D and Pucci P 2005 Functional proteomics *Clin. Chim. Acta* **357** 140–50

[10] Heng L and Homer N 2010 A survey of sequence alignment algorithms for next-generation sequencing *Brief Bioinform.* **11** 473–83

[11] Bhagwat M and Aravind L 2007 PSI-BLAST tutorial *Comparative Genomics* ed N H Bergman1,2 (Totowa, NJ: Humana) ch 10
[12] Gabanyi M J and Berman H M 2015 Protein structure annotation resources Methods *Mol. Biol.* **1261** 3–20
[13] Vonderviszt F and Namba K 2000–2013 Structure, function and assembly of flagellar axial proteins. In: Madame Curie Bioscience Database (Austin, TX: Landes Bioscience)
[14] Magdeldin S 2014 Basics and recent advances of two dimensional–polyacrylamide gel electrophoresis *Clin. Proteomics* **11** 16
[15] Date S V 2008 The Rosetta stone method *Bioinformatics. Methods in Molecular Biology*™ vol 453 ed J M Keith (Totowa, NJ: Humana)
[16] Wasinger V C, Pollack J D and Humphery-Smith I 2000 The proteome of *Mycoplasma genitalium.* Chaps-soluble component *Eur. J. Biochem.* **267** 1571-82
[17] Hinshaw J E, Carragher B O and Milligan R A 1992 Architecture and design of the nuclear pore complex *Cell* **69** 1133-41
[18] Oliver S G 2002 Functional genomics: lessons from yeast *Philos. Trans. R. Soc. Lond.*. B **357** 17–23
[19] Feezor R J, Cheng A, Paddock H N, Baker H V and Moldawer L L 2005 Functional genomics and gene expression profiling in sepsis: beyond class prediction *Clin. Infect. Dis.* **15** 41
[20] Brown T A 2002 Transcriptomes and proteomes *Genomes* 2nd edn (Oxford: Wiley-Liss, Garland Science) ch 3
[21] Wojtowicz W M, Flanagan J J, Millard S, Zipursky S L and Clemens J C 2004 Alternative splicing of *Drosophila* Dscam generates axon guidance receptors that exhibit isoform-specific *Homophilic Binding Cell* **118** 619–33
[22] Cutler D J, Zwick M E and Carrasquillo M M *et al* 2001 High-throughput variation detection and genotyping using microarrays *Genome Res.* **11** 1913–25
[23] Yan P S, Chen C M and Shi H D *et al* 2001 Dissecting complex epigenetic alterations in breast cancer using CpG island microarrays *Cancer Res.* **61** 8375–80
[24] Pollack J R, Perou C M and Alizadeh A A *et al* 1999 Genome-wide analysis of DNA copy-number changes using cDNA microarrays *Nat. Genetics* **23** 41–6
[25] Relogio A, Ben-Dov C and Baum M *et al* 2005 Alternative splicing microarrays reveal functional expression of neuron-specific regulators in Hodgkin lymphoma cells *J. Biol. Chem.* **280** 4779–84
[26] Wang D, Coscoy L and Zylberberg M *et al* 2002 Microarray-based detection and genotyping of viral pathogens *Proc. Natl Acad. Sci. USA* **99** 15687–92
[27] Conejero-Goldberg C, Wang E and Yi C *et al* 2005 Infectious pathogen detection arrays: viral detection in cell lines and postmortem brain tissue *Biotechniques* **39** 741–51
[28] Peregrín-Alvarez J M and Ouzounis C A 2007 The comparative genomics of protein interactions *Genome Inform.* **19** 131–41
[29] Brückner A, Polge C, Lentze N, Auerbach D and Schlattner U 2009 Yeast two-hybrid, a powerful tool for systems biology *Int. J. Mol. Sci.* **10** 2763–88
[30] Boireau W, Rouleau A, Lucchi G and Ducoroy P 2009 Revisited BIA-MS combination: entire "on-a-chip" processing leading to the proteins identification at low femtomole to sub-femtomole levels *Biosens. Bioelectron.* **24** 1121–7
[31] Natsume T, Nakayama H, Jansson O, Isobe T, Takio K and Mikoshiba K 2000 Combination of biomolecular interaction analysis and mass spectrometric amino acid sequencing *Anal. Chem.* **72** 4193–8

[32] Phizicky E M and Fields S 1995 Protein-protein interactions: methods for detection and analysis *Microbiol. Rev.* **59** 94–123

[33] König H, Frank D, Heil R and Coenen C 2013 Synthetic genomics and synthetic biology applications between hopes and concerns *Curr. Genomics* **14** 11–24

[34] Gibson D G 2011 *A8 Creation of a Bacterial Cell Controlled by a Chemically Synthesized Genome* Institute of Medicine (US) Forum on Microbial Threats (Washington, DC: National Academies Press)

[35] Gibson D G *et al* 2010 Creation of a bacterial cell controlled by a chemically synthesized genome *Science* **329** 52–6

[36] Itaya A, Bundschuh R, Archual A J, Joung J G, Fei Z, Dai X, Zhao P X, Tang Y, Nelson R S and Ding B 2008 Small RNAs in tomato fruit and leaf development *Biochim. Biophys. Acta* **1779** 99–107

[37] Gibson D G *et al* 2008 Complete chemical synthesis, assembly, and cloning of a *Mycoplasma genitalium* genome *Science* **319** 1215–20

[38] Itaya M, Fujita K, Kuroki A and Tsuge K 2008 Bottom-up genome assembly using the *Bacillus subtilis* genome vector *Nat. Methods* **5** 41–3

[39] Kulkarni M M and Perrimon N 2013 Analyzing the structure, function and information flow in signaling networks using quantitative cellular signatures *Handbook of Systems Biology* (Amsterdam: Elsevier), 5 pp 89–113

[40] Dick D M 2011 Gene-environment interaction in psychological traits and disorders *Annu. Rev. Clin. Psychol.* **7** 383–409

[41] Lodish H, Berk A and Zipursky S L *et al* 2000 Gene replacement and transgenic animals *Molecular Cell Biology* 4th edn (New York: Freeman) Section 8.5

[42] Rogge L *et al* 2000 Transcript imaging of the development of human T helper cells using oligonucleotides arrays *Nat. Genet.* **25** 96–101

[43] DeRisi J, Penland L, Brown P O, Bittner M L, Meltzer P S, Ray M, Chen Y, Su Y A and Trent J M 1996 Use of a cDNA microarray to analyse gene expression patterns in human cancer *Nat. Genet.* **14** 457–60

[44] Fodor S P, Read J L, Pirrung M C, Stryer L, Lu A T and Solas D 1991 Light-directed, spatially addressable parallel chemical synthesis *Science* **251** 767–73

[45] Pease A C, Solas D, Sullivan E J, Cronin M T, Holmes C P and Fodor S P 1994 Light-generated oligonucleotide arrays for rapid DNA sequence analysis *Proc. Natl Acad. Sci. USA* **91** 5022–6

[46] Lipshutz R J, Fodor S P, Gingeras T R and Lockhart D J 1999 High density synthetic oligonucleotide arrays *Nat. Genet.* **21** 20–4

[47] Chee M, Yang R, Hubbell E, Berno A, Huang X C, Stern D, Winkler J, Lockhart D J, Morris M S and Fodor S P 1996 Accessing genetic information with high-density DNA arrays *Science* **274** 610–4

[48] Lockhart D J *et al* 1996 Expression monitoring by hybridization to high-density oligonucleotide arrays *Nat. Biotechnol.* **14** 1675–80

[49] Wodicka L, Dong H, Mittmann M, Ho M H and Lockhart D J 1997 Genome-wide expression monitoring in *Saccharomyces cerevisiae Nat. Biotechnol.* **15** 1359–67

[50] Wallraff G, brock P, Dipietro R, Nguyen T, Huynh T, Hinsberg W and McGall G 1997 DNA sequencing on a chip *Chemtech.* **27** 22–32

[51] Nuwaysir E F *et al* 2002 Gene expression analysis using oligonucleotide arrays produced by maskless photolithography *Genome Res.* **12** 1749–55

[52] Ferguson J A, Steemers F J and Walt D R 2000 High-density fiber-optic DNA random microsphere array *Anal. Chem.* **72** 5618–24

[53] Michael K L, Taylor L C, Schultz S L and Walt D R 1998 Randomly ordered addressable high-density optical sensor arrays *Anal. Chem.* **70** 1242–8

[54] Steemers F J, Ferguson J A and Walt D R 2000 Screening unlabeled DNA targets with randomly ordered fiber-optic gene arrays *Nat. Biotechnol.* **18** 91–4

[55] Walt D R 2000 Techview: molecular biology. Bead-based fiber-optic arrays *Science* **287** 451–2

[56] Kettleborough R N, Bruijn E D, Eeden F V, Cuppen E and Stemple D L 2011 High-throughput target-selected gene inactivation in zebrafish *Methods Cell Biol.* **104** 121–7

IOP Publishing

Introduction to Pharmaceutical Biotechnology, Volume 2
Enzymes, proteins and bioinformatics
Saurabh Bhatia

Chapter 9

Bioinformatics

9.1 Introduction

According to the Oxford Dictionary, bioinformatics involes conceptualizing biology in terms of molecules and applying informatics techniques to understand and organize the information associated with these molecules, on a large scale. In brief, bioinformatics is a management information system for molecular biology and has many practical applications.

Bioinformatics analysis involves [1]:

- the assembly of sequenced data (directly from environmental samples) in order to construct contiguous sequences (contigs and scaffolds);
- the prediction of genes (and putative proteins) based on the assembled data; and
- prediction of domains, functions and pathways for the putative proteins.

Bioinformatics is a discipline related to the exploitation and application of computer hardware and software to the procurement, storage, investigation and imaging of biological material (figure 9.1) [1]. Scientific developments in recent years have encouraged marked progress in understanding the genetic basis of phenotypes. With these advances, genomics has tremendously transformed the scope of biological problems at a genome-wide scale, exploring vast data and opening numerous possibilities [2]. However, the enormous amount of information that has been produced increases the challenges that must be overcome for storage (Moore's law) and the processing of biological information. Bioinformatics and computational biology have been pursued to overcome such challenges [2]. A general scheme of the process of bioinformatics is shown in figure 9.2.

Bioinformatics has the following three main components:

- New algorithms and statistics production for evaluating the connections among large sets of biological data, e.g. DNA sequence data.

doi:10.1088/978-0-7503-1302-5ch9

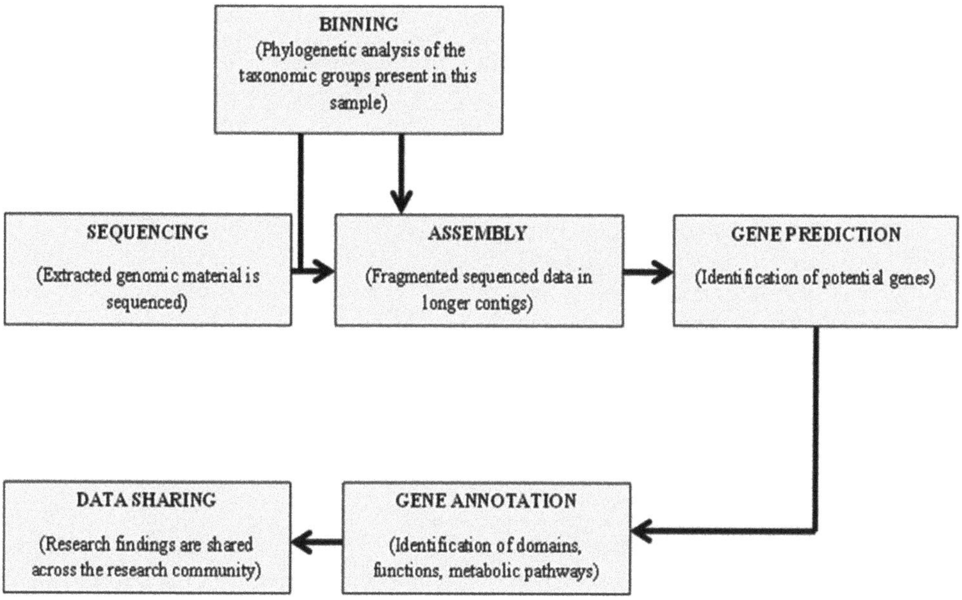

Figure 9.1. Bioinformatics: genomic material is sequenced and processed using assembly, gene prediction and gene annotation tools. Finally, the results are pooled by scientific groups around the world [1].

- Application of these techniques for the examination and understanding of different biological data, including nucleotide sequences, amino acid sequences, etc.
- The production of systems for effective storage, access and management of a large body of varied biological information.

The field of bioinformatics first developed for the production of amino acid sequences of proteins and nucleotide sequences of DNA. Zuckerkandl and Pauling (1962) predicted that the amino acid sequences of proteins could be employed to investigate the evolutionary relationship between different organisms. This suggestion was based on the evidence that the amino acid sequences of homologous proteins, i.e., proteins having similar functions, were related. This started a new area of investigation known as 'molecular evolution'. Subsequent studies have allowed inferences of evolutionary relationships from the relative analysis of amino acid sequences of functionally related proteins. These investigations became possible due to the development of quantitative procedures for sequence comparisons.

9.2 History of bioinformatics

The first complete assortment of amino acid sequences was collected in the *Atlas of Protein Sequence and Structure* by the National Biomedical Research Foundation (USA). This assortment was modified by Margaret Dayhoff in 1965–78 [3]. Dayhoff and colleagues also made considerable contributions to the assessment of amino acid sequences by means of exploring applications of computer software in the

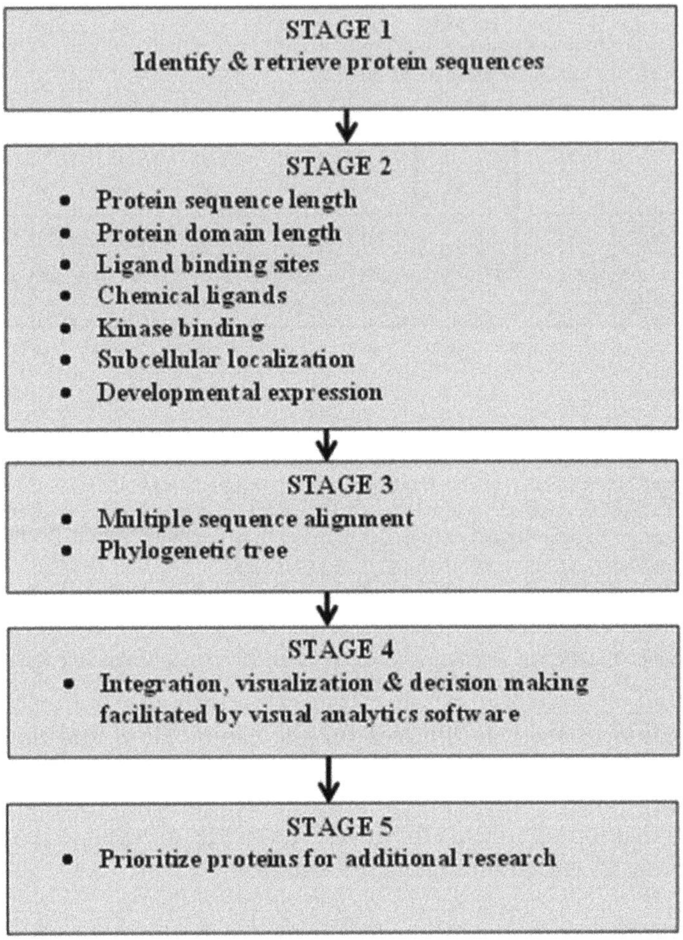

Figure 9.2. Basic process of bioinformatics.

identification of remotely related sequences, deducing evolutionary relationships, etc. In 1990, the European Molecular Biology Laboratory established their data library to collect, arrange and allocate nucleotide sequence data and associated information.

This task is now carried out by the European Bioinformatics Institute (Hinxton, UK). In the early 1980s, the National Centre for Bioinformatics Information (NCBI) was established in the USA. NCBI works as a main information databank and source for other information. It is one of the leading databases and provides search engines for diverse information in the field. The DNA Data Bank was later established by Japan. In 1984, the National Biomedical Research Foundation established the Protein Information Resource. This assists investigators in the detection and elucidation of protein sequence information. These large databases function in close association with each other and frequently exchange data. These databanks are a key resource for all scientists involved in the study of biological phenomena, predominantly the molecular aspects of biologic sciences. The organization and investigation of the rapidly accumulating sequence data require advanced computer software and statistical procedures. This has resulted in the involvement of teams of researchers from computer science and mathematics in the development of the discipline of bioinformatics. Diverse procedures and techniques have now been established that allow the organization, utilization and dissemination of biological information. In bioinformatics, we can design compounds that bind specifically with an expressed protein, or perhaps more importantly, a transcription regulator can cause changes in expression levels, as shown in figure 9.3.

9.3 Sequences and nomenclature

The incorrect use of sequence analysis techniques may result in numerous errors in genome annotation. In bioinformatics, sequence examination is the process of exposing a DNA, RNA or peptide sequence to any of an extensive range of analytical approaches to recognize its features, function, structure or evolution. The procedures employed in sequence analysis include sequence alignment and searches in biological databases. With the development of approaches of high-throughput production of gene and protein sequences, the rate of addition of new sequences to databases has increased exponentially. Such a collection of sequences does not, by itself, increase the scientist's understanding of the biology of organisms. However, equating these new sequences to those with known functions is an important approach to understanding the biology of an organism. Therefore, sequence examination can be employed to allocate functions to genes and proteins through the study of the similarities between the compared sequences. Currently, there are numerous tools and techniques that offer sequence comparisons (sequence alignment) and examine the alignment of the product to understand its biology. One of the major difficulties of bioinformatics is

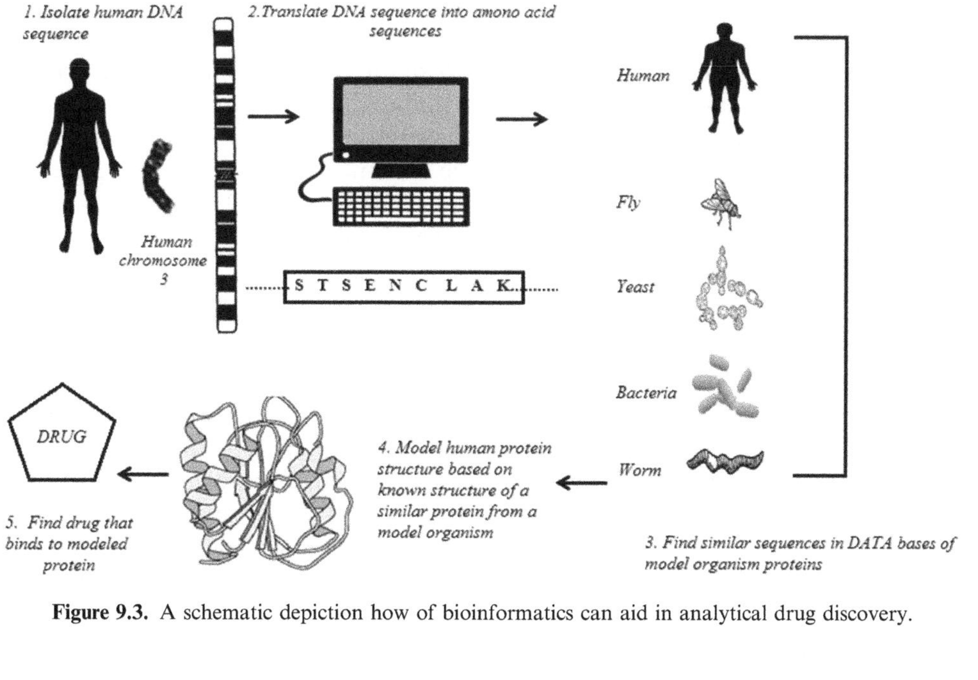

Figure 9.3. A schematic depiction how of bioinformatics can aid in analytical drug discovery.

the arrangement of the vast data in a well-organized and easily comprehensible format. The nucleotide and amino acid sequences are easily reduced to digital data by using single letter codes. The nomenclature system accepted in bioinformatics is founded on the protocols set out by IUPAC, which allow the understanding and utilization of the data produced by different individuals or research groups.

9.3.1 DNA sequences

The key symbols employed to signify DNA sequences are denoted by single letters A (adenine), C (cytosine), G (guanine) and T (thymine). However, sequence data often encompass doubts as to which of the four bases is present at several locations. These doubts in DNA sequences are addressed by frequent sequencing of the related DNA segments. The base sequences of the two complementary strands of a DNA molecule are signified by means of identical symbols. Even those locations that display uncertainty can be represented by this system of symbols. The base sequence of only one strand is registered in databanks. This sequence runs from the 5′ to the 3′ direction, i.e., the 5′-end is at the left-hand extreme and the 3′-end is at the right-hand extreme of the sequence. The base sequence of the complementary strand is effortlessly obtained either manually (for short sequences) or by means of a suitable software package. Exclusively in the case of RNA sequences, the symbol U (for uracil) takes the place of T.

9.3.2 Amino acid sequences of proteins

Amino acids are usually represented by three-letter symbols, e.g. Ala (alanine), Val (valine), etc. However in bioinformatics, they are represented by single letters, such as A (alanine), C (cysteine), D (aspartic acid), etc. However, a number of locations in protein sequences have uncertainties; this situation is comparable to that for DNA sequences. For example, it can be unclear whether a site has glutamine or glutamic acid; such a site is denoted by the symbol Z. Similarly, B represents either asparagine or aspartic acid. The character X is used to show that the position may have any amino acid. The protein synthesis usually initiates at the N-terminus and ensues to the C-terminus. The amino acid sequences in databanks are thus listed from the N-terminus (at the extreme left of the sequence) to the C-terminus (at the extreme right) of the polypeptide.

9.3.3 Types of sequences in nucleotide sequence databases

The databanks of DNA sequences comprise a variety of sequence types. A short explanation of each of these sequence types is provided in the following.

cDNA sequences. A cDNA molecule is derived by reverse transcription of an RNA molecule. The cDNA sequences thus represent that part of the genome that is transcribed into RNA. If the cDNA is derived from mRNA, it will represent only the exon sequences of the genes expressed in the cell/tissue/organism of interest.

Genomic DNA sequences. These sequences signify the complete genome of the organism, regardless of whether it is expressed or not. Once the genome sequence is complete, it will encompass the sequence of the complete genome of the organism. In prokaryotes, the genome typically entails a single chromosome, whereas in the case of eukaryotes the genome comprises the nuclear DNA.

Expressed sequence tag (EST) sequences. ESTs are mRNA sequence fragments resulting from single sequencing reactions executed on randomly selected clones from cDNA libraries [4]. To date, almost 45 million ESTs have been produced from over 1400 different species of eukaryotes. Usually EST projects are used to either match existing genome projects or as low-cost options for purposes of gene discovery. However, with developments in accuracy and coverage, they are starting to find applications in fields such as phylogenetics, transcript profiling and proteomics [4]. These sequences are derived by sequencing only a portion of the cDNA molecules made using mRNA [4]. These sequences are dubbed 'tags' as they can be utilized as probes for the isolation of the related genes from the genomic DNA. This strategy was used by Venter and colleagues for deriving the sequence of the expressed portion of the human genome. The EST sequences method produced massive amounts of sequence data that allowed the assembly of a preliminary transcript map of the human genome. Large numbers of EST sequences have been assembled in an EST databank, dbEST. One of the challenges with ESTs is duplication; long genes may be represented by two or more ESTs. For example, one major databank has over 1300 ESTs for a single gene. This results as an EST has to be short enough to signify a single exon or its part, and long genes have many exons.

Genome sequence tag (GST) sequences.
A genes-first approach to genome sequencing has been described which efficiently generates GSTs from genomic DNA [5]. GSTs were first produced to recognize the genes of *Plasmodium falciparum*. It was noticed that the enzyme mung bean nuclease (Mnase) cuts *P. falciparum* genomic DNA between genes. GSTs are produced by sequencing the DNA fragments on either side of the points of cuts generated by Mnase [5]. Examination of gene sequence tags prepared from mung bean nuclease-digested *P. falciparum* DNA demonstrates that this technique has numerous advantages over the popular cDNA expressed sequence tag approach. So far, 673 sequence tags containing over 215 kb of sequence have been generated from 400 clones [5].

Organellar DNA sequences.
Mitochondria and plastids are membrane-bound organelles that translate energy from foodstuffs or sunlight into cellular energy. Organelles have their own independent genome that encodes a range of genes directly related to producing energy for the cell. Organellar DNA is the DNA existing in mitochondria (mtDNA) and chloroplasts (cpDNA). The sequences of these are stored in databanks. Mitochondria and chloroplasts are thought to have developed from bacteria that formed a symbiotic relationship with ancestral cells containing a eukaryotic nucleus. The majority of the genes initially within these organelles have been transferred to the nuclear genome over evolutionary time, leaving different genes in the organelle DNAs of different organisms.

Sequences of other molecules.
In addition to the DNA sequence databases, sequences of molecules such as tRNA, small RNAs, etc, are also collected in databanks.

9.3.4 Databases

Various database and software resources have reported and applied within the fields of medicine, biology and bioinformatics [6, 7]. Up-to-date information on existing databases and software would be a valuable resource [8]. Previous efforts have been made to preserve accurate lists of available bioinformatics resources, however, most have not been adequately completed due to the slow process of manual curation, or specialized requirements for resource inclusion. There are three public domain bioinformatics services:

- The National Centre for Biotechnol Information (NCBI), located in the USA.
- The European Bioinformatics Institute (EBI), located in the UK.
- GenomeNet (the Japanese Bioinformatics Service), located in Japan.

These organizations develop databases as well as suitable analysis techniques. These computational tools and databases are essential for the organization and management of the vast amounts of biological data. A database/databank is a huge assortment of information relating to a specific topic, such as nucleotide sequence, protein sequence, etc, maintained in an electronic environment. Databanks are at the core of bioinformatics. The amount of data is growing rapidly.

9.3.4.1. Nucleotide sequence databases

The EMBL Nucleotide Sequence Database (http://www.ebi.ac.uk/embl/) incorporates, organizes and distributes nucleotide sequences from all available public sources [9, 10]. The database is located and maintained at the European Bioinformatics Institute (EBI) near Cambridge, UK. In an international collaboration with DNA Data Bank of Japan (DDBJ) and GenBank, data are exchanged amongst the collaborating databases on a daily basis to achieve optimal synchronization [9, 10]. The main nucleotide sequence databanks are thus GenBank, maintained by NCBI, the DDBJ and the Nucleotide Sequence Database maintained by EMBL. These, and several other nucleotide sequence databases that have been created, are listed in the following:

- E. coli: This databank, established by NCBI, has nucleotide sequences of the *E. coli* genome.
- Mito: This databank contains sequences of mitochondrial genomes and is held at NCBI.
- GenBank: This is the key nucleotide sequence databank created by NCBI. This databank comprises nucleotide sequences of genomic DNA.
- dbEST: This database contains EST entries from the GenBank, EMBL and DDBJ databases. This databank is likely to have sequencing errors, contamination by heterologous sequences and the presence of transcribed repetitive elements. The database is maintained by NCBI.
- EMBL: This is a complete nucleotide (DNA and RNA) sequence database held at EMBL and compiled from numerous sources. It is in partnership with the GenBank and the DDBJ. EMBL connects with GenBank and DDBJ on a regular basis and to exchange data and update its contents.

- Kabat: This databank is maintained at NCBI. It has nucleotide sequences that are useful from the immunological point of view.
- Yeast: This NCBI database comprises the nucleotide sequence data of the yeast genome.
- The International ImmunoGenetics Database (IMGT): This databank holds the nucleotide sequences of immunologically important genes, such as T-cell receptors, B-cell receptors, etc.

9.3.4.2 Protein databases

Protein databases have now become an important part of modern biology. Vast amounts of information on protein structures, functions and, mainly, sequences are being generated. Examining databases is often the initial step in the study of a new protein. Comparisons between proteins or between protein families offer information about the association between proteins within a genome or across various species, and therefore offer much more information than can be acquired by reviewing only an isolated protein [11]. Moreover, secondary databases obtained from new databases are also extensively available. These databases rearrange and annotate the data or provide predictions. The utilization of multiple databases often helps scientists to understand the structure and function of a protein [11]. Although certain protein databases are well known, they are far from being fully utilized in the protein science community:

- The Protein Data Bank (PDB): This databank has protein sequences whose three-dimensional structures are already identified. It is maintained at Brookhaven National Laboratory, USA. The records are predominantly nonredundant. The databank is also held ay NCBI's PDB mirror site MMDB, and at the PDB mirror site of EBI, UK.
- Swiss-Prot Database: This is a nonredundant protein sequence databank, held at the University of Geneva, Italy, and maintained by EBI, UK. It is an adaptation of the Protein Identification Resource (PIR) databank of NCBI. NCBI also offers Swiss-Prot, which has the most releases of protein sequence entries from the Swiss-Prot Database of EBI.
- Yeast: This NCBI databank contains yeast protein sequences.
- Kabat: This databank is maintained by NCBI. It has protein sequences that have immunological relevance.

9.3.4.3 Other databases

A number of other types of databanks have been generated, such as the Radiation Hybrid Database (RHdb), which comprises experimental conditions and radiation hybrid records from the human and mouse. The radiation hybrid score vectors can be employed to create chromosome maps that are an alternative to the genetic map. The Kyoto Encyclopedia of Genes and Genomes (KEGG) comprises data on metabolic pathways in numerous micro-organisms. It is part of the GenomeNet (Japan) database system.

9.3.5 Search engines and analysis tools

The exploitation of several databases requires the use of suitable search engines and analysis tools. These tools are often called database mining tools and the process of database utilization is known as database mining.

9.3.5.1 The Basic Local Alignment Search Tool (BLAST)

A method for rapid sequence comparison, BLAST directly estimates alignments that optimize an amount of local similarity, the maximal segment pair (MSP) score [12]. Current mathematical findings on the stochastic characteristics of MSP scores permit an examination of the performance of this technique as well as the statistical implications of the alignments it generates. The basic algorithm is simple and robust; it can be applied in different ways and in a variety of contexts comprising straightforward DNA and gene identification searches, protein sequence database searches, motif searches and in the examination of multiple regions of similarity in long DNA sequences [12]. BLAST [13, 14] is a sequence similarity search program that can be employed to rapidly search a sequence database for matches to a query sequence. A number of variants of BLAST exist to compare all combinations of nucleotide or protein queries against a nucleotide or protein database. In addition to performing alignments, BLAST provides an 'expect' value, statistical information about the significance of each alignment.

BLAST is a family of easy to use sequence similarity search techniques available on the Internet. BLAST is maintained by NCBI. This technique is intended to recognize possible homologs for a given sequence. It can examine both DNA and protein sequences. Documentation of homologs permits the prediction of possible roles and demonstration of the three-dimensional structure. A local arrangement finds the ideal alignment between subregions or local regions of particular sequences. A local alignment search engine discovers the sequence motifs, domains, etc, in the database that are homologous to the submitted sequence motif, domain, etc. Various BLAST programs have been explored. Each of them functions for specific purpose. The newest BLAST programs are known as BLAST 2.0. A variety of BLAST programs are listed in the following:

- BLASTn: This program links a sample nucleotide sequence with a nucleotide sequence database.
- BLASTp: This program matches a sample protein sequence to a protein database.
- BLASTx: This program interprets a sample nucleotide sequence into an amino acid sequence and compares the latter with a protein database.
- tBLASTn: This program translates a sample protein sequence into a nucleotide sequence and compares it with a nucleotide sequence database.
- tBLASTx: This program interprets a sample nucleotide sequence as well as the nucleotide sequence database into amino acid sequences and examines for homology between the two.

As an example, a scholar has created a nucleotide (DNA/RNA) or amino acid (protein) sequence and needs to compare this sequence with those contained in a database with a view to detect homologous sequences. The BLAST search engine carries out the assessment by means of algorithms. Simply, the logic used by BLAST programs is as follows:

- The sequence submitted by researcher is compared base-per-base or amino acid-per-amino acid with the database sequences.
- A replacement scoring matrix is employed by the BLAST programs. Each counterpart is conferred a specified score, while each divergence is penalized by a particular negative score.
- The sequence alignment is then allocated a complete score, which is the summation of scores allocated to each of its paired amino acids/nucleotides.
- Top scoring alignments are rated according to set standards. These measures differentiate between a similarity due to an ancestral relationship and that due to random chance.
- Discovered homologies or counterparts are further studied by means of data available through ENTREZ and other search engines.

The BLAST engine can be accessed online (www.ncbi.nlm.nih.gov). The steps users have to take are as follows:

- The sequence for which homology is to be examined is initially submitted into the 'input sequence' box of the BLAST interface. This sequence has to be in a appropriate format, i.e., the FASTA format.
- A suitable BLAST program is designated depending on the type of assessment to be made.
- The suitable database from which homologous sequences are to be searched has to be selected. The default database used by BLAST is the NR database of NCBI. The NR protein database maintained by NCBI as a target for their BLAST search services is a composite of SwissProt, SwissProt updates, PIR, PDB.
- The sequence is now submitted to the BLAST server.
- The outcomes of the search can be accessed either by email or via the BLAST interface.

9.3.5.2 ENTREZ search and Taxonomy Browser

The NCBI Taxonomy database is a curated set of names and classification for all the organisms represented in the gene bank. There are two main tools for viewing the information, the Taxonomy Browser and Taxonomy ENTREZ. Both systems allow the searching of the databases for names and links to relevance sequence data.

ENTREZ is the text-based search and retrieval system used at the NCBI for all of the major databases, including PubMed, nucleotide and protein sequences, protein structures, complete genomes, taxonomy and many others [15]. ENTREZ is simultaneously an indexing and retrieval system, a collection of data from many sources and an organizing principle for biomedical information [15].

ENTREZ is one of the most standard search tools in the field. The ENTREZ engine explores bibliographic citations and biological information from a range of reliable databases, e.g. Swiss-Prot, PDB, GenBank, EMBL, etc. It uses PubMed's bibliographic database for bibliographic or citation search. ENTREZ provides a range of criteria for searching, and is an extremely useful and compliant search engine. It can be used to explore diverse information, e.g. all potential citations from a specific author in a given area, standard names for given genes, a specific sequence in the databases, etc. There are several database retrieval tools such as ENTREZ, LocusLink, Taxonomy Browser, etc.

The NCBI Taxonomy database (www.ncbi.nlm.nih.gov/taxonomy) is the standard nomenclature and classification repository for the International Nucleotide Sequence Database Collaboration (INSDC), comprising the GenBank, ENA (EMBL) and DDBJ databases. The diversity of organisms is such that millions of species are known. This search engine offers taxonomic information on different species. The Taxonomy databank of NCBI has data (including scientific and common names) on all organisms for which some sequence data are known (over 79 000 species). The engine offers genetic data and the taxonomic relations of the species in question. The Taxonomy Browser has links with the other servers of NCBI, e.g. Structure and PubMed. The browser supports two different kinds of web page hierarchies, which present the familiar indented view of the taxonomic classification, and taxon-specific pages, which summarize all of the information that we associate with a particular taxonomic entry in the database [16].

9.3.5.3 NCBI's LocusLink

The LocusLink and RefSeq databases were started to report data-access issues ensuing from significant increases in both sequence data and the number of web sites relating to information about genes [17]. LocusLink offers a single point-of-access to a variety of gene-specific information sources including web resources [17].

LocusLink is an NCBI project to link information applicable to specific genetic loci from several disparate databases. LocusLink covers data about genes, comprising their official names. Moreover, it permits one to search for genes homologous to a specific gene, and to derive data about these genes. For example, one can effortlessly obtain data about mouse genes that are homologous to given human genes. One can also search for homologs of a specific gene in numerous other organisms. The LocusLink website is www.ncbi.nlm.nih.gov/LocusLink/.

9.3.5.4 PROSITE

PROSITE contains entries describing protein domains, families and functional sites, as well as associated patterns and profiles to identify them [18–21]. It is complemented by ProRule, a collection of rules based on profiles and patterns, which increases the discriminatory power of these profiles and patterns by providing additional information about functionally and/or structurally critical amino acids [18–21]. PROSITE is mainly used for the annotation of domain features of UniProtKB/Swiss-Prot entries. Among the 983 (DNA-binding) domains, repeats and zinc fingers present in Swiss-Prot (release 57.8 of 22 September 2009), 696

(~70%) are annotated with PROSITE descriptors using information from ProRule. In order to allow better functional characterization of domains, PROSITE developments focus on subfamily specific profiles and a new profile building method giving more weight to functionally important residues [18–21]. In other words the PROSITE database consists of a large collection of biologically meaningful signatures that are described as patterns or profiles. Each signature is linked to documentation that provides useful biological information on the protein family, domain or functional site identified by the signature. The PROSITE web page has been redesigned and several tools have been implemented to help the user discover new conserved regions in their own proteins and to visualize domain arrangements [18–21].

PROSITE has an assortment of active sites and sequence patterns present in various proteins. Entries in PROSITE are usually associated with Swiss-Prot and other significant databases. The PROSITE file comprises the sequence entries that share the matched sequence motif-of-interest. The characterized motifs are well documented to minimize redundancy. PROSITE has search engines for comparing patterns/motifs. The PROMOT search engine can be employed to compare a sequence against the PROSITE database. PROSEARCH is another search engine to explore the Swiss-Prot and Tremble databases for a specific motif.

9.3.6 Various Indian databases

9.3.6.1 GM crops database

This database, based at the National Research Centre on Plant Biotechnology, New Delhi, is an active web resource storing data on the biosafety of transgenics (released in India), comprising over 800 publications on the subject. For example, a total of 139 transgenic lines using four genes (crylAb, crylAb, cry2Ab and vip3A) and a single promoter (CaMV 35S) have been established.

9.3.6.2 Vanshanudhan

The Vanshanudhan database (http://125.18.242.23:8080/genome/Login.jsp) allows searches based on complete genome data comprising genes, cDNA and protein sequences. Currently, the Vanshanudhan data are based on rice pseudomolecule version 3.0, released from Michigan State University with a unique gene nomenclature, e.g. 01-0001, which defines the chromosome number as well as gene number. This database has been established by the NRC on Plant Biotechnology researchers as an result of an Indian rice genome initiative [22]. It covers information on the 56 298 rice genes.

9.4 Investigation by means of bioinformatics tools

Bioinformatics based engines have been established for distribution of the large amount of biological data being produced at a very rapid rate. For example, when a long piece of DNA has been sequenced, the first objective is to regulate if it contains any genes. This objective is comparatively simple in the case of prokaryotes, but the search for eukaryotic genes is quite difficult.

9.4.1 Identification of genes

With the increase in genomic and functional genomics information, approaches for disease gene identification are rapidly being developed [23]. Databases are now indispensable to the process of selecting candidate disease genes. Relating positional information with disease characteristics and functional information is the normal approach by which candidate disease genes are selected. Enrichment for candidate disease genes, however, depends on the skills of the operating researcher [23].

Prior to genome sequencing, gene identification used approaches based on transcript mapping. The genomic clone of interest was exposed to zoo blot hybridization, i.e., hybridization with the whole genomic DNA of a range of species. It has been discovered that coding sequences are strongly conserved during development. Thus, a clone that tested positive in a zoo blot is possible to characterize as a gene. The clone of interest can be hybridized with cDNA libraries or employed for northern blot or reverse northern blot assays. A constructive assay identifies a gene as only genes are transcribed. CpGProD is an application for identifying mammalian promoter regions associated with CpG islands in large genomic sequences [24]. Detection of CpG islands in the clone suggests it to be a gene as about 50% of human genes have related CpG islands. CpG islands are short stretches of G. C-rich DNA often found in association with vertebrate genes. Other methods of gene identification included cDNA selection, cDNA capture and exon trapping. These methods provide research that is appropriate for individual gene identification but unsuitable for genome annotation. Consequently, the correct reading frame is identified by carrying out a six-frame translation of the DNA sequence. The accurate reading frame is predicted to be the longest frame uninterupted by a stop codon (TGA, TAA or TAG); this type of reading frame is known as an open-reading frame (ORF). The longer this ORF, the better is the probability that it represents a gene. Locating the 3′-end of an ORF is relatively easier than locating its 5′-end, although the 5′-end may be indicated by an ATG. However, more tools are required to locate the 5′-end of an ORF, such as the presence of a Kozak sequence (CCGCCA UGGG) flanking the AUG codon. Investigation of codon usage may also be supportive, and the presence of CpG islands may designate the 5′-end of numerous vertebrate genes. However, determination of ORFs may be hindered by sequencing errors. The most significant single progress in genome annotation is the use of PCs to predict genes from DNA sequences. Identification of genes may achieved using one of the following two approaches. In the first technique, sequences of known genes, cDNAs, ESTs and proteins present in databases are equated with the genome sequence; this is achieved by homology search techniques, such as BLAST. In the second method, specific software is employed to identify genes.

In prokaryotes, programs such as GenMark (a modified GeneScan algorithm) and Glimmer can identify all the genes, including overlapping genes, present in the genome [25].

Numerous sophisticated software programs have been developed for gene prediction in eukaryotes, e.g. Genie, GeneScan, Grail, GeneFinder, HMM Gene, etc [26]. These programs required training with information or need to be provided with a set of rules. Grail can be employed with human, mouse, *Arabidopsis*, etc, genome sequences, but Genie has only been trained on human and *Drosophila* sequences. However, no single program is 100% perfect at identifying genes. An algorithm trained with the DNA sequence of one organism, *Caenorhabditis elegans*, will not execute suitably with that of another organism, e.g. a plant, without being retrained.

The gene prediction programs search for gene-specific characteristics, e.g. promoters, splice sites and polyadenylation sites, or for pertinent gene content such as ORFs. The currently available gene search programs are associated with different search criteria and their sensitivities vary widely. The identification of ORFs, usually exceeding 300 nucleotides, is sufficient to find most genes in prokaryotic genomes. However, such a simple search criterion will miss smaller genes and overlapping genes. These problems are resolved by using algorithms that consider differences in base composition between genes and noncoding DNA, e.g. in GenMark. The gene prediction programs used in eukaryotes use the output of several algorithms to generate a whole gene model. In this model, a gene is defined as a series of exons that are coordinately transcribed. The various features of eukaryotic genes recognized during gene detection include transcriptional and translational controls, e.g. the TATA box, cap site, Kozak consensus and poly-adenylation sites. But problems arise as the TATA box is missing in 70% of human genes, and polyadenylation signals can differ considerably from the consensus sequence AATAAA. Moreover, these sequences recognize only the first and the last exons of a gene. Thus, additional features have been encompassed in modern gene search tools; these features include 5′- and 3′-splice sites, differences in base composition between coding and noncoding DNA (typically, comparison of hexamer base composition), etc.

9.4.2 Identification of the function of a new gene

A convenient approach to detect the function of a new gene is as follows. The gene sequence is translated into the amino acid sequence and thus the protein it is anticipated to encode. This protein sequence is then compared with a protein database. A program such as tBLASTx will execute both these functions. If the encoded protein is homologous to a protein in the databank, it allows identification of the new gene and also suggests the function of the new gene. The FASTA and HMMER programs are slower than BLAST, but they are more sensitive.

9.4.3 Identification of functional domains

Several bioinformatics techniques for the detection of protein motifs and protein domains have been explored. Among these tools are PRINTS, PROSITE, SMART, BLOCKS, etc.

9.4.4 Detection of noncoding RNA

A number of RNAs are noncoding, such as rRNA, tRNA, small RNAs, etc. Of these, rRNAs are the easiest to find; this is done by sequence similarity search. The program tRNAScanSE searches for aRNA sequences.

9.4.5 Genome annotation

After the detection of a gene and prediction of its functions, suitable experiments have to be devised to authenticate these findings. This information is then employed for genome annotation. Standard genome annotation languages have become extensively accepted. GAME is a program for describing investigational confirmation to support annotation. Likewise DAS (the distributed annotation system) is mainly useful for indexing and visualization. Several software programs such as BioPerl 2001, BioJava 2001, etc, are employed for storing, manipulating and imagining the genome annotations.

9.4.6 Molecular phylogenetics

DNA and protein sequence statistics can be employed to examine the development of genes and their protein products; this is called molecular phylogeny. Bioinformatics techniques are employed to define phylogenetic relationships, which may be accessible in the form of either a phylogenetic tree or a dendogram. While the two depictions look different, they depict exactly the same relationship.

References

[1] Roumpeka D D, Wallace R J, Escalettes F, Fotheringham I and Watson M 2017 A review of bioinformatics tools for bio-prospecting from metagenomic sequence data *Front. Genet.* **8** 23

[2] Diniz W J and Canduri F 2017 Bioinformatics: an overview and its applications *Genet. Mol. Res.* **15** 16

[3] Strasser B J 2010 Collecting, comparing, and computing sequences: the making of Margaret O. Dayhoff's atlas of protein sequence and structure, 1954-1965 *J. Hist. Biol.* **43** 623-60

[4] Parkinson J and Blaxter M 2009 Expressed sequence tags: an overview *Methods Mol. Biol.* **533** 1–12

[5] Reddy G R, Chakrabarti D, Schuster S M, Ferl R J, Almira E C and Dame J B 1993 Gene sequence tags from *Plasmodium falciparum* genomic DNA fragments prepared by the "genease" activity of mung bean nuclease *Proc. Natl Acad. Sci. USA* **90** 9867–71

[6] Cannata N, Merelli E and Altman R B 2005 Time to organize the bioinformatics resourceome *PLoS Comput. Biol.* **1** e76

[7] Wren J D and Bateman A 2008 Databases, data tombs and dust in the wind *Bioinformatics* **24** 2127–8

[8] Gilbert D 2004 Software review: bioinformatics software resources *Brief. Bioinform.* **5** 300–4

[9] Kulikova T *et al* 2004 The EMBL nucleotide sequence database *Nucleic Acids Res.* **32** D27–30

[10] Stoesser G *et al* 2002 The EMBL nucleotide sequence database *Nucleic Acids Res.* **30** 21–6

[11] Xu D and Xu Y 2004 Protein databases on the internet *Curr. Protoc. Protein Sci.* **33** 2.6.1–2.6.15

[12] Altschul S F, Gish W, Miller W, Myers E W and Lipman D J 1990 Basic local alignment search tool *J. Mol. Biol.* **215** 403-10

[13] Camacho C, Coulouris G, Avagyan V, Ma N, Papadopoulos J, Bealer K and Madden T L 2009 BLAST+: architecture and applications *BMC Bioinform.* **10** 421

[14] Altschul S, Madden T, Schäffer A, Zhang J, Zhang Z, Miller W and Lipman D 1997 Gapped BLAST and PSI-BLAST: a new generation of protein database search programs *Nucleic Acids Res.* **25** 3389–402

[15] Ostell J 2013 The ENTREZ Search and Retrieval System *The NCBI Handbook* 2nd edn (Bethesda, MD: National Center for Biotechnology Information)

[16] Federhen S 2012 The NCBI taxonomy database *Nucleic Acids Res.* **40** D136–43

[17] Maglott D R, Katz K S, Sicotte H and Pruitta K D 2000 NCBI's LocusLink and RefSeq *Nucleic Acids Res.* **28**(1) 126–28

[18] Bairoch A 1992 PROSITE: a dictionary of sites and patterns in proteins *Nucleic Acids Res.* **20** 2013–18

[19] Christian J A 2010 PROSITE, a protein domain database for functional characterization and annotation *Nucleic Acids Res.* **38** D161–6

[20] Hulo N *et al* 2008 The 20 years of PROSITE *Nucleic Acids Res.* **2008** D245–9

[21] Nicolas H *et al* 2004 Recent improvements to the PROSITE database *Nucleic Acids Res.* D134–37

[22] Jayaswa P K, Singh B P, Sharma T R and Singh N K 2012 Vanshanudhan: a rice gene database for geneticists and breeders *3rd World Congress on Biotechnology. Agrotechnol. (13–15 September, Hyderabad, India)*

[23] Driel M A and Brunner H G 2006 Bioinformatics methods for identifying candidate disease genes *Hum. Genomics* **2** 429–32

[24] Ponger L and Mouchiroud D 2002 CpGProD: identifying CpG islands associated with transcription start sites in large genomic mammalian sequences *Bioinformatics* **18** 631-3

[25] Aggarwal G and Ramaswamy R 2002 *Ab initio* gene identification: prokaryote genome annotation with GeneScan and GLIMMER *J. Biosci.* **27** 7-14

[26] Rogic S, Mackworth A K and Ouellette F B F 2001 Evaluation of gene-finding programs on mammalian sequences *Genome Res.* **11** 817–32

IOP Publishing

Introduction to Pharmaceutical Biotechnology, Volume 2
Enzymes, proteins and bioinformatics
Saurabh Bhatia

Chapter 10

Protein and enzyme engineering

10.1 Protein and enzyme engineering

Protein engineering is the design of new enzymes or proteins with new or desirable functions. It is based on the use of recombinant DNA technology to change amino acid sequences. Many different protein engineering methods are available today, owing to the rapid development in biological sciences, more specifically, recombinant DNA technology. In an earlier chapter, we discussed the principles, uses and techniques of biocatalysis and enzyme technology. In the past, the industry was largely dependent on these naturally occurring enzymes. However, naturally occurring enzymes and their related bioprocesses are often limited by a set of conditions. With advancements in biotechnology, it has been possible to plan and engineer novel enzymes and other macromolecules with predicted characteristics. This field of research also gained more support from computer aided molecular modeling (CAMM), which has contributed considerably, not only to protein and enzyme engineering, but also to drug design. Different characteristics of protein and biocatalyst engineering will be discussed in this chapter. To advance the production of a particular macromolecule or to produce novel products, metabolic pathways are also being designed in different ways. The various steps involved in protein engineering are shown in figure 10.1.

10.2 Designing macromolecules

Proteins are versatile macromolecules with various important functions, including structural, catalytic, sensory and regulatory functions. The rational design of enzymes is a great challenge to our understanding of protein structure and physical chemistry and has numerous potential applications. Protein design algorithms have been useful to design or engineer proteins that fold, fold faster, catalyze, catalyze faster, signal and adopt preferred conformational states. The field of *de novo* protein design, although only a few decades old, is beginning to produce exciting results. Advances in this field are already having a considerable influence on biotechnology

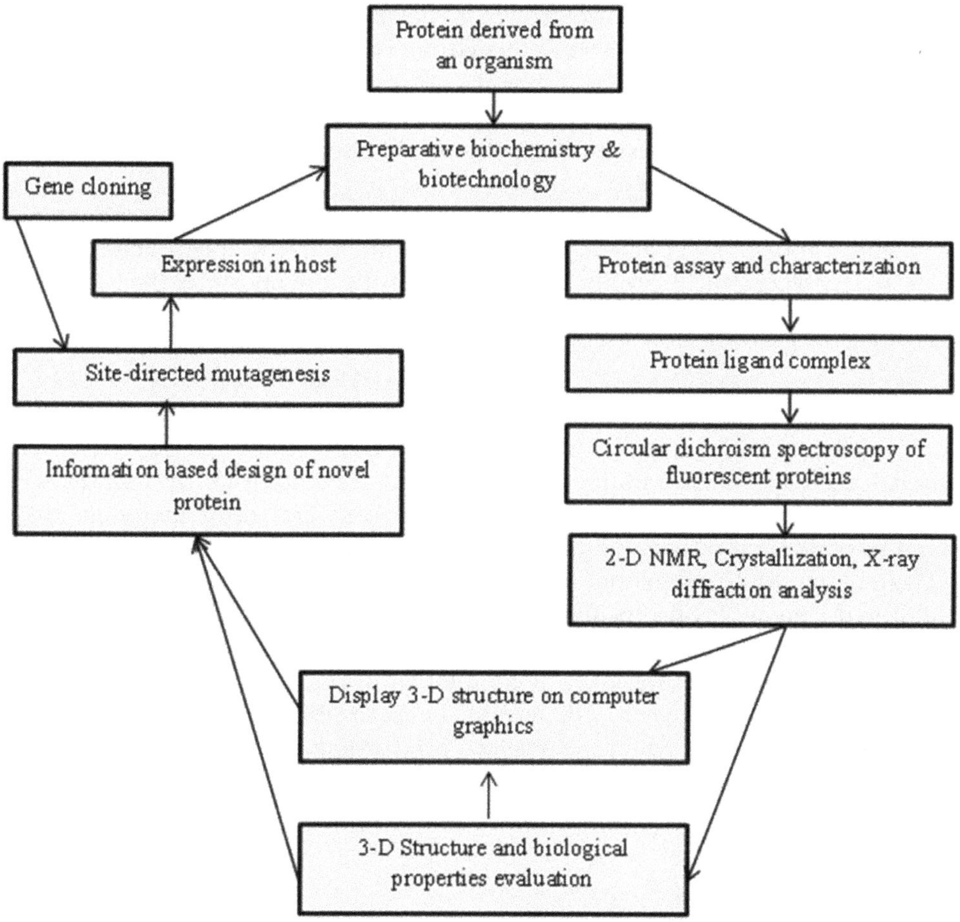

Figure 10.1. The various steps involved in protein engineering.

and chemical biology. The application of powerful computational approaches for functional protein design has currently succeeded in engineering target activities [1].

In the last quarter of the twentieth century, various major discoveries have made it possible to modify/alter the structure of novel proteins/molecules/products in a deliberate way leading to a predicted functional role. Site-directed mutagenesis allows changes in sequences at identified sites, resulting in modification of proteins in a deliberate way. Moreover, computational and graphical developments have allowed access to the three-dimensional structure of proteins. These two developments allow us to understand protein–drug interactions more deeply. The steps involved in protein engineering are as follows:

- The protein sample is characterized with a ligand (drug, enzyme substrate, receptor hormone or antigen–antibody interaction).
- The three-dimensional structure is investigated using nuclear magnetic resonance (NMR) and x-ray diffraction patterns of crystals.

- The three-dimensional structure is displayed with interactive computer graphics and the available information is used to suggest a novel design suiting the requirements.
- The desired DNA sequence that is expected to give the novel designed protein is then either synthesized or obtained by site-directed mutagenesis of an available gene.
- The novel gene is incorporated into an appropriate expression system and the gene product is extracted, purified and characterized biochemically.
- If the biochemical profile does not fulfill the previously earlier anticipated structure, the cycle may be repeated again, until the anticipated structure is achieved.

It must be noted that the aforementioned steps of protein engineering share steps employed in the engineering of drugs, insecticides, herbicides and peptide vaccines, although there are other additional methods that are employed for drug discovery. The various cycles involved in genetic engineering and drug designing display common steps, as shown in figure 10.2.

Drug design and protein engineering are different from each other in the sense that drug design only concerns the ligand, which can be modified more efficiently by chemical synthesis, whereas in protein engineering rDNA technology is usually used. Three-dimensional structures for a small number of proteins that are interrelated to the novel protein of interest will usually be accessible to a protein engineer in developing a procedure for novel protein production as per the desired features. Consequently, a process is required for production of the three-dimensional structure of the protein of interest. This can be achieved *de novo* through modeling

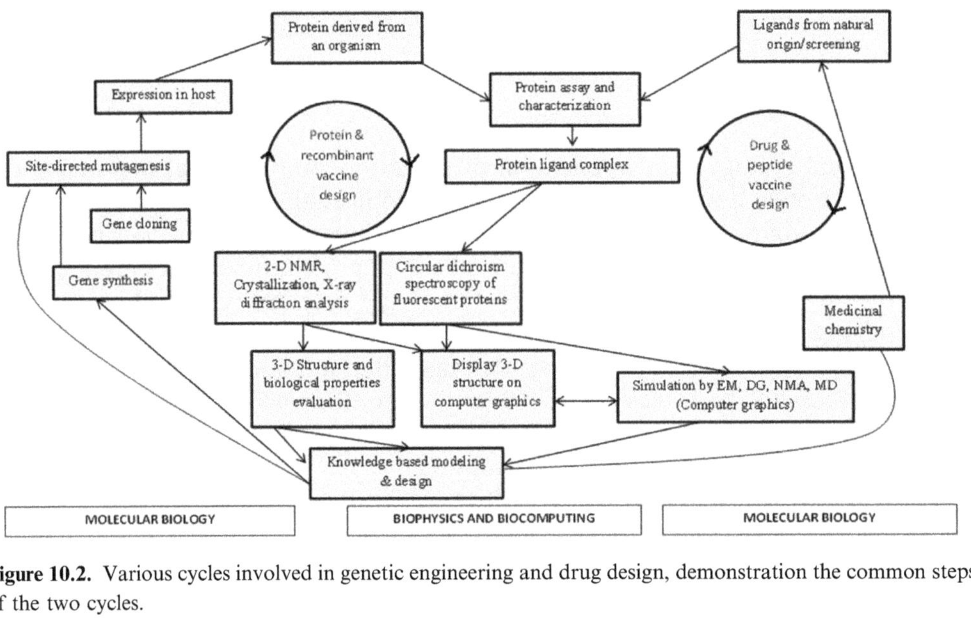

Figure 10.2. Various cycles involved in genetic engineering and drug design, demonstration the common steps of the two cycles.

from homologous or analogous proteins, if they are accessible. This is defined as knowledge based modeling and involves design by means of computer software, comprising computer graphics, computer simulations and databases. Computer software comprises the complete set of methods and programmes used on the computer to rectify problems and provide operational aids. The various components of computer software used in knowledge based molecular modeling are illustrated in figure 10.3.

10.3 Protein engineering versus enzyme engineering

Protein engineering involves changing the structure of a protein to enhance or amend its properties. Protein engineers are progressively able to rely on structure–function insights, computational methods and deeper understanding of natural biosynthesis processes, to streamline the design and applications of enzymes [2, 3]. Enzyme engineering is experiencing the most intense and exciting transformation in its history [2, 3]. This promises extraordinary expansion in the scope and applications of modified or improved enzymes with the desired physical and catalytic properties. Two complementary approaches are currently available: rational redesign and directed evolution.

Although the terms protein engineering and enzyme engineering may often be used interchangeably, there is an indirect difference between the two. Protein engineering allows structural modifications of the function and selectivity of enzymes, mainly in an aqueous medium.

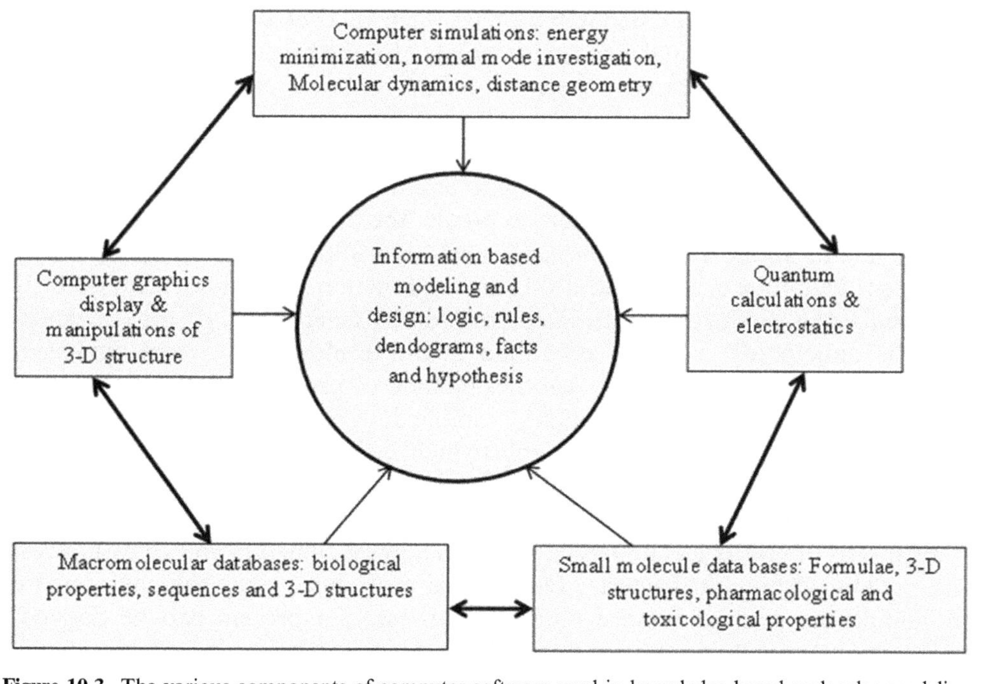

Figure 10.3. The various components of computer software used in knowledge based molecular modeling.

Enzyme engineering, in contrast, sometimes also includes engineering of the enzyme micro-environment, therefore offering applications in non-aqueous environments, as enzymes can function in organic solvents. Such non-aqueous environments offer benefits such as:

- higher substrate solubility;
- reversal of hydro reactions; and
- modified enzyme specificity.

These advantages offer new enzyme activities that can be achieved by means of genetic modifications or using complex multistep pathways within the cells.

10.4 Protein engineering

Protein engineering is the design of new enzymes or proteins with new or desired functions. It is based on the use of rDNA technology (to change amino acid sequences). The first report on protein engineering dates back to the early 1980s, in a publication by Ulmer (1983). Currently, due to the advancements in rDNA technology and high-throughput screening approaches, protein engineering methods and applications are becoming increasingly important and widespread [4].

In the last two decades rapid development has been made in the examination of protein structure and function. Amino acid sequences are now accessible for as many as 8000 proteins, however, the three-dimensional structures of only about 400 proteins have been determined using the x-ray crystallography technique. It has been observed that the investigation of protein structure and function has now reached the level which for DNA existed in the 1970s. Such an assessment is supported by the fact that only one center in the world is available for the procurement of protein data (the Brookhaven Data Bank), whereas a number of such centers are available for DNA sequence data. From the three-dimensional structure of the 400 proteins examined so far, it is likely that if two proteins are comparable in their amino acid sequence, they will be inclined to fold into similar three-dimensional structures, so that it will be possible for researchers to predict the three-dimensional structure of a protein and its amino acid sequence. This will then allow researchers to identify the perfect structure for a protein and forecast its function. This model can be further utilized for the production, identification and characterization of a gene that will give the anticipated sequence of amino acids, resulting in a three-dimensional structure for a specified protein. Another method of protein engineering may be to alter a protein by suitable reaction to make it more appropriate for desirable function. This area of protein research (which includes the modification/development of protein either by recombinant technology or by any chemical reaction) is currently growing considerably.

By means of rDNA technology, a gene can now be cloned in an expression vector and made to express in bacteria. During this procedure, using the considered cell containing the gene of interest cloned in a vector, a protein can be derived in sufficient quantity. As a gene can be artificially incorporated and altered by means of rDNA technology, novel proteins can be achieved. The field of protein engineering includes enzymes, synthetic peptides, storage proteins and drugs to be used in medicine, industry and agriculture. The goals of protein engineering are as follows:

- To generate superior enzymes to further catalyze the production of high value synthetic chemicals.
- To synthesize enzymes for large-scale use in the chemical industry and to produce biological compounds (including peptides, storage proteins and specific proteins) that are more therapeutically active/functional than natural ones.

10.5 Foundation of protein (enzyme) engineering

Many proteins have been characterized in prokaryotes and eukaryotes, but only a few have become commercially significant. This is because of the high cost of purifying enzymes in sufficient quantities. The cost factor can be overcome by producing an enzyme in sufficient quantities from bacteria, but for its commercial usage, an enzyme (under *in vitro* conditions) should also have some features in addition to those typical of enzymes in cells. These features are as follows:

- The enzyme should be vigorous with a long life.
- The enzyme should be able to use the substrate provided by industry even if it differs slightly from that in the cell.
- The enzyme should remain active under the conditions (e.g. extremes of pH, temperature and concentration) of the industry, even if these conditions differ from *in vivo* conditions.

As mentioned above, an enzyme should be designed to meet the specific needs. Thus, efforts have been made to modify the properties of enzymes. The following properties need to be modify to meet the requirements with the help of protein or enzyme engineering:

- allosteric regulation;
- cofactor requirements;
- the kinetic properties of enzyme-turnover and the Michaelis constant, K_M;
- the molecular weight and subunit structure;
- optimum pH;
- protease resistance;
- the stability and activity of the enzyme in non-aqueous solvents;
- substrate and reaction specificity; and
- thermostability and the optimum temperature for the enzyme.

For a specific category of enzymes, differences may be present for each of the above properties, so that one may like to combine the ideal characteristics to obtain the most effective form of enzyme. Occasionally, however, it may not be possible to obtain a combination of ideal properties. For an example, an enzyme with maximum activity may not be the most stable. Consequently, a compromise among properties might have to be made to produce an appropriate enzyme from the existing variability or even from inducing variability by mutagenesis. However, if the structure–function relationship of an enzyme is known, the structural properties for anticipated function may be combined and protein engineering procedures may then be employed to create a novel enzyme presenting a combination of all necessary functional properties. This part of protein engineering can be demonstrated by means of the example of glucose isomerases, which convert glucose into other isomers such as fructose, and is used to make high fructose corn syrup, which is vital for the soft drink industry. It displays wide variation in its properties. Glucose isomerases from the TIM barrel family of enzymes look similar, with a highly characteristic domain called a TIM barrel with an active site for catalytic action at one end. This TIM barrel can be present in enzymes that may vary in sequence and may catalyze diverse reactions. As discussed, a resemblance in structure of a protein should suggest resemblance in function, thus the TIM barrel offers a challenge to this concept. However, it is interesting that certain enzymes in this family show similarity in their metabolic pathways, so that they catalyze two successive steps, consequently presenting coupling of their functions. As an example of two enzymes in the TIM barrel family, triose phosphate isomerase is one of the most efficient catalysts, while glucose isomerase is very inefficient. Consequently, if the glucose isomerase enzyme is restructured to use the highly effective domain of the TIM barrel family, it will be an extraordinary achievement for the soft drink industry.

10.6 Basic assumptions for protein engineering

In protein engineering, one should recognize the following characteristics of enzymes:

- Several amino acid substitutions, deletions or additions result in no enzyme activity, so they are silent mutations.
- Proteins have a limited number of basic structures and only minor changes need to be made to them to lead to variations.
- Related patterns of chain folding and domain structure can arise from different amino acid sequences, which show little or no homology (while similar amino acid sequences never give different folding or domain structures).

The above characteristics suggest that many base changes may occasionally result in modification in function; a number of modifications at particular positions may result in desired favorable modifications. For example, single amino acid replacement (glycine to aspartic acid) in *E. coli* aspartate transcarbamylase results in:

- loss of activity, and
- a modification in the binding of catalytic and regulatory subunits.

Another example is the engineering of a single biosynthetic antibody binding site, which is only 1/6 of the size of the complete antibody but preserves its antigen-binding specificity. A heavy and light chain variable region present in this synthetic fragment is linked by a 15-amino acid linker. For this fragment, a synthetic gene has also been prepared, which is allowed to express in *E. coli*. This fragment binds to digoxin, a cardiac glycoside. These types of single amino acid replacements in BABS fragments have occasionally resulted in major modifications in its binding affinity. Consequently it is essential to examine not only the crystal structure but also the active sites therein, so that the gene can be modified or artificially produced for protein engineering to meet the desired requirements.

10.7 Steps involved in protein engineering

Due to the rapid development in biological sciences, more specifically rDNA technology, different protein engineering methods are currently available. Figure 10.4 shows the steps involved in protein engineering.

10.7.1 Studying three-dimensional protein structure

Proteins are molecular devices (nanometer scale), where biological function is exercised [5]. There are 20 natural amino acids, whose occurrence is greater than that of other special ones with particular functions. These 20 amino acids can be grouped together forming polypeptide chains, or proteins, in different ways determined by the genetic code and limited by stereochemical properties. These proteins may have a constitutive or transient cell expression with regard to its functions. It is worth mentioning that efforts are underway to make proteins of unnatural amino acids as well [6, 7].

An investigation of the three-dimensional structure of a protein (including examination of active sites) is the first and foremost step in any application of protein engineering. For such an investigation, the protein should be present in crystal form. X-ray diffraction procedures have also been established to produce data at a rapid rate, to allow us to forecast the three-dimensional structure of proteins by means of model building. Experimental differences in the structures of

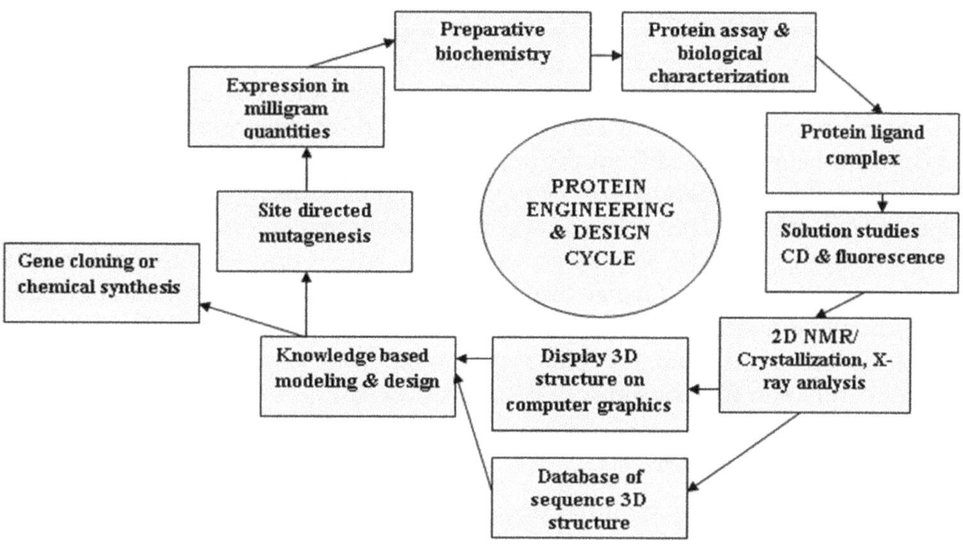

Figure 10.4. The steps involved in protein engineering.

various enzymes can be connected with the differences in their function, so that the information from such investigations can be further employed for protein engineering. X-ray diffraction information at altered temperatures also permits researchers to understand the modifications anticipated in protein structure due to changes in temperature, signifying that protein structure is dynamic. Likewise, the NMR method will permit researchers to understand the structure of enzymes in solution rather than in crystals, so that researchers can avoid the time-consuming process of crystallization, thus speeding up the study of protein structure. The information generated from all these techniques is utilized in protein modeling, which is then further employed for an exercise in protein engineering. When crystals of a protein are not accessible for investigation, but amino acid sequence data exist, identical sequences in any two proteins are matched and the effects of substitution are studied. This is occasionally called 'sequence garing' and permits researchers to understand which amino acids are involved in catalytic binding at active sites.

10.7.2 Protein modeling

The tertiary structure of proteins can disclose information that is difficult to detect in a linear sequence. Understanding the tertiary structure is important when generating hypotheses and interpreting data. Unluckily, the gap between the number of known protein sequences and their associated structures is widening. One way to bridge this gap is to use computer-generated structure models of proteins [8]. By using information produced by x-ray diffraction and NMR investigations, models can be created with the help of computer graphics. Advanced computer programmes are available (interactive color graphics programmes) by means of which a protein structure can be fitted to the electron density map (obtained from x-ray diffraction) through simultaneous display on a computer monitor. Likewise, van der Waals surfaces for the specific protein can be presented and interaction between several molecules simulated. Other interactive molecular graphics can also be employed (with the help of computer software) to discover the perturbations (disturbances) in protein structure that will result from specific modifications of amino sequences. It was also observed that to a certain extent the three-dimensional structure of a protein can be anticipated from the amino acid sequence, however for more clarity researchers are still dependent on x-ray diffraction patterns for elucidating the three-dimensional structure. It is well known that accurate prediction of three-dimensional structure by amino acid sequences will rectify many problems in protein based therapeutics, which may further result in long term success in protein engineering. Once the three-dimensional structure is accessible, different protein models can be established to investigate the effect of amino acid sequence alterations on its structure–function relationships.

10.7.3 Perturbation theory

The theory of perturbation allows researchers to accurately calculate the effects of small changes in amino acid sequence. So, according to this theory, there will be no prerequisite to perform experiments at each step of the protein engineering process to evaluate the outcome (chat function) of recommended modifications [9]. Such a theory will result from protein refinement and calculation of the theoretical structure of a protein which will be homologous to other proteins of known structure. This theory can be tested by using families of monoclonal antibodies. When established for prediction, it can be used to engineer a variety of proteins [9].

10.8 Methods of protein engineering

Various different protein engineering approaches are available today, due to the rapid advancements in biological sciences, more specifically, rDNA technology. The most traditional approach in protein engineering is the so-called rational design approach, which encompasses site-directed mutagenesis of proteins [10]. Site-directed mutagenesis permits the introduction of specific amino acids into a target gene. The popular approach for site-directed mutagenesis is called the overlap extension approach. This approach includes two primer pairs, where one primer of each primer pair contains the mutant codon with a mismatched sequence.

Different methods have been introduced for future use in protein engineering (figure 10.5). In this context, mutagenesis, selection and recombinant DNA technology are being used and will be utilized more widely in the future.

10.9 Mutagenesis and selection of mutant enzymes

Mutagenesis is performed for the improvement of a specific property of an enzyme. The following are some examples of the selection of mutant enzymes:

- *Anthranilate synthetase*: Physiological studies were executed under nutritional stress and nonstress conditions to measure the relative importance of the various regulatory mechanisms that *Escherichia coli* can use to modify its rate of tryptophan synthesis [11]. The enzyme *E. coli* anthranilate synthetase is normally sensitive to tryptophan inhibition due to feedback inhibition. After mutation, an MTR 2 mutation mutant strain of *E. coli* was found to possess a different form of anthranilate synthetase that is insensitive to tryptophan inhibition. This will allow the continuous synthesis of tryptophan without any inhibition accumulating as a product.
- *Xanthine dehydrogenase*: A point mutation in the structural gene for purine hydroxylase I (xanthine dehydrogenase) of *Aspergillus nidulans* results in

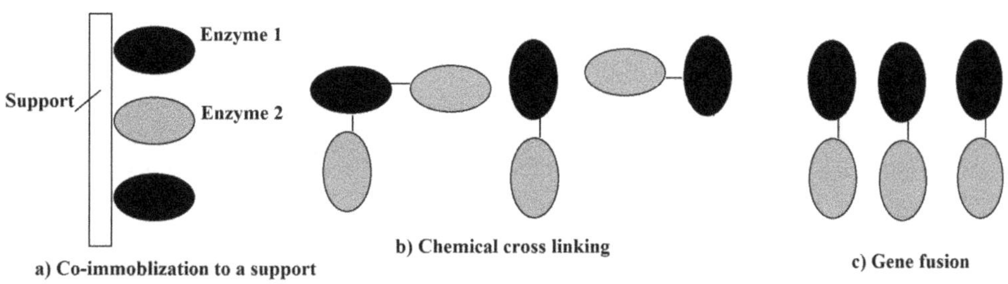

Figure 10.5. Basic methods for generating the proximity between two enzymes.

several dramatic pleiotropic effects [12]. This enzyme oxidizes 2 hydroxy-purine at position 8, but a mutant has been isolated which oxidizes 2 hydroxy-purine at position 6.
- *Lactate dehydrogenase*: The mechanism of catalysis of malate dehydrogenase is similar to that of lactate dehydrogenase, an enzyme with which it shares a similar three-dimensional structure. This enzyme, present in bacteria, was altered to malate dehydrogenase by a natural mutation resulting in amino acid substitution. Substitution of a single amino acid residue of a lactate dehydrogenase changes the enzyme specificity to that of a malate dehydrogenase, but a similar substitution in a malate dehydrogenase resulted in relaxation of a high degree of specificity for oxaloacetate [13].

As mentioned above, single amino acid modification or addition/deletion is possible, however, if further improvement requires modifications in several amino acids, such a mutant will be rare or nonexistent and this type of alteration can only be achieved by gene modifications.

10.10 Gene modifications or gene synthesis for protein engineering

In gene modifications some oligonucleotides can be synthesized by means of automatic DNA synthesizers with solid supports or on microarrays. Two procedures are available for gene modification using synthetic oligonucleotides.

In vitro mutagenesis using synthetic oligonucleotides. In this procedure synthetic oligonucleotides can be employed for *in vitro* mutagenesis of genes of interest. This method involves synthesis of small synthetic oligonucleotide primers containing the desired modifications. Then it is hybridized to a suitable position in a cloned gene and the rest of the gene is then replicated by means of the polymerase enzyme, so that the rest of the gene remains intact. This strategy was employed to alter the active site of tyrosyl-tRNA synthetase, whose structure was already known. In this enzyme at position 35, cysteine was substituted by serine, with the expected effect of reducing K_M for adenosine triphosphate. In 1983, this was defined as the main step forward in protein engineering. PCR can also be employed for inducing mutations in known genes for the purposes of protein engineering.

The synthesis of complete modified genes (de novo). Complete genes can in some cases be chemically synthesized in the form of oligomers. For instance insulin, somatostatin and interferon are common examples in which complete genes have been synthesized in the form of oligomers. To introduce restriction enzyme sites at suitable positions, the sequence of the synthetic gene can be designed in a modular fashion. The cleaving gene at suitable positions allows the incorporation of desirable fragments, resulting in modification of function. As various oligomers are involved, modification can be incorporated in different oligomers and these oligomers can be further utilized in ligation mixtures for the synthesis of a complete gene, which will have a number of modifications. This procedure can be employed for making extensive modifications in the amino acid sequence of the protein. For example, the interferon gene which was artificially synthesized opened the way to modify the portions of the interior so that a variety of artificial interferon genes can be prepared, which can be later allowed to express in *E. coli.*

10.11 Multi-enzyme systems

In biotechnological production, multi-enzyme systems have been artificially synthesized. These enzymes are capable of catalyzing sequential reactions. By using a number of techniques such as co-immobilization, chemical cross linking and gene fusion, the proximity of more enzymes can be achieved (figure 10.6). During gene fusion, the structural genes of two or more enzymes are joined together to prepare bi- and polyfunctional enzymes.

At the 3′-end the translational stop signal of the first gene is deleted and ligated in frame to the ATG start codon of the second gene. Short linkers can also be used for this purpose. A number of examples of multienzymes prepared by gene fusion *in vitro* are mentioned below:

- β-galactosidase–galactokinase (tetramer) for the sequential hydrolysis of lactose to glucose.
- β-galactosidase–galactose dehydrogenase (the hybrid enzyme is a dimeric enzyme) catalyzes the hydrolysis of lactose to galactose which is oxidized to galactolactone.
- β-galactosidase–galactose dehydrogenase–galactokinase (tetramer or an octamer).
- Galactose dehydrogenase–luciferase (a bifunctional enzyme).

10.12 Chemical modification of enzyme

Proteins produced by genes undergo post-translational modifications which lead to stability, structural integrity, altered solubility and viscosity of individual proteins. This may also affect their chemical reactivity. These modifications can be achieved

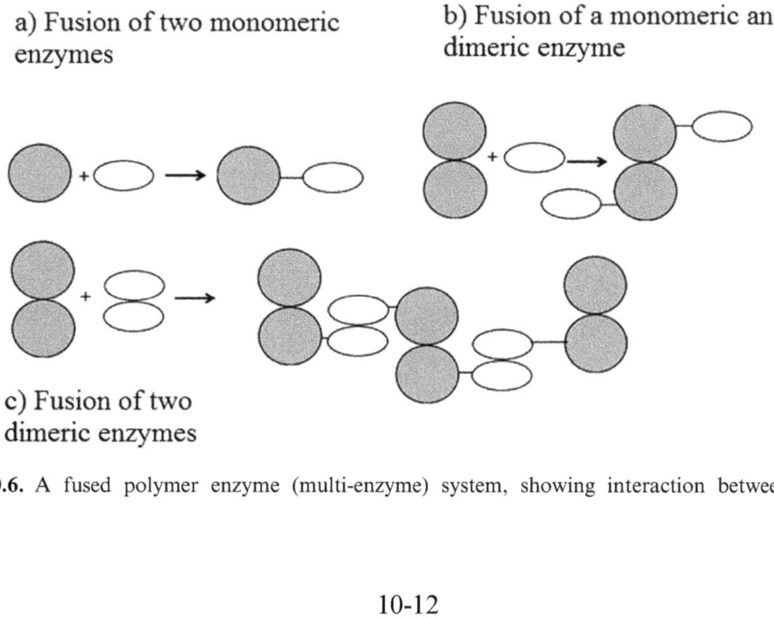

Figure 10.6. A fused polymer enzyme (multi-enzyme) system, showing interaction between component enzymes.

under laboratory conditions and occasionally it is even possible to produce entire new enzymes by exploring new active sites or altering old ones.

Enzyme–PEG conjugates. The enzyme called L-asparaginase is usually isolated from microbes and anticancer properties, however, its toxicity and half-life (less than 18 h) limit its utilization. It was observed that this enzyme can be modified by polyethylene glycol derivatives to form PEG–asparaginase conjugates. These conjugates differ from the native enzyme in the following characteristics:

I. It preserves only 52% of the catalytic activity of native enzyme.
II. It becomes resistant to proteolytic degradation.
III. It does not cause allergy.

Currently, these types of enzyme conjugates (e.g. uricase, catalase, etc) are prepared and utilized at the industrial scale to treat various types of tumors in rats and mice.

Modification of proteases into peptide ligases. To achieve high specificity and stereoselectivity, and to suppress side reactions, peptide ligation to the native enzyme is one of the best alternative procedures. Consequently, production of any enzyme that may catalyze peptide ligation will be most encouraged. Modification of protease subtilisin into thiol and selenolsubtilisin (two semisynthetic enzymes) can catalyze peptide ligation. Both of these modified proteases efficiently catalyze peptide ligases [14].

Production of site specific nucleases. In this procedure chemical cleavage agents are used to combine DNA recognition and the binding properties of proteins.

Production of artificial semisynthetic oxidoreductases. By covalently attaching redox-active prosthetic groups to existing sites, artificial oxidoreductases can be prepared. Joining 10-methylisoalloxazine derivatives to specific sites of several proteins has been achieved. The ability of these semisynthetic enzymes compares favorably with that of naturally existing flavoenzymes.

10.13 Some early achievements of protein engineering

To determine the effects of site specific mutagenesis involving substitution of one or more amino acids, a number of proteins have been examined. Attempts have also been made to examine the function of different regions of a protein. Certain early achievements of protein engineering are mentioned as follows:

- *Acetylcholine receptor*: This protein is involved in delivery of acetylcholine across the membrane. The exact regions of this involved in acetylcholine binding and formation have been explored.
- *Cytochrome C*: A phenylalanine residue recognized to be non-essential for electron transfer, however, it is involved in determining the reduction potential of the protein.
- *Dihydrofolate reductase*: In this enzyme, substitution of a single amino acid, aspartic acid (Asp) by asparagine (Asn), results in a decrease in specific activity by a thousand fold, signifying that aspartic acid is significant for the active site. Other similar modifications were also examined.

- *Human-13 interferon*: Removal of the three cysteine residues led to an improvement in stability of the enzyme.
- *Insulin*: Contains A and B chains connected by a C-peptide of 35 amino acids. It was observed that a sequence of six amino C-peptides was satisfactory for the connecting function.
- *Lactose permease* (product of 'y' 'lac' operon): This enzyme is involved in delivery of lactose and, a cysteine to glycine replacement showed that this amino acid was vital for delivery. Further, of four histidine residues, two at positions 35 and 39 do not play any essential role in transport while the mutation in any of the other two at positions 208 and 322 result in a transport function.
- *T4 lysozyme*: A mutation of isoleucine to cysteine in this enzyme leads to formation of a disulfide bridge, leading to thermal stability and a 200-fold increase in enzyme activity even at 67 °C.
- *Trypsin*: This can be restructured to have modified substrate specificity.
- *β-lactamase*: This bacterial derived enzyme hydrolyzes and inactivates the *β*-lactam ring of penicillin derivatives and allows delivery across the inner membrane. In the course of transport a polypeptide (23 amino acids) is cleaved off. Comprehensive examination suggests that transport and processing are not influenced by this polypeptide of 23 amino acids alone. An active site comprising the amino acid serine has also been recognized, as its substitution by cysteine results in a reduction in the activity of this enzyme.
- *λ repressor*: This protein could be engineered to develop a specific site for cro-protein, since the alteration led to the development of a cro recognition site.
- *Subtilisin*: The serine protease subtilisin is an important industrial enzyme as well as a model for understanding the enormous rate enhancements effected by enzymes. Mutations in well over 50% of the 275 amino acids of subtilisin have been reported in the scientific literature. Most subtilisin engineering has involved catalytic amino acids, substrate binding regions and stabilizing mutations. Additional effective modification of substrate specificity includes the subtilisin.
- *Lactate dehydrogenase*: A lactate dehydrogenase isolated from *Bacillus stearothermophilus* was modified individually by each of the three substitutions of amino acids.

References

[1] Tiwari M K, Ranjith S, Singh R K, Kim I W and Lee J K 2012 Computational approaches for rational design of proteins with novel functionalities *Comput. Struct. Biotechnol. J.* **2** e201209002

[2] Chen R 2001 Enzyme engineering: rational redesign versus directed evolution *Trends Biotechnol.* **19** 13-4

[3] Li Y and Cirino P C 2014 Recent advances in engineering proteins for biocatalysis *Biotechnol. Bioeng.* **111** 1273-87

[4] Ulmer K M 1983 Protein engineering *Science* **219** 666-71

[5] Lesk A M 2001 *Introduction to Protein Architecture* (Oxford: Oxford University Press)

[6] Xie J and Schultz P G 2005 An expanding genetic code *Methods* **36** 226–38

[7] Anthony-Cahill S J, Griffith M C, Noren C J, Suich D J and Schultz P G 1989 Site-specific mutagenesis with unnatural amino acid *Trends Biochem. Sci.* **14** 400–3

[8] Malmström L and Goodlett D R 2010 Protein structure modeling *Methods Mol. Biol.* **673** 63-72

[9] Kokh D B, Czodrowski P, Rippmann F and Wade R C 2016 Perturbation approaches for exploring protein binding site flexibility to predict transient binding pockets *J. Chem. Theory Comput.* **12** 4100–13

[10] Arnold F H 1993 Engineering proteins for non-natural environments *FASEB J.* **7**(9) 744-9

[11] Yanofsky C and Horn V 1994 Role of regulatory features of the trp operon of *Escherichia coli* in mediating a response to a nutritional shift *J. Bacteriol.* **176** 6245–54

[12] Scazzocchio C and Sealy-Lewis H M 1978 A mutation in the xanthine dehydrogenase (purine hydroxylase I) of *Aspergillus nidulans* resulting in altered specificity. Implications for the geometry of the active site *Eur. J. Biochem.* **91** 99-109

[13] Goward C R and Nicholls D J 1994 Malate dehydrogenase: a model for structure, evolution, and catalysis *Protein Sci.* **3** 1883-8

[14] Bryan P N 2000 Protein engineering of subtilisin *Biochim. Biophys. Acta* **1543** 203-22

Lightning Source UK Ltd.
Milton Keynes UK
UKHW051627270219
338125UK00003B/60/P

9 780750 313032